9/30/92

Augustin-Louis Cauchy

Painting of Augustin-Louis Cauchy by J. Roller ($\simeq 1840$).

Bruno Belhoste

Augustin-Louis Cauchy

A Biography

Translated by Frank Ragland

With 34 Illustrations

Springer-Verlag
New York Berlin Heidelberg London
Paris Tokyo Hong Kong Barcelona

Bruno Belhoste
Service d'histoire de l'éducation
Institut national de recherche pédagogique
Paris 75005
France

Mathematics Subject Classification (1980): 01A70, 01A55, 01A50

Library of Congress Cataloging-in-Publication Data

Belhoste, Bruno.
[Cauchy, 1789–1857, English]
Augustin-Louis Cauchy: a biography/Bruno Belhoste: translated by Frank Ragland.
p. cm.
Translation of: Cauchy, 1789–1857.
Includes bibliographical references.
ISBN 0-387-97220-X (alk. paper)
1. Cauchy, Augustin Louis, Baron, 1789–1857. 2. Mathematicians—France—Biography, 3. Mathematics—France—History—19th century.
I. Title.
QA29.C36B4513 1991
510'.92—dc20
[B] 89-26329

Printed on acid-free paper.

Typeset by Thomson Press (India) Limited, New Delhi, India.
Printed and bound by Edwards Brothers, Inc., Ann Arbor, Michigan.
Printed in the United States of America.

9 8 7 6 5 4 3 2 1

ISBN 0-387-97220-X Springer-Verlag New York Berlin Heidelberg
ISBN 3-540-97220-X Springer-Verlag Berlin Heidelberg New York

Foreword

To write a biography about a leading scientific figure is admittedly an ambitious undertaking for a historian, since, in addition to the usual difficulties presented by biographical studies, one must now contend with the strange intricacies of scientific research and thought. This difficulty is compounded considerably in at least two ways for one who would write a biography of Augustin-Louis Cauchy (1789–1857). First of all, the vast output of Cauchy's creative genius can be overwhelming on its own terms—even for mathematicians. Second, mathematical notation, concepts, and terminology were far from logical or uniform in Cauchy's day, and thus the historian faces a double challenge in pursuing the development of scientific ideas and the relationships between them. An equally difficult (but no less important) task facing a biographer of Cauchy is that of delineating the curious interplay between the man, his times, and his scientific endeavors. I think Professor Belhoste has succeeded admirably in meeting all of these challenges and has thus written a vivid biography that is both readable and informative.

Professor Belhoste's subject stands out as one of the most brilliant, versatile, and prolific figures in the annals of science. Cauchy was a man who, though a creator of our times, was nevertheless very much a creation of his own age, a particularly dynamic period in Western history that has come to be known as the Age of Revolutions (1789–1848). Although the revolutions of that era may have been primarily political upheavals, they were also scientific in the important sense that it is during this period that science and scientific scholarship passed from the control of a few enlightened despots and aristocrats into the hands of the state and its appendages, the universities and other forerunners of today's research institutions. It can, of course, be debated whether this professionalization of science was really revolutionary or merely evolutionary in the sense of a clearly discernable shift in how, where, by whom, and under whose auspices systematic scientific investigations should be carried out. However, what Professor Belhoste's work makes admirably clear is that by the time Cauchy became established in the world of science, these issues had already been fairly well decided.

Nearly two hundred years have now passed since the young Cauchy set about his task of clarifying mathematics, extending it, and applying it (whenever possible) and placing it on a firm theoretical footing. Even as he reacted to his society's sudden shifts and turns—political, religious, and intellectual—in ways which may seem to us today as hasty and ill-considered, he doggedly pursued this youthful vision. In doing so, he made many fundamental contributions not only to mathematics but to physics and astronomy as well. The true measure of Cauchy's success in the grand undertaking of his youth must necessarily be sought in the standards and methods that he bequeathed to modern science. If he was not wholly successful in the task he originally set for himself, his lack of success would seem to have less to do with any shortcomings or defects on his part than with the stupendous vastness and subtlety of mathematics and its related disciplines. And this, too, is yet another level on which Professor Belhoste's work achieves practical importance; for here we are afforded a detailed, rather personalized picture of how a first-rate mathematician worked at his discipline—his strivings, his inspirations, his triumphs, his failures, and above all, his conflicts and his errors. In this respect, then, this study should be of signal interest to young students of the mathematical sciences, since, all too often, they only catch a glimpse of Cauchy as a supremely confident and creative genious and not as a human being endowed with his share of the errors, weaknesses, and shortcomings that are such an important part of human nature.

In translating this work, I have tried to strike a balance between the flavor of the original French study and requirements of readability. To this end, in the body of the text I have given the French titles of Cauchy's works (and for those of his contemporaries also). Similarly, the original French titles are used exclusively in the bibliographical notes, and this applies also to the works of other authors mentioned in the course of the text. In this way, any reader desiring to consult the originals will have little difficulty in doing so. Similarly, I have taken care to maintain the French system of citation used by the author.

Finally, I would like to thank Mss. Connie Burgess and Joan Passalacqua for their patience and steadfast devotion in typing the manuscript.

Frank Ragland
New York
Winter, 1989

Preface

Cauchy was the true heir of the great mathematical tradition of the 18th century, the heir of Euler, Legendre, Lagrange, and Laplace; and yet he was both a leading standard-bearer and an active creator of an essentially new approach to mathematics. Indeed, Cauchy and Karl Friedrich Gauss (1777–1855) may rightly be called the first truly modern mathematicians. The sheer bulk of his scientific productivity is immense. The *Oeuvres Complètes*, a publication which covers almost an entire century, from 1882 to 1975, fills more than 27 large volumes and contains, in addition to five complete textbooks, nearly 800 research articles and treatises. Cauchy's work does contain an element of redundancy. But, on balance, such an enormous scientific creativity is nothing less than staggering, for it presents research on all the then-known areas of mathematics: arithmetic, algebra, geometry, statistics, mechanics, real and complex analysis, and mathematical physics. Nevertheless, in spite of its vastness and rich multifaceted character, Cauchy's scientific works possess a definite unifying theme, a secret wholeness. This, at least, is the essential point of departure of the present study.

There are several ways in which a study of this type might have been developed. A historian by training and by temperament, I have chosen the most natural method: the biography. To be sure, there already exists a biography of Cauchy. In 1868, eleven years after Cauchy's death, C.A. Valson, a mathematician of Catholic persuasion who was preparing the publication of the *Oeuvres Complètes*, published a biography.[1] Published under the auspices of the Cauchy family, this two-volume study is not without interest today. This is mainly so because Valson had at his disposal certain of Cauchy's personal papers—documents which have now been completely destroyed—as well as the family's archives and the actual testimony of many of Cauchy's associates. In spite of this, however, Valson's study fails to meet the standards of rigor and scholarship demanded by modern historical studies. Replete with irrelevant

[1] *La Vie et les Travaux du Baron Cauchy* (2 volumes, Paris, 1868).

moral judgements, and frequently succinct on many aspects of the great mathematician's life and works, Valson's work would seem to be more hagiography than history.

In the present work, I have earnestly sought to present a portrait of a mathematician and the age in which he lived. In this way, I hope to underscore, to highlight, as it were, the essential coherence of an active, productive career in the sciences which stretched over nearly half a century. Certainly, Augustin-Louis Cauchy was a far cry from the romantic heros of that era. His long and active life lacks the ephemeral sparkle and brilliance of that of his contemporary Evariste Galois. A devout Catholic and close associate of the Jesuits as well as a strict royalist, Cauchy quickly became a recognized mathematician who, at a very early age, took a place in the leading scientific institutions of his day. On the other hand, however, he was a proud man, a man of passionately held convictions who, whenever occasions arose for him to defend or to explain those things that he regarded as "the truth," consistently refused to allow himself to be swayed by considerations of personal convenience or self-interest. Thus, for example, in the political sphere his adherence to the Bourbon cause, was, for better or worse, absolute and unyielding. Caught up in the political purges of the 1830s, though he himself had earlier profited from similar procedures in 1816 after the fall of Bonaparte, Cauchy chose exile over perjury. In fact, from 1830 until his death in 1857, he steadfastly refused to swear loyalty oaths to any of the regimes that governed France. In a similar way, Cauchy's belief in Catholicism was uncompromising and apparently untarnished by any doubts whatsoever. Indeed, throughout his life, Cauchy practiced his faith with all the zeal of a new convert and with all the feverish passion of a missionary.

Cauchy's view of mathematics sprang from the same deeply felt need for absolutes, for certainty. If, in politics, this need manifested itself in uncompromising adherence to the Bourbon cause and in religion to strict Roman Catholic orthodoxy, then in mathematics its expression was reflected in a demand for rigor and irrefragable proof. To Cauchy, the real work of a scholar, of a scientist, must necessarily be a quest for truth. "Truth," he wrote in 1842, "is a priceless treasure which, whenever we manage to acquire it, cannot bring us remorse and sorrow; it cannot disquiet and distress our soul. The mere thought of its heavenly attributes, of its divine beauty suffices to replenish us for all the sacrifices we may have made in discovering it. Indeed, the joy of heaven itself is but the full and complete possession of immortal truth." Feeling uncomfortable in the age in which he lived and often misunderstood by his contemporaries, Cauchy found a refuge in mathematics. Here, in a world far removed from the everyday one, his creative genius could thrive, expand, and reach its full measure of expression.

Still, in a sense, his view of mathematics represents a certain expression of the concerns of the era in which he lived. Thus it was that Cauchy early rejected the optimistic rationalism of the Age of the Enlightment in no uncertain terms even though Laplace, his mentor and protector, was ever faithful to the credo of the philosophes. As it specifically concerns mathematics, it should be noted that analysis had experienced tremendous growth all during the 18th century. Unfortunately, this growth had come about at the expense of mathematical rigor; and Cauchy, like the more able of his associates and contemporaries, set about strengthening the theoretical foundations of mathematics and restructuring the entire edifice. His position as a professor of mathematics at the École Polytechnique provided him with an excellent forum for working towards these ends. Certainly, his demand for rigor in mathematics proved to be exceedingly fruitful, since it opened up wholly new rich fields of mathematical research.

But Cauchy's creative genius found broad expression not only in his work on the foundations of real and complex analysis areas to which his name is inextricably linked, but also in many other fields. Specifically, in this connection, we should mention his major contributions to the development of mathematical physics and to theoretical mechanics, two fields which experienced rapid growth during the 19th century. Along these lines we mention, among others, his two theories of elasticity and his investigations on the theory of light, research which required that he develop whole new mathematical techniques such as Fourier transforms, diagonalization of matrices, and the calculus of residues.

It should be observed that as to his mathematical talents, Cauchy—like Euler and Gauss—was a universalist in the fullest sense of the term. His few works on the theory of numbers, algebra, and geometry make us deeply regret that he did not devote more effort to these areas. It can hardly be doubted that had he done so, he would have obtained results of signal importance. Cauchy's creativity, however, bore the stamp of his training as an engineer. Accordingly, Cauchy, like other great mathematicians of the French School, always gave priority to questions about applied mathematics which were a great source of inspiration for him.

Today, two centuries after Cauchy's birth, his work has been completely integrated into the austerely beautiful and elegant structure of mathematics, a structure which is constantly changing. And thus it is that Cauchy's theorems and theories have been reformulated in newer, more modern terms and the memory of the scholar himself has been progressively weakened. Only his name attached to a few outstanding results remains to give testimony to future generations of the importance of his works. It is the fate of all mathematicians to see their individual contributions become quickly absorbed into the great common structure. And while this is hardly the place to dwell on this simple fact, we should nevertheless pause and consider another simple fact: mathematics—regardless of how impersonal it may be in its formal beauty and power, is not something handed down complete and perfect from heaven. Rather it is the cumulative result of the work (and, quite frequently, of the sufferings also) of many individual human beings. If, somehow, this biography of Cauchy should help in establishing this belief, then I will have attained my goal.

This book was the outcome of several years of work, but it could not have been completed without the help of a number of persons. First of all, I am deeply grateful to Professor René Taton, who directed my initial research with great care and kindness. I must also express my gratitude to those who were kind enough to provide me with often rare documents and information. In this respect, I am particularly indebted to Professors Dugac, Grattan-Guinness, Ross, Russo, and Yuschkevich. I also recall with special gratitude and fondness the very fruitful discussion that I had with my good friends Amy Dahan-Dalmedico and Jesper Lützen. Nor can I forget the warm reception I was always accorded at the archives and at the libraries. Special mention must be given to the Secretariat of the Academy of Sciences where M. Berthon and Mme Pouret kindly placed at my disposal their profound knowledge of the archives. Similarly, I was greatly assisted by the cogent advice of Mlle Billoux at the Central Library of the École Polytechnique. I also want to express my gratitude to Professor Frank Ragland of the City University of New York who translated this work into English with great skill and care. Finally, I should like to thank Jeremy Gray and John Greenberg who carefully read the manuscript. Their comments enabled me to make a number of changes in the final text.

Paris, France
Autumn, 1989

Bruno Belhoste
INRP, Service d'histoire
de l'éducation

Contents

Chapter 1

The Formative Years

Augustin-Louis Cauchy was born on August 21, 1789 in Paris and was baptized in the parish church of Saint-Roch. He was christened Augustin after the month of his birth and Louis after his father, Louis-François. Louis-François Cauchy was the principal commis of the Lieutenant Général de Police of Paris. By an unfortunate coincidence, his firstborn came into the world at the moment when his career became tragically compromised by the Paris insurrection. Disturbed by a fear of shortages, and cheered on by a bourgeoisie that was worried about its incomes and investments, the people of Paris rioted following the dismissal of the principal minister Necker. But, this was to be no ordinary riot; it would soon become outright revolution.

A revolutionary council was installed by the citizens of Paris during the night of July 12–13, with the astronomer Bailly as president. On the next day, July 14, the insurgents, searching for powder and weapons, attacked the Bastille, the hated symbol of absolutism. Flesselles, the Provost, was slaughtered. A few days later, the royal intendants Foullon and Bertier de Sauvigny fell to the same fate. Thereafter, the royal administration was powerless in Paris. Louis Thiroux de Crosne, the Lieutenant Général de Police, was deprived of all authority during the first hours of the insurrection and withdrew discredited from office; on July 15, de Crosne handed his command of the municipal police over to the new mayor of Paris and on the 23rd fled to England where he remained for the time being. The events also meant the end of the career of his principal commis, Louis-François Cauchy.

It was one of fate's strange quirks that Augustin-Louis should be born at the time his father lost his position and his protector, and the coincidence could not have gone unnoticed by the family. Indeed, one is inclined to think that this particular coincidence had an indelible subconscious effect on young Cauchy. In later life, he worked unceasingly to undo the results of the Great Revolution of 1789. He regarded it as a fatal disaster and fought against the ideas that had triggered it. His hatred of the Revolution was so intense and so uncompromising that when the events of July 1830 occurred, he simply could not bear to watch the reemergence of the Revolution. It is likely that he always felt a secret guilt vis-à-vis his father, a guilt for having dared to come into the world at so unpropitious a time.

A few words about Louis-François are in order. The elder Cauchy was born on May 27, 1760 in Rouen to Louis-Charles Cauchy, a Rouen businessman (1). Louis-Charles, who, according to Théodore Lebreton (2), was a master locksmith and metal worker, had only one sister, Marie-Madeleine, who married Laurent Larsonnier. Larsonnier was employed at the customs house at Versailles on the eve of the Great Revolution and became treasurer of the Chamber of Peers during the Restoration with the support of his brother-in-law. Louis-François received a good education, a fairly common practice among the bourgeoisie of that era. Sent to the Collège de Lisieux in Paris, he was awarded honors in 1777 by the University of Paris for his performance in the Concours général, (a celebrated competition between the students of the Parisian colleges). Afterward, he returned to his hometown, where he worked as a lawyer near the Parliament of Normandy. Soon, in 1783, he was appointed Secretary-General of the Intendance (Administration) of Rouen (3), the intendant Louis Thiroux de Crosne apparently recognized that this young lawyer was uncommonly able. Thiroux, himself a gifted administrator, had been intendant at Rouen since 1767. He was a very cultured man, a member of the Academy of Rouen since 1771. Imbued with modern ideals, he was capable of recognizing talent when he saw it. Thus, when he was appointed Lieutenant Général de Police of Paris in August of 1785, he took his esteemed secretary along, naming him his principal commis.

During the final years of the Old Régime, the government made extensive, elaborate plans for the rebuilding of Paris. At the time, Paris was a city of more than 600,000, among the largest cities in Europe—second in size only to London. In spite of its size, however, it was still essentially a medieval town: overcrowded, dirty, dangerous, and unhealthy. Its narrow, dusty streets, open drains and ditches, and cemeteries in the heart of town distressed and shocked the enlightened minds of polite society, a society that prized order, balance, and reason. The government, urban designers, and engineers looked forward to the building of a new city: a city that would be healthy, well lighted, well policed, and beautiful. At the time, however, the city districts were hemmed in by tollgates and stations, and traffic swarmed across the bridges. But, new metropolitan areas and avenues were in the planning.

Louis-François took part in these projects and discussions, which were soon to be interrupted by the onslaught of the Great Revolution, but which ultimately would be carried through to fruition under successive regimes during the 19th century. In 1786, he supervised the removal of the remains from the Cemetery of the Innocents with Thouret and Fourcroy (4). He was also in charge of controlling theaters and the book trade and managing state prisons such as the Bastille (5).

Louis-François was still unmarried, living with his widowed mother, perhaps at the home of a cousin, Antoine Thibaut Beauvais, a middle-class

Parisian, in the rue Saint-Honoré. In October 1787, he married a 20-year-old Parisienne, Marie-Madeleine Desestre. By his marriage to Marie-Madeleine, Louis-François became part of a family that was well established in the lower ranks of Paris officialdom; Marie-Madeleine's father, Louis-Jacques Desestre, was the Elder Dean of Bailiffs of the King's Conseil d'État; her uncle, Jean-Baptiste Desestre, was employed as Inspector of Revenues at the Hôtel de Ville de Paris (Town Hall). Similarly, her mother, Madeleine Paupelin, came from a solidly bourgeois Parisian family, a family containing lawyers, merchants, and even some members of the bureaucracy, who were well on their way to ennoblement.

Marie-Madeleine was a good match for Louis-François, who was not well off; he owned a small family plot at Frettecuisse in the Picardie, and he had an income of 150 pounds, 5000 pounds in notes and coin, 8000 in other movable property, and a life annuity valued at 800 pounds. Present worth aside, he now had an important position and a very bright future indeed. As for his wife, she had brought a dowry of 40,000 pounds, 15,000 in cash, and an income of 1250 pounds per annum to be paid in perpetuity, as well as a life annuity worth 500 pounds (6). Shortly after their marriage, the young couple bought a few acres and a spacious country house in Arcueil, in the rue des Carnaux (7).

The Revolution interrupted his climb upward, and the following years were more difficult for Louis-François. He was compromised; he had to make himself forget. According to Valson, he took a position as Chief of the Bureau of Almshouses and Charity Workshops, a post that he kept until the Reign of Terror (8). But, finally, when Louis Thiroux returned from abroad and was arrested and quickly condemned to death on the same day, April 28, 1794, Louis-François began to fear for himself and his family and decided to leave Paris where, he realized, he stood in danger of being denounced by the revolutionary authorities. Fleeing the dangerous turmoil that now gripped the capital, he took his wife and two children, Augustin-Louis and the baby Alexandre-Laurent, who had been born on March 12, 1792, to their country house in Arcueil, where they remained until the Reign of Terror had passed.

But, if Arcueil was a haven from denunciation and execution, life there was hardly easy; sufficient food was a major concern, as the following remarks by Louis-François, which are cited in Valson, attest:

> We never have more than a half pound of bread—and sometimes not even that. This we supplement with the little supply of hard crackers and rice that we are allotted. Otherwise, we are getting along quite well, which is the important thing and which goes to show that human beings can get by with little. I should tell you that for my children's pap I still have a bit of fine flour, made from wheat that I grew on my own land. I had three bushels, and I also have a few pounds of potato starch. It is as white as snow and very good, too, especially for very young children. It, too, was grown on my own land (9).

During this period, Augustin-Louis contracted smallpox (10); he did not seem to have adjusted easily to these stressful conditions, to the privation and material uncertainties, and most importantly, to the threats of arrest that tormented his father. We know very little about these early years, which must have been crucial in the development of his personality. Although we have no direct reports left, we may suppose that henceforth he found refuge in thought and quiet study, a refuge from the agonies and fears of an uncertain period. A timid, frail boy, withdrawn and pensive, he had no liking for sports and games. His love for purposeful work, so unusual in a boy his age, was in striking contrast to the carefree openness of his younger brother Alexandre.

The fall of Robespierre on July 27, 1794 was greeted with relief: Louis-François and his family could now return to Paris unafraid. In the autumn of that year, he became Assistant Director of the Division of Crafts and Manufacturing in the Commission of Arts and Crafts, which was housed in the Hotel Molé in the rue Dominique. The following spring, in April 1795, while the repression was striking telling blows against the last radicals, he moved up to the post of director in that bureau, replacing Pierron. From the reestablishment of the ministries on October 2, 1795 until the coup d'état of 18 Brumaire (November 10, 1799), Louis-François remained Director of the Bureau of Crafts and Manufacturing in the 4th Division of the Ministry of the Interior [this was the new name of the former Division of Crafts and Manufacturing since the time of the Convention (11)]. Carrying a salary of 5000 to 6000 francs, this position was certainly not minor. According to the Almanach National, the Bureau had authority over all industry and manufacturing, useful arts, handicrafts, patents, and certificates of invention, as well as the administration and supervision of the Conservatoire des Arts et Métiers and the national manufacturing enterprises at Sèvres, Chaillot (the Savonnerie), Beauvais, Aubuisson, etc.

During the forced leisure at Arcueil, Louis-François made good use of his time by undertaking the education of his children, and he continued this for several years after their return to Paris. Louis-François was extremely solicitous of his children's education. He himself had always been a model student: was it not knowledge, the fine fruit of education, that had enabled him to rise to the upper level of government under the Old Régime on the eve of the Great Revolution? Socially, he belonged to the bourgeoisie of talent, a kind of meritocracy, that had begun, little by little, to occupy important state positions even before the Revolution. No doubt he envisioned his sons as following in his footsteps by entering government service.

Louis-François wrote 'little didactic compositions' for his sons' educations (12). These writings were usually written in the orderly, balanced style that was then characteristic of French verse, and they dealt with subjects such as grammar, history, and ethics. He later introduced the children to Latin and Greek, requiring that they study the works of the ancient writers. Also he did not neglect their religious education. Indeed, from an early age, Augustin-Louis familiarized himself with biblical texts. A sincere Christian, Louis-François was particularly devoted to the 'humanities', the term being used here in its 18th-century sense, and wrote poetry in French and Latin. Indeed, as we will see, he even acquired a certain reputation as a poet by publishing verses during the years of the Consulate, the Empire, and the Restoration. While these verses were mediocre pieces, flattering and praising the current rulers, it is interesting to note that his three sons, particularly Augustin-Louis, picked up the habit of writing French or Latin poems that dealt with diverse topics and were periodically published.

By all indications, Louis-François did not neglect his children's education in the sciences. A letter dated February 18, 1799 from Louis-François Cauchy to the Central Bureau of Correspondence of Le Mans, of which he was, as Director of the Bureau of Manufactures, a nonresident affiliate, shows that he sought diversion from administrative routines by studying nature:

> The sciences are sisters of the arts, and anyone interested in the latter cannot be unfamiliar with the former... I myself have always been especially fond of natural history, and, although conchology (the study of shells) may not be the most attractive part of it, even it seemed to me to present, in various respects, a particular charm. From a practical point of view, it might present useful applications that have not been discerned heretofore. I think, for example, that conchology can do more than any other branch of zoology to acquaint us with the way in which sun and climate affect animal species. Consequently, the thought occurred to me to form the most complete collection possible of indigenous testaceans in France. I hope, citizens, that some of you will want to show me everything that the department of Sarthe possesses... Good health and fraternity! (13)

Again, Louis-François' intellectual interest stirred similar interests in his oldest son, because, later on, at Cherbourg, Augustin-Louis enjoyed collecting plants in his spare time

The coup d'état of 18 Brumaire overthrew the Directory and established the Consulate with Bonaparte as First Consul. Louis-François supported the new regime enthusiastically and praised its virtues in several writings (14). On January 1, 1800, he was elected to the post of Secretary-General by the newly created Senate. In this capacity, he was responsible for the transcription and editing of that body's proceedings. On the same occasion, he became Archivist and Keeper of the Seal of the Senate; it was his duty to countersign and release all Senate dispatches after they had been authorized by its president. He worked directly under the Chancellor of the Senate, Count Laplace. For Louis-François, his election to a post at the Senate represented an extraordinary advancement, a promotion that not only doubled his salary to more than 12,000 francs, but also brought him influence and prestige: he was in daily contact with the senators, most of whom were men of considerable standing. Exactly how Louis-François came to receive the nomination to so prestigious a position is not known. But, it can hardly be doubted that it was through the good offices of some influential person, such as Fontanes, a habitué of the reactionary and staunchly catholic Elisa salon (15), or François de Neufchâteau, a senator and former Minister of the Interior. The Senate sat in the Palais du Luxembourg; Louis-François Cauchy and his family lived nearby in the rue de Tournon, and Augustin-Louis frequently came over to work in his father's office.

By that time, he had developed an interest in mathematics, since, according to Valson, who studied Cauchy's school notebooks, 'it was not an infrequent thing to find a paper on a literature assignment suddenly interrupted: A mathematical idea would have crossed the youngster's mind and so absorbed him that he would be forced to translate the compelling notion into numbers and figures' (16). On several occasions, Louis-François presented his son to Laplace and Lagrange, both of whom were mathematicians of the first rank as well as senators (17). Lagrange seems to have taken some interest in young Cauchy, for, according to Valson, one day in 1801, Lagrange made the following remark to several members of the Senate who were meeting in Louis-

Pierre-Simon Laplace (1749–1827), in the dress of Grand Chancellor of the Senate. Cauchy was his protégé until 1816.

François' office, among them Lacépède:

> Now you see that little fellow there, don't you? Well, one day he will replace all of us simple geometers (18).

Even if we should doubt the remark Valson attributed to Lagrange, it is likely that he did advise Louis-François about to his son's education in mathematics:

> Do not allow him even to open a mathematics book nor write a single number before he has completed his studies in literature (19).

On Lagrange's advice, Louis-François enrolled his son in the École Centrale du Panthéon in the fall of 1802. There he was to complete his studies in the humanities.

The École Centrale du Panthéon was located not far from the Palais du Luxembourg, on the Montagne Sainte-Geneviève, housed in the buildings of the Abbey of the Génovéfains (20). This institution was perhaps the best of the 3 écoles centrales that had been established in Paris after 1795. A little more than 300 students attended it. In the first section, which was open to boys of more than 12 years of age and which was the section that Augustin-Louis entered, 2 years of study were devoted to ancient languages, drawing, and natural history, subjects that were taught by Bachelier, Maherault, and Cuvier (21). Augustin-Louis attended the École Centrale du Panthéon during a time when France's educational structure was in a state of transition, between the creation of the Lycées under the law of May 1, 1802 and the establishment of the Lycée Napoleon, which replaced the École Centrale du Panthéon, in September 1804. Students at the École Centrale were not grouped into classes, but rather could select the courses that interested them, passing freely from one course to the other. The code of discipline of these institutions was very liberal, a far cry from the quasi-military code that would soon be applied to the imperial Lycées.

Augustin-Louis worked particularly hard in ancient languages during the two years he spent at the École Centrale du Panthéon and showed himself to be a very bright young scholar. As early as 1803, he took first place in the competitive examinations in Latin composition, which had been instituted the preceding year by Napoleon. In 1804, he won second place in the Latin oratory competition for advanced students and first place in Greek and Latin poetry. Most important, however, he won the Grand Prize for the Humanities, an award that was bestowed on the student who ranked highest in the most competitions (22). He worked unremittingly in preparation for this examination, even though he was to take first communion on Easter Sunday, 1804 (23).

Encouraged by Louis-François, the young scholar was already so fired by an ambition to always be first that, writing in his resolutions for first communion, he modestly pledged that:

> I shall never flaunt the little learning that I have acquired through the care and help my father has given me. If I have learned anything, it is only

> because he took care to teach me. Had he not taken upon himself the trouble of instructing me, I would be as ignorant as many other children (24).

He was already reaping the fruits of his increasing reputation: receiving his awards at the Institut; dinner at the home of the Minister (25). These were the things that could easily turn the head of a teenager, who, in his mother's words, 'then had many faults of character' (26).

In spite of his successes in the humanities, Augustin-Louis decided to prepare himself for entrance into the École Polytechnique the following year. In doing so, he was rejecting a family tradition: Louis-François had studied law, and whatever interest he had in natural history seems rather to have come from a respect for learning in general rather than from a strong inclination toward the sciences per se. Augustin-Louis' younger brothers, Alexandre and Eugène, were to follow in their father's footsteps, and both would enjoy distinguished careers in law (27). Only Augustin-Louis chose to become an engineer. His decision seems to have been in response to a kind of peculiar personal bent, which could be discerned in his early liking for mathematics and which would be confirmed later. Meanwhile, we should not completely disregard the influence of the family on his decision. It should be noted, first of all, that the École Polytechnique had been created in response to the requirements of the public sector. By preparing to enter into a major institution of the state, Augustin-Louis was acting in a way that was wholly consistent with his father's views and plans.

As for Louis-François, far from opposing the choice his oldest son had made, he actively endorsed it, using his influence to advance his son in his chosen career and providing him with moral and material support when, later on, Augustin-Louis began to work on his first scientific projects. Moreover, for the rest of his life, Augustin-Louis retained a deep filial respect for his parents, particularly his father. Each week he, along with his sisters and brothers, would go to the Palais du Luxembourg to enjoy a family luncheon, to sit at the paternal table 'like a young olive tree', and to realize the blessings promised by the prophet to 'the man who feared the Lord' (28). Later on, father and son would study Hebrew together; and, on March 11, 1842, they would even present a report on biblical prosody before the Académie des Inscriptions et Belles-Lettres (29).

Augustin-Louis' intellectual personality, which is so strongly stamped on all of his scientific works, was nurtured in an intimate family circle, in close contact with a very strict and pious mother and a very open and hard-working father. It was in this circle that he developed his exceptional capacity for hard work and his curiosity and interest in learning that, as time passed, became an almost exclusive passion for truth. He also inherited a certain stubbornness and rigidity of character that his contemporaries frequently mistook for narrow-mindedness—and this was particularly the case in political matters. Still, if stubbornness and rigidity were indeed basic features of his character, it is hardly likely that without them he would have been able to persist in attacking and solving so many fundamental and difficult research questions, some of which did not yield to his efforts for long periods of time.

Without overly simplifying, we can explain Cauchy's peculiar mathematical qualities in terms of his family: an ability to formalize situations and manipulate abstractions and a conceptual and logical rigor, as well as clarity and precision in exposition. Such qualities are, of course, of great value in the theory as well as the practice of law. Not surprisingly, then, one of Cauchy's major projects was to put on a firm, rigorous footing many mathematical methods and procedures that had up to then been used without sufficient theoretical justification. In other words, his aim was to establish clearly defined guidelines as to what could and could not be done in mathematics. Could he pay a more worthy homage to his family's tradition in law than this? Seen in this light, then, the Cauchy family's propensity for law was not far removed from the concerns of the mathematician of the family.

Starting in the autumn of 1804, Augustin-Louis attended mathematics classes given by Dinet, professor at the Lycée Napoleon and an examiner for admission to the École Polytechnique (30). Cauchy made very rapid progress, and in 1805, he took the competitive entrance examinations at the École Polytechnique. Examined on October 30, 1805 by J.-B. Biot, he was second out of the 293 applicants and the 125 admitted. According to the rules he now had to choose the field of public service that he would enter once his studies at the École had been completed. In satisfaction of this requirement, then, he presented the following choices, in order of preference: (1) Ponts et Chaussées (highways and bridges), (2) Génie maritime (maritime engineering), (3) Mines (mining engineering), (4) Génie militaire (military engineering), (5) Corps des ingénieurs géographes (topography), (6) Artillerie de terre (land artillery), and (7) Artillerie de marine (naval artillery). He put at the top of the list the Ponts et Chaussées service, a select one, to which almost all the Polytechniciens aspired.

The École Polytechnique had opened its doors at the end of 1794. The school gave the future civil and military engineers a high-level scientific education. Once they had completed work at the École Polytechnique, the students rounded out their education with more specialized training in an école d' application, such as the École des Ponts et Chaussées or the École de l'Artillerie et du Génie at Metz, which prepared them for a specific public service. At the beginning of October 1805, an important reform instituted by Napoleon went into effect. The École Polytechnique was now transferred from the Palais-Bourbon to the Collège de Navarre on the Montagne Sainte-Geneviève. Henceforth, students were to be organized into military corps and quartered in barracks. For this, they had to pay a fee of 800 francs, provide their own uniforms and other personal necessities, and pay for their own books and equipment. During his first year at the École, Augustin-Louis belonged to the 4th squadron of the 1st company, which was under the command of Charles-Émile Laplace, the son of the great Laplace.

In general, the mathematical sciences occupied the largest portion of the scheduled class time at the École. Indeed, in 1806, the distribution of time among the various disciplines during the first and second years was as follows: for analysis, 29% during the first year and 18% during the second; for mechanics, 17% and 22%; and for descriptive geometry, 26% and 3%, respectively; the physical sciences, on the other hand, were allotted considerably less time in the academic program. Indeed, in 1806, 5% was allotted to physics in the first year and 7% in the second; the figures for chemistry are 9% for each of the first two years (31). The balance of the schedule at the École was devoted to literary studies and to drafting and drawing during the first year and to studies in applications (fortifications, construction, and mines) during the second. Basically, at this time, the students were required to read a number of books and other materials that had been selected by the professors. We can get a view of the readings and course requirements from the following list for 1805, which has been supplied by A. Fourcy:

Analysis instruction: the *Cours d'Analyse Algébrique* by Garnier and the *Traité Élementaire de Calcul Différentiel et Intégral* by Lacroix;
Mechanics instruction: the *Traité de Mécanique*, using the methods of Prony and edited by Francoeur and the *Plan Raisonné du Cours de Prony*;
Descriptive geometry instruction: the *Géométrie Descriptive* by Monge;
Applied analysis instruction: the *Feuilles d'Analyse Appliquée à la Géométrie* by Monge and the *Application de l'Algèbre à la Géométrie* by Monge and Hachette.

These works constituted the core of the required texts for the mathematical sciences. In addition to these, there were numerous works that were required in chemistry, physics, and other areas.

The faculty at the École was indeed an illustrious group: Hassenfratz in physics, Fourcroy in general chemistry, and Guyton in applied chemistry; Durand in architecture; Neveu in drawing and drafting; Andrieux in grammar and fine arts; Sganzin in civil engineering problems; and Duhaÿs in fortifications and topography. In the mathematical sciences, the core discipline at the École, two professors worked in each area: Lacroix and Poisson in analysis, Monge and Hachette in applied analysis and descriptive geometry, and Prony and Labey in mechanics. In April 1807, Poisson and Labey exchanged professorial chairs. Cauchy studied analysis under Lacroix, whose course closely mirrored his *Traité Élémentaire de Calcul Differentiel et Intégral*; descriptive geometry and applied analysis were studied under Hachette; and mechanics under Prony (32). Much of the student's work was done under répétiteurs (tutors). Ampère was Cauchy's tutor in analysis and mechanics, with Teysseyrre and later Bazaine as répétiteurs adjoints (assistant tutors); Livet and later Paul Binet were his tutors in descriptive geometry.

As we have seen, Cauchy began his work at the École Polytechnique with great dash, although his successes during the first year were not especially brilliant. In July 1806, he progressed from the second to the first division, being ranked thirteenth of the 25 students who had chosen the Ponts et Chaussées as their public service speciality. Given the military discipline under which the École functioned and the promiscuity in the barracks, Augustin-Louis no doubt experienced some difficulty in adjusting to life at the École. Everything was in sharp contrast to the warm, easy comfort of life in the rue de Tournon. Here, the young man, who had been brought up so carefully and in such a devout way, now found himself among other young people who were not only older but also loud, flashy, libertine, and irreligious. Fortunately, the second year was better (33); and Augustin-Louis asserted himself, so that in October 1807 he was admitted to the École des Ponts et Chaussées. He now ranked first of the 17 students entering the Ponts et Chaussées and was ranked third in the entire student body. Even though he showed himself to be a very good technician per se, he was, nevertheless, not a very exceptional all-around student of engineering sciences. In fact, his special talent for solving geometric problems caused some comment among his professors. For example, in July 1806, he solved a problem requiring the determination of the lines of maximal slope (34). That same month, Hachette published in the *Correspondance sur l'École Polytechnique*, a periodical he edited, an elegant proof by Cauchy of an important theorem by Monge: 'If a surface whose equation is of degree m is touched by a cone, the curve of contact of these two surfaces lies on another surface curve of degree $m-1$' (35). In the same periodical, Hachette also published a resume of a report by Cauchy on the problem of constructing a circle tangent to three given circles (36).

When he left the École Polytechnique in October 1807, Cauchy was barely 18 years old. To judge by the comments of the administration, he was of average height and had light brown hair and gray eyes (37). On several different occasions, his fragile health had been grounds for various special permissions. It would appear that Augustin-Louis was the victim of illnesses that doubtlessly resulted from overwork, but the real source of the problem seems to have been a kind of nervous tension (38). At the end of 1807, he enrolled in courses at the École des Ponts et Chaussées. This école was housed in the outbuildings of the Palais-Bourbon, occupied by the École Polytechnique until its move to the Collège de Navarre in 1805. Its students were commuters. They were divided into three classes according to their degree of knowledge and not to a circle of study. Each year, they submitted their works—construction projects and theoretical memoirs—for several competitions, which were solemnly judged in the presence of the Minister of Interior. As a result of these competitions and their practical work during the summer months, the students obtained marks, called degrés, on which depended their passage from a given class to a more advanced one. The best of the students could expect to be named engineers-aspirants after two years of study.

There were, at that time, three major professorial chairs at the École des Ponts et Chaussées: a chair for applied mechanics held by Prony (actually, Eisenmann substituted for Prony, who was in service elsewhere), one for applied stereotomy and construction practices held by the engineer Louis Bruyère, and one for civil architecture held by the architect Charles-François Mandar. The students could also take courses in modern languages, first in English and German only, and then, from 1806 on, in Italian, also (39). Cauchy studied these languages and learned them sufficiently well to be able to read scientific treatises in all of them. Courses at the school lasted for four months each year, from December to March, the major portion of the students' instructional time being spent on fieldwork in the countryside. During the warm season, students in the second and third classes were often sent on field expeditions far away from the capital.

Augustin-Louis proved himself to be a very brilliant student at the École des Ponts et Chaussées. In the competitions of 1807, he won four first prizes (40). The library of the École still keeps an elementary study on wheels written by Cauchy during his first year. Its contents and objectives are indicated by the title: 'Mémoire sur les roues des voitures et les perfectionnements dont elles sont susceptibles, sur les roues à voussoirs connues sous le nom de roues à la Duboville, les vélocifères, etc'. Written before he was 20 years old, this work is far removed from the future concerns of the great mathematician (41).

When he left the École Polytechnique, Cauchy lived near his parents, at the Palais du Luxembourg and in Arcueil. It is likely that he attended some meetings of the Société d'Arcueil, a reunion of young scientists studying about Laplace and Berthollet, in their country houses in Arcueil, in 1808 and 1809. In the spring of 1808, he was sent on assignment to the works then in progress in the Ourcq Canal and the Paris Water System. By not sending him far from the capital at this time, the administration was granting him an exceptional favor.

Cauchy was sent to a vast construction site that was under the direction of the engineer Pierre Simon Girard, of the Ponts et Chaussées. Girard had been on Napoleon's expedition to Egypt and was author of several treatises and scientific works. His most important work was a treatise on the strength of materials (42). Upon his return from Egypt, Girard had been appointed director of the Ourcq Canal project by the First Consul. From the Ourcq Canal, as it then stood, Girard had conceived the idea of constructing an aqueduct, as well as a navigational canal. Though these twin projects seemed impossibly contradictory to many, Girard had the Emperor's confidence and in October 1807 was appointed director of the project with full authority to proceed with the works. Napoleon attached great importance to the Ourcq Canal project: this undertaking would modernize and beautify Paris, the city that had now become the capital of Europe. Not surprisingly, then, Bonaparte intervened directly on several occasions to ensure that the project was carried on efficiently.

In 1808, work on the project was started inside Paris. Cauchy was closely associated with this phase of the construction effort. First, Girard placed him for four months under the direction of the engineer Pierre-Marie Thomas Égault, responsible for the general survey of Paris (43). Later on, in December 1808, Cauchy took part in the construction of the Saint-Denis Aqueduct. On this project, he supervised the unearthing of an ancient Roman highway, which had been hidden for centuries beneath the cobble stones in the rue Saint-Denis. At the end of this field assignment, Girard sent the Directeur Général of the Ponts et Chaussées a highly complimentary report on the young Cauchy:

> The laying-out of the circle-aqueduct inside Paris was the first operation on which this student worked. He was employed on the leveling work for this aqueduct. He later worked on the drawing of land over which the arched Saint-Laurent gallery would pass, from the circle-aqueduct to the large drain. He prepared various plans under orders from and with the direct instruction of M. Égault, and he did so with great enthusiasm, as well as with all the accuracy that could have been desired.
>
> The theoretical knowledge and understanding that he acquired at the École made it easy for him to carry out these operations, and I believe that the exactness of these past performances is indicative of the performance that he will give on similar assignments in the future.
>
> After having assigned him, for nearly four months, to work on surveying, laying out and leveling, I thought it would be to his advantage to assign him to one of the large building sites. He was personally charged with overseeing the construction of a portion of the rue Saint-Denis Aqueduct and surveying for work done. In this assignment, he again gave proof of his aptitude and enthusiasm. I should add in this connection that he has made some useful observations on the procedures that are ordinarily followed for quantity surveying of materials.
>
> Finally, I directly charged him with the responsibility of raising the original land of the site and an ancient Roman highway that we found some two meters beneath the present paving stones. This operation, which in a way was foreign to our work, has nevertheless inspired in him the degree of interest that it deserved, and he has achieved a satisfying result.
>
> I now turn to your question as regards his conduct and use of time while he is not on duty. It is enough that I tell you that as M. Cauchy's family resides in Paris he has spent with them the little free time that he has from his duties. It has seemed to me that his education and principles have made it unnecessary until now that he be accorded the surveillance and advice that must ordinarily be given to young people (44).

Cauchy's second year at the school proved to be as remarkable as the first one had been. During this year, Cauchy was sent on his second field assignment, again to the Ourcq Canal project, and then later on to the Saint-Cloud bridge project, the latter under the direction of the engineer Beautemps-Beaupré. Unfortunately, no record exists of Cauchy's service on the bridge project (45). However, we know something about his achievements on the competitive examinations of 1807.

In these competitions, Cauchy took first place in rational mechanics and in mechanical engineering; he took second place in bridges and street engineering and in wooden bridge engineering (46). Meanwhile, on two other examinations, navigation and stone bridges, Cauchy's studies were lost, and he could not compete. These studies, rediscovered in Prony's private papers, are kept today in the library of the École des Ponts et Chaussées (47). The paper on navigation was divided into 2 distinct parts. The first part contains 11 statements and observations, accompained by various calculations in which the statements and observations were subjected to a mathematical treatment, and the second part applied the analysis given in the earlier part to the particular problems of navigating the river Marne. The most impressive feature of this paper is the quality of its exposition (48).

Two other papers discuss problems involved in constructing stone bridges. The first of these studies presents an elementary discussion of construction problems (49). The second study, which is the more interesting, investigates the problems connected with constructing bridges with vaulted arches (50). In the first part, Cauchy examined the relevant theories that had been developed by various authors, such as La Hire, Couplet, Coulomb, and Bossut. In the second part, he compares these theories and the results they gave rise to with experience. The third part of the paper is analytical in its approach. Here, Cauchy sets forth the general equation for the equilibrium of an arbitrary arch and then applies his general results to several particular types of arches. Of all the papers written in 1809, this paper on vaulted arches was by far the most indicative of the future. Indeed, while he was at Cherbourg in 1811, Cauchy briefly considered resuming his research on this topic (51).

Given his record and achievements, Cauchy only needed two years at the school to complete his studies. Thus, in January 1810, he left the institution, brilliantly ending his apprenticeship in engineering. When he took leave of the school, he not only had a solid scientific background, but he had also developed a definite personal, religious, political, and intellectual outlook.

The Cauchy family was very pious, and we have seen the care that Augustin-Louis took in preparing himself for First Communion. On the other hand, the vast majority of the students at the École Polytechnique were fairly indifferent to religion, to faith per se no less than to practices; there, the dominant attitude among the students was one of liberalism and anticlericism. In the meantime, however, the Catholic Renaissance had taken roots among the urban middle classes, which were still largely devoted to the philosophical ideas of the preceding century. This movement had slowly and almost imperceptibly begun to make itself felt at the École.

During the summer of 1805, a few weeks before Augustin-Louis entered the École Polytechnique, a young student at the École des Ponts et Chaussées was appointed répétiteur adjoint (assistant tutor) in analysis and mechanics, replacing Matthieu, the incumbent holder of that position. The newly appointed répétiteur adjoint, a certain Paul-Emile Teysseyrre, was an ardent Catholic member of the Congrégation de la Sainte Vierge.

The Congrégation, founded four years earlier, in 1801, by a Jesuit priest named Father Jean-Baptiste Bourdier-Delpuits, would bring 'young people of

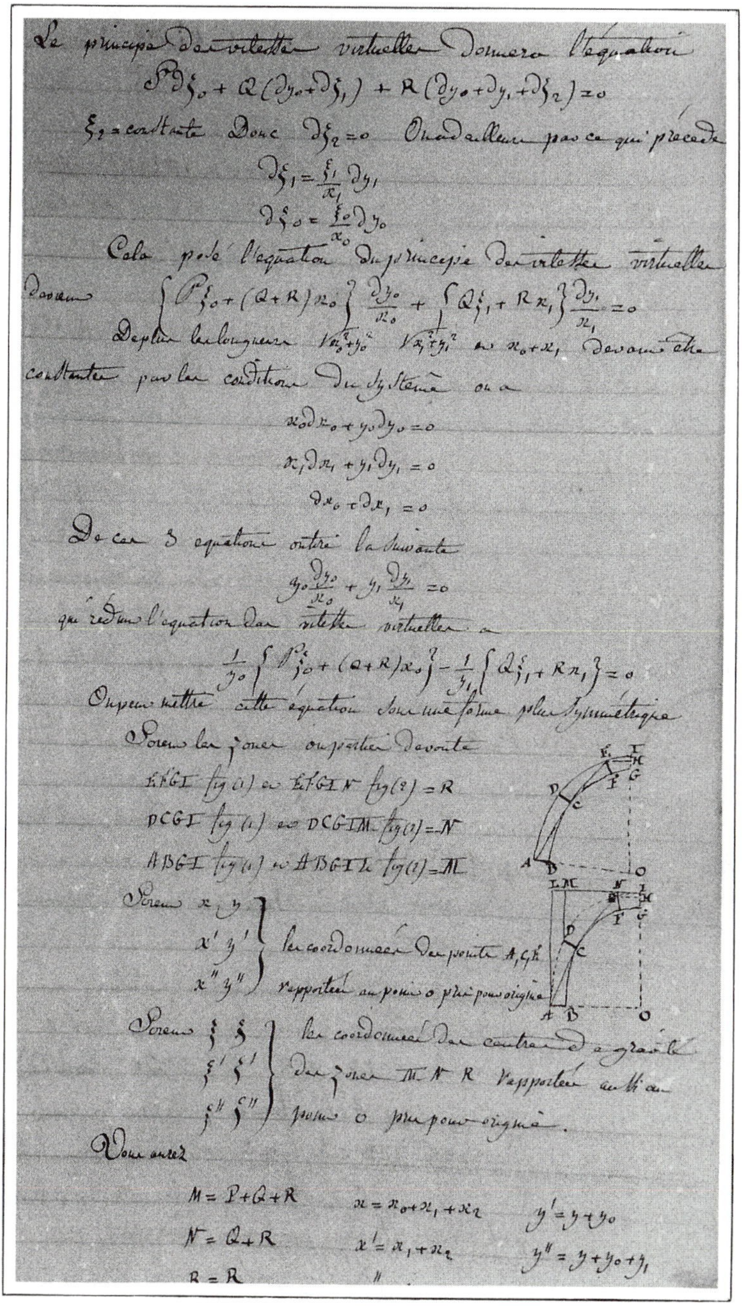
Le principe des vitesses virtuelles donnera l'equation

$$P\,\delta\xi_0 + Q(\delta y_0+\delta\xi_1) + R(\delta y_0+\delta y_1+\delta\xi_2)=0$$

ξ_2 = constante donc $\delta\xi_2=0$ On a d'ailleurs par ce qui précède

$$\delta\xi_1=\frac{\xi_1}{x_1}\delta y_1$$
$$\delta\xi_0=\frac{\xi_0}{x_0}\delta y_0$$

Cela posé l'equation du principe des vitesses virtuelles devient

$$\{P\xi_0+(Q+R)x_0\}\frac{\delta y_0}{x_0}+\{Q\xi_1+Rx_1\}\frac{\delta y_1}{x_1}=0$$

De plus les longueurs $\sqrt{x_0^2+y_0^2}$ $\sqrt{x_1^2+y_1^2}$ et x_0+x_1 devant être constantes par les conditions du système on a

$$x_0\delta x_0+y_0\delta y_0=0$$
$$x_1\delta x_1+y_1\delta y_1=0$$
$$\delta x_0+\delta x_1=0$$

De ces 3 equations on tire la suivante

$$y_0\frac{\delta y_0}{x_0}+y_1\frac{\delta y_1}{x_1}=0$$

qui reduit l'equation des vitesses virtuelles a

$$\frac{1}{y_0}\{P\xi_0+(Q+R)x_0\}-\frac{1}{y_1}\{Q\xi_1+Rx_1\}=0$$

On peut mettre cette équation sous une forme plus symmetrique

Soient les zones [illegible]

EFGI fig(1) ou EFGIN fig(2) = R
DCGI fig(1) ou DCGIM fig(2) = N
ABGI fig(1) ou ABGIL fig(2) = M

Soient x, y / $x'\ y'$ / $x''\ y''$ les coordonnées des points A, C, E rapportées au point O pris pour origine

Soient $\xi\ \zeta$ / $\xi'\ \zeta'$ / $\xi''\ \zeta''$ les coordonnées des centres de gravité des zones M N R rapportées aussi au point O pris pour origine.

Nous aurons

$M = P+Q+R \quad x = x_0+x_1+x_2 \quad y' = y+y_0$
$N = Q+R \quad x' = x_1+x_2 \quad y'' = y+y_0+y_1$
$R = R \quad x'' \ldots$

Second Mémoire sur les Ponts en Pierre. Théorie des Voûtes en Berceau, by Cauchy. Manuscript, 1809, ENPC Library, ms 19. Published by permission of the École Nationale des Ponts et Chaussées

good families' together in prayer and unite them against the menace of the current faithlessness, irreligion, and secularism (52). Its members met every two weeks in the rue Saint-Guillaume. They were under a duty to render each other mutual assistance and charity according to the Congrégation's motto: *cor unum et anima una.* In the beginning, the membership of this society numbered only six youths, but it grew rapidly, so that by 1804 it could claim 198 members. Young ecclesiastics flocked to its meetings, as did many young aristocrats, often scions of the best families, and a considerable number of young students, such as Hennequin in law, Laënnec in medicine, and Teysseyrre in the sciences. Father Delpuits' aim was to have members of the Congrégation scattered liberally about in positions of importance in society and in government. He was particularly desirous that his youthful adherents would be among the intellectual elite.

In a very natural way, then, the Congrégation undertook to infiltrate the École Polytechnique, a veritable hotbed of secular minded scientists and liberals. The infiltration of the Ecole was Teysseyrre's mission. Teysseyrre was as busy trying to convert the souls of his young charges as he was trying to educate their minds. He gathered about himself a number of Catholic youths, among them Augustin-Louis Cauchy. Teysseyrre was an unusual person with a strong, ardent personality, who soon converted young Jacques Binet into a Congregationalist. Jacques Binet, who was Cauchy's personal friend and who would later become a notable mathematician in his own right, had heretofore been a complete atheist. Similarly, Teysseyrre exerted a strong influence on the young Lamennais and left a lasting impression on Cauchy, even though Teysseyrre would be at the school for only one year. For his own part, Cauchy became a Congregationalist a few months after he left the École on April 3, 1808.

Little by little, this ostensibly religious society spread its influence and, in due time, became suspect in the eyes of the authorities. As earlier noted, the Congrégation had begun as no more than a prayer group, a purely religious organization that no doubt would have been left unmolested by the State had it remained what it was at its inception. But, that was not to be, and the Congrégation increasingly became a gathering of young people who harbored royalist notions and who were, therefore, opposed to the Empire. The growing quarrel and bitterness between the Papacy and Napoléon Bonaparte pushed the Congregationalists into determined opposition to the Empire. Cauchy's father, of course, had been a devoted royalist. But once the Great Revolution had swept away the Old Régime, Louis-François had consistently played a very prudent hand under the Directorate, and when Bonaparte staged his stunning coup d'état of 18 Brumaire, Louis-François had promptly rallied to Bonaparte's cause. Furthermore, his position as Secretary of the Senate and the verses he had occasionally written in praise of the Emperor now bound the elder Cauchy more closely than ever to the present regime. However, being billeted at the École Polytechnique had the effect of freeing the younger Cauchy from any such concerns. Furthermore, the repressive measures taken by the government against the Papacy and against anyone in France who dared support it made the Empire, the reign of the Usurper, become more and more odious to Augustin-Louis. By the time the Congrégation managed to resume its meetings in 1813—this time very discreetly—Cauchy had returned to Paris from Cherbourg and was once again able to take part in these almost clandestine activities.

The four years spent at the École Polytechnique were undoubtedly of decisive importance to the intellectual development of young Augustin-Louis Cauchy. Not only had he mastered a very impressive amount of mathematics, but he had also acquired a set of political and religious convictions to which he faithfully adhered for the rest of his life. Unfortunately, we know little about the influential events that took place during this period. For example, did he already have plans to embark on a career in science? We do not know for sure. However, the enthusiasm he showed upon leaving Paris to work in Cherbourg at the beginning of 1810 rather suggests that even then he had dreams of a brilliant career in engineering, a career that had been made altogether possible by his successes at the two engineering institutions.

Chapter 2

Sojourn at Cherbourg

Having completed the required two years at the École des Ponts et Chaussées, Augustin-Louis Cauchy was appointed a junior engineer (aspirant-ingénieur) on January 18, 1810, by Count Molé, the Directeur Général of the Ponts et Chaussées (1). Chosen from among the engineering students who had completed their program of study, the 15 appointees were given all the duties and responsibilities of full-fledged engineers, except for the title and the pay of an engineer. A few weeks after his appointment, in February 1810, Cauchy was assigned to Cherbourg, where he was to assist the engineers responsible for the excavation and construction of Port Napoléon (2). Placing a copy of Laplace's *Mécanique Céleste* and a copy of Lagrange's *Théorie des Fonctions Analytiques* at the bottom of his trunk, he barely had time to pack before he had to leave on the stagecoach for Cherbourg. Copies of *Virgil* and the *Imitation of Christ* under his arm, he left for his first real position in March (3).

There was the excitement of new experiences and a sense of expectation: the reading of the two classics that he had brought along; the meditations and profound thinking to which he could abandon himself with such ease; the pleasant excitement of discovering the countryside, and the positive, confident reflections stirred his naturally inquisitive mind. The conversation he had with his fellow travelers during the long stagecoach ride and the exciting diversity of France's regional cultures and architectural styles could not but have helped to relieve the sadness of leaving home. Perhaps he would be away from his family, from his parents, for a long time. Now, for the first time in his life, he was leaving the city of Paris for the provinces.

Could he measure up to the confidence that Count Molé had shown in him by assigning him to Cherbourg, the site of the largest and most important construction project in all the Empire? A vast engineering effort under the direction of some of the most outstanding engineers that the École des Ponts et Chaussées had produced! For more than 30 years, there had been a fierce desire to construct a first-rate naval facility there. As the inspecteur général of the Ponts et Chaussées, Joseph-Marie-François Cachin, noted in 1803:

> There has always been a deep-felt need for a Channel port that would guarantee a place of safety for our naval vessels. But, it was especially during the disastrous era of the Battle of La Hougue that the government realized the necessity of establishing a great naval facility for France on this region of the coast. Since then, the roadstead of Cherbourg, which is at the edge of the Peninsula known as Cotentin, has seemed to be the most favorable place for such a facility. The fact is that its forward location on the sea lanes through the Channel offers everything we need either for maximum surveillance of the enemy or for harassing its convoys or, finally, for assembling all the details necessary for a major military expedition. The Cherbourg roadstead, with its excellent anchor hold, is equally favorable for the arrival and departure of vessels, no matter the winds and tides. This location has vast anchoring and docking space and is susceptible to all kinds of methods of attack and of protection and defense. Finally, by all military and maritime accounts, this location has all the advantages that can influence the fortunes of our naval forces and commercial relations (4).

On June 23, 1786, in the presence of King Louis XVI, construction began on an enormous dike enclosing the roadstead from the seashore. This project, begun some three years before the outbreak of the Revolution, was completed in 1806. During the Revolution, an arrangement was envisioned whereby within the Cherbourg roadstead there would be a port of careenage and vessel refitting separate from the commercial harbor; and, under this arrangement, there was even the possibility that there would be an arsenal. However, it was not until the establishment of the Consulate that it was decided on April 15, 1803 to construct a naval base at Cherbourg, an entirely new port that would be distinct from the old city. The various projects as originally outlined by Joseph-Marie-François Cachin comprised 'a forward port, as well as a port capable of containing a dozen warships and a proportional number of frigates, along with the dry docks, no less than all the fortifications necessary to shelter the port from the enemy' (5). A little later, on September 26, 1804, the First Consul personally decided that 'the plans be modified so that a huge naval facility capable of holding thirty front-line vessels that would be kept in a constant state of readiness' could be built around the inner basin, according to Cachin's plans. However, this phase of the work was not completed until after 1830.

Cachin began construction of the new harbor, which was first called Port Bonaparte and then Port Napoleon, at the end of April 1803. There was much to be done. In fact, it was necessary to dig into rock in order to construct the forward harbor and refitting docks with sufficient depth to maintain a fleet at low tide. At any given time, there were 2000 to 3000 laborers working on this construction site; some were soldiers, but by far the greater number were prisoners of war. Conditions were so bad at the Cherbourg project that some

20,000 prisoners died during the 15 years of work; and, according to Cachin, about 2000 died on the job (6). Clearly, so vast an undertaking required that new procedures be developed and original techniques devised. Thus, for example, two steam machines were installed to work continuously at draining off the seepage (7).

By the time Augustin-Louis was assigned to Cherbourg, work on the forward port had already been carried forward to an advanced state. Cauchy had barely arrived in Cherbourg before he became involved in the direction of the construction site; and, it would seem that he found this responsibility to be in no way disagreeable. In fact, on June 8, 1810, he wrote to his father that 'the project at Port Napoleon is more and more important and my job assignments are very instructive' (8). For his own part, Louis-François, from afar, remained ever watchful of his son's future and, on July 9, 1810, asked Count Molé to allow Augustin-Louis to remain at Cherbourg since '[he] found the work there to be very instructive' (9). On December 13, 1810, an imperial decree appointed Augustin-Louis a second-class engineer-ordinary attached to the project at the port of Cherbourg.

Throughout his sojourn at Cherbourg, Augustin-Louis worked on almost all phases of the project: the excavation of the basins for the military port by blasting through schistose rock (10); the construction of the two piers that formed the entrance to the port and the construction of the encircling dikes joining the one on the north to Fort du Homet; the draining of the water that seeped into the harbor across the cofferdam that temporarily barred its way; the construction of a row of signal posts along the shore; and the building of a lock for the dike enclosing the roadstead. These were all considerable undertakings for the young engineer. But, in addition to these, he devoted time to various other construction projects, such as shelters for the prisoners; barracks for the troops on the site, forges, and covered buildings for the granite piers on which the future arsenal would be located. Cauchy described his work during those days at Cherbourg in a letter cited by Valson. Dated July 3, 1811, the letter reads:

> I get up at 4 o'clock each morning and I am busy from then on. My usual work load has been increased over the past month by the unexpected arrival of Spanish prisoners of war. We were told of their coming only a week before they arrived, and within those few days' time, we had to build lodgings and bunks for 1200 men. The buildings in which these prisoners are housed consist of two structures, each containing 19 rooms for a total of 38. When the prisoners arrived, 12 rooms had already been prepared, and we had to build another 12, the walls of which had not even been completely raised. We had no roof tiles for these buildings, and I had to visit the local quarries to try and get some. We had no kitchens, and I had to put together some temporary stoves. At last, two days ago all the Spaniards were set up in lodgings. They have bunk beds and straw mattresses and are given food. They count themselves as being quite

well-off. I had no sooner finished the job of lodging the prisoners of war than I had to get busy on another assignment. Today, I drew the plans for forges that I am to have built in granite. I am also constructing two lighthouses, one on each of the two piers that are located at the entrance to the harbor.

I do not get tired of working; on the contrary, it invigorates me and I am in perfect health...(11).

Augustin-Louis' performance and enthusiasm for his work certainly pleased his superiors. His real scientific investigations—investigations that were begun at Cherbourg and as we will see, represent his first breakthroughs in this area—had the effect of raising him in the esteem of the other engineers on the Cherbourg project. He seems to have been presented to the Emperor, with a very flattering introduction, when Napoleon and the Empress Marie-Louise visited the Cherbourg construction site in May 1811. Be that as it may, it is certain that after Cauchy had left the project Cachin addressed his former subordinate in particularly warm terms when he informed him of the grand opening of the forward port by the Empress on August 1, 1813.

So it is done, my young friend. The ocean has taken possession of our works... and, just think of it—you were not even here! Your absence made me feel rather out of sorts; however, hopes for your happiness lifted my spirits. I hope that soon, when I get a chance to see you, it will be possible to tell you of my wishes that you will never stop being a member of a group that regards you as one of its most distinguished members, a group among which you will always find true friends. I myself am happy to be among that group and to offer you, no matter the circumstances, testimony of my unalterable devotion (12).

At Cherbourg, Augustin-Louis not only had to supervise teams of men, make decisions, endure the hardships of the construction site, and spend long hours outside in all kinds of weather, he also had to learn how to move about in polite society, how to form and cement the types of social relations his new position demanded. Certainly, being barely 20 years old and unmarried, he would have to establish a good, solid reputation for himself. But this would not be easy for this rather frail young man, a youngster who, though he might have the perfect drawing-room politeness and polish he had learned in the salons of the Palais du Luxembourg, was nevertheless formal and too cold. On the recommendation of his parents, he had been received at the home of Marie-Charles-César Fay, Count of Latour-Maubourg, then the leading citizen of Cherbourg. The count had been appointed to the Senate by Napoleon, with whom he had allied himself upon returning from his stay abroad as an 'émigré'. Returned to France and in the Emperor's good graces, Latour-Maubourg was now commander of the Cherbourg Military District. Thus, without any effort on his own part, Augustin-Louis was introduced into the small, tightly knit circle of notables in the port city. This privilege, an honor that many others might have been very happy to have, soon began to appear to Augustin-Louis as tedious. A friend of Lafayette, the Count was a liberal and a 'philosophe', a man who was indifferent to religion and whose company offered no comfort to the young Congregationalist. Cauchy's interest soon flowed toward people who shared his piety, his studiousness, and his general serious-mindedness:

> The homes I visit in Cherbourg, and which I naturally should visit, are those of my superiors: the homes of M. Cachin, of M. Franqueville, and of MM. Duparc and Vallot. And I am well received at each of them. I cannot get out of seeing M. L..., to whom I am so obliged. At these homes, particularly that of M. Franqueville, one meets people from all the various groups in the city. Among them are some in which religion is honored and respected; there are others, however, in which the only thoughts are of amusement. I am very closely connected with some of the former and the time I spend there is not wasted. I visit some four or five of these homes. They are very closely connected to each other and to those of MM. Cachin, Franqueville, and Duparc. I also associate with a few other persons whose acquaintance should prove useful, such as M.D...., a cleric now returned from England; M.G...., the headmaster of the high school; and M.V...., who is very able in mathematics (13).

Quite a few people were antagonized by Cauchy's austere behavior and his cool exposure of his religious sentiments. Moreover, the slowness he showed in calling on those whose ideas he judged to be different from his own was especially offensive. No young man should be so pensive and rigid. So, the gossiping began. Finally, echoes of the criticisms that were leveled against him reached his parents, probably through the offices of Count Latour-Maubourg, and Augustin-Louis had to defend his conduct:

> So they're claiming that my devotion is causing me to become proud, arrogant, and self-infatuated. Exactly who is making these claims? Not people who have much religion themselves; for the people that I have talked with have all urged me to follow my usual code of conduct, and everything that has been reported to me (about these people) tells me that they do not blame me. However, a few days ago, a certain person in the... Society told me in a friendly way that religion often makes young people self-infatuated. I talked with her for a while on the subject and showed her that I was not self-infatuated. Now, as for those people who have no religion themselves, I have decided never to discuss it at all with them; and I will only reply (to them) when they attack me on the matter. Thus, when I arrived in Cherbourg, M... decided one day to tell me, while we were talking about religious obligations, that I would soon get over all such notions. Without getting upset, I told him that when a man was engaging in wickedness and wrongdoing, he would do well to get over it quickly; and then I asked him if he found anything in my conduct

> that was wicked or harmful. Another person who was present overheard what was being said and, being of my mind on the matter, took my part. After making a bit of uneasy small talk, the gentleman ended his remarks with a great deal of politeness and said nothing more to me about religion. I am now left alone about religion, and nobody mentions it to me anymore except to encourage me to continue with what I believe in. There are a few philosophes who say that religion has made me self-infatuated. But, for my own part, I am truly happy that in a country where one hears so much gossip and where certain people keep busy slandering and picking from dawn to dusk, they have not found other grounds upon which to reproach me (14).

Augustin-Louis was basically a solitary person, a man who was more attracted by nature's spectacle than by the doings of human beings. Whenever his work left him time to do so, he would devote himself to such thoughts, thoughts that would later prove to be the source of his most abstract mathematical discoveries. But, right now, of course, Cauchy's meditations on nature were quite far removed from any kind of learned scientific investigations, rather they were indicative of his own personal style. In a letter to his parents, he stated:

> From time to time I get a moment to relax, and I use it to make little walks around Cherbourg. There are some very scenic spots hereabouts, one of the most picturesque being the Quincampoix Valley. There is a small river that twists its way through this valley and on whose banks are several factories. The river winds its way across a magnificent prairie, which is filled with flowering apple trees. The clear waters of this little stream can only be seen through a dense growth of green bushes and shrubs, which seem to form a kind of cradle about the river. On one side, the meadow is bordered by great patches of fern and plants of every kind; on the other, by immense rocks and cliffs whose peaks rise sharply into the air or hang suspended over the dale.

Cauchy, of course, was not content to observe the sights of the surrounding countryside in a passive way. He wanted to understand, to explain, to classify. So, since his father had taught him much about plants during his childhood, he began to collect various plants. Changing its tone, Cauchy's letter describing the countryside around Cherbourg continued:

> Among the plants that are found in abundance in this part of the country, I have noticed foxglove, white orchids with tiger stripes, orchids with red petals, various thick-leaved plants, and a great deal of ferns, called polypode, which cover all the walls of the city like a hood. There is a treelike fern with shiny, varnished leaves that split up into many smaller ones; there is another fern that looks much like the one I just described, except that it is not treelike; there is still another species that resembles the two preceding ones except that it is rather short; then another kind of

> fern grows in a rampant, random way; and, finally, there is the type a few leaves of which I sent to you in an earlier letter. I must also tell you that while walking I have come across some uncommon insects, such as the 'cardinale' [red dragon fly?] and the blue stag beetle. So, you see, I hardly need leave the city in order to have all the delights of the countryside (15).

When he described to his parents how the sea looked during a storm, perhaps thinking of the awful tempests that had twice damaged the great dike at Cherbourg—once on November 2, 1810, and then shortly thereafter on the night of November 10–11, 1810—he sought to give an analytical explanation of how the billows swelled and rolled as they approached the shore rather than to present a strictly visual portrayal of the motion.

> Vernet's pictures can give some idea of how it looks when it is calm, but cannot possibly describe it when it is disturbed. They cannot show how the wave, after having dashed against the rocks, angrily withdraws, receding only to come again to the point whence it had departed, more furious and more terrible than before. They cannot show how, at the meeting of swells that have been hurled back by the rocks with those rolling in from the open sea, each particular wave rises up and then smashes against the shore, where it leaves a long trail of foam (16).

This way of looking at things, essentially the scientist's way of viewing nature, was presented more fully in a lecture that Cauchy gave before the Cherbourg Academic Society on November 14, 1811 (17). In this address, he used bold strokes to paint the portrait of a thinker, a scholar–scientist, who is trying to penetrate the secrets of nature:

> He fixes his gaze on the earth on which he stands and maps out a design of it with the same ease that he would use in mapping out a design for his own garden. With a sure and steady hand, he sketches the courses of the rivers and the contours of the oceans. He measures the heights of the mountains and the depths of the abysses. He notes the plants that cover the globe and the beasts that dwell among them, from the moss that grows on stones to the Cedars of Lebanon, and from the delicate little shellfish scampering along beneath the seas to the gigantic elephant possessed of a tremendous strength. Even the bowels of the earth seem no longer to posseses anything that can be kept hidden from him. He questions. Armed with mining tools, he sets forth to discover the hitherto unheard-of minerals that lie within the bowels of the earth. He lifts his gaze from the land to the sky. His thoughts soar and embrace the general system of the universe. He marvels so as thus to articulate the secrets of nature. From the spot where he is placed in the great vastness of the world, he measures distances, though he himself cannot run their course. He plots the paths of the stars that surround him and studies their influence on the seasons and on the climes and on the tides. On his scales,

> he weights the moon and the planets and then tells the comets the days on which they shall once more visit them.

Clearly, as the little of his address suggests, he was trying to give a precise definition of the 'view of things', which 'measures' and 'marvels so as thus to articulate the secrets of nature' in order to determine their proper limits:

> If we observe that all our intelligence and all our skills and devices are contained within certain bounds that cannot be cast aside, we can convince ourselves without too much trouble that our knowledge is as limited as our senses.

Although the bare words of Cauchy's address might ostensibly suggest that he was giving an impersonal detached description of the 'general' scholar–scientist, we suspect that he, a heretofore almost unknown young engineer, had completely dedicated himself to science and was, in fact, describing himself to his audience.

Whatever may have been his interest in the natural sciences, he spent his spare time at his desk studying the exact sciences. As earlier noted, Cauchy had taken Laplace's *Mécanique Céleste* and Lagrange's *Théorie des Fonctions Analytiques* to Cherbourg with him. Other works were available to him either through his father who sent them from Paris or by borrowing them from various libraries in Cherbourg. At first, he had planned to 'make a coherent study of all branches of mathematics, starting with arithmetic and proceeding to astronomy, clearing up obscure points as well as possible, working on simplifying proofs, and trying to discover some new propositions' (18). Soon he became involved in his own research. On Lagrange's advice (19), he first attacked a problem in pure geometry that had been posed two years earlier by Poinsot.

In a paper presented to the Institut on July 24, 1809, Poinsot had established the existence of three new nonconvex regular polyhedra: two dodecahedra and one icosahedron. These were now added to the five regular convex polyhedra that had been known since antiquity and to Kepler's star-shaped dodecahedron. But Poinsot had been unable to show that no other regular polyhedra existed. 'This', he wrote, 'is a question that deserves to be investigated; and it seems to me that it is one that will not be easy to solve in a fully rigorous way'. The Institut received Cauchy's study, 'Recherches sur les polyèdres', on February 11, 1811 (20). Employing a development that generalized the methods Poinsot had used in constructing all the star-shaped regular polygons and his three nonconvex regular polyhedra, Cauchy dealt with Poinsot's problem in the first part of his study. The second part was devoted to the Euler Formula

$$S + F = A + 2 \tag{2.1}$$

on the number of faces F, edges A, and vertices S of a polyhedron. Cauchy pointed out the equivalence of the Euler formula Eq. (2.1), to the relation

$$S + (F - 1) = A + 1 \tag{2.2}$$

between the number of sides A and vertices S of the network of $F - 1$ polygons, which can be obtained by cutting out one face of the polyhedron and projecting the other faces on the cut face. Thus, from a proof of Eq. (2.2), he deduced a proof of the Euler formula Eq. (2.1). Then, Cauchy generalized the Euler formula to the case of a network of polyhedra. He obtained the relation

$$S + F = A + P + 1 \tag{2.3}$$

on the number of polyhedra P, faces F, edges A, and vertices S of the network; the Euler formula is given by the special case $P = 1$ in Eq. (2.3). The commission whose responsibility it was to evaluate the study was composed of Legendre and Malus, and its report on Cauchy's research was very favorable. 'The proofs', observed Malus' report on May 6, 1811, 'are rigorous and developed in an especially elegant way' (21).

Acting on the advice of Legendre and Malus, Cauchy undertook further studies on polygons and polyhedra, and on January 20, 1812, he presented another paper to the Institut (22). In the first part of it, he established eight theorems on the variations of the angles of rectilinear and spherical convex polygons. These basic results and the Euler formula, Eq. (2.1), were used in the second part of the study to give a proof by reductio ad absurdum of theorem 11 in Book 9 of Euclid's *Elements*. According to this theorem—a result that had not theretofore been proved—two convex polyhedra whose faces are equal and are similarly placed are necessarily equal, either by symmetry or by superposition; consequently, a polyhedron with rigid faces is completely rigid (23).

This last result made a strong impression. At first, Cauchy had a somewhat difficult time of it in trying to convince Malus of the validity of his proof by reductio ad absurdum.

> If M. Malus seemed not satisfied with the proof I sent to you [wrote Cauchy to his father who was acting as intermediary] it probably has to do with the fact that you did not advise him of what I had taken care to tell you; namely, that my proof rests on several lemmas that are easy to prove. It does not, therefore, surprise me in the least that M. Malus has concluded that I assumed things that could not be assumed. But, that is not the question: if I had the time, I should have sent you the proofs of the lemmas I used. Today, I will reduce the question down to the matter of knowing whether or not my proof is acceptable, assuming the lemmas are established. As to the form of proof that I used, I think it would be not only difficult to change it, but downright impossible. The reason is that until now a geometric argument has not been given, except in terms of reductio ad absurdum, of the theorem that in 2-dimensional geometry is analogous to the one in 3-dimensional geometry that I dealt with: I

Title page of the manuscript of the paper on the polygons and polyhedrons. January 20, 1812. Published by permission of the École Polytechnique.

> mean the theorem by which it is proved that two triangles are equal if their three sides are equal. If one should establish this latter theorem without using either trigonometry or reductio ad absurdum, I would agree that my proof ought not be admitted. It thus seems impossible to banish the reductio ad absurdum proof from geometry; and this is particularly true in the present case. In fact, in order to prove that under certain conditions only one polyhedron can be constructed, it is necessary to see that after the first figure has been constructed subject to the given conditions then one cannot construct a second figure without encountering a contradiction. I insist on this argument because the type of proof I gave seems to me to be inherent in the nature of the theorem in question. Moreover, it is precisely what M. Legendre used in establishing several particular cases of the same theorem (24).

Malus died on February 24, 1812, and thus did not participate in the commission that evaluated the study. This commission, composed of Legendre, Carnot, and Biot, gave a very glowing report on February 17, 1812. Written by Legendre, the report concluded:

> We wanted to give only an idea of M. Cauchy's proof, but have reproduced the argument almost completely. We have thus furnished further evidence of the brilliance with which this young geometer came to grips with a problem that had resisted even the efforts of the masters of the art, a problem whose solution was utterly essential if the theory of solids was to be perfected (25).

Cauchy's first research on polyhedra demanded more mathematical virtuosity than mathematical knowledge per se. Strangely enough, he seems not to have attached much importance to mathematical works of this kind, works that were destined to make him famous. At his lecture on November 14, 1811, he declared:

> What can I say about the exact sciences? Most of them seem to have already been carried forth to their highest stage of development. Arithmetic, geometry, algebra, and higher mathematics are sciences that can rightly be regarded as having been completed, as it were; and nothing more remains to be done with them except to find new areas of useful applications (26).

This opinion was widely held in France's scientific community at that time. Most decidedly, the future seemed to belong to the engineering sciences (27).

Such pessimism could hardly have fired Cauchy with a desire to persevere in the mathematical sciences. However, his research on polyhedra had given him a reputation in Paris. On February 22, he was nominated a corresponding member of the Société Philomatique. Founded in Paris in 1788, the Société Philomatique was patterned after the old Académie des Sciences, with regular members and correspondants. The society published the *Bulletin*, which contained scientific information and summary papers of new works communicated by its members. On March 21, Poisson, who was the mathematical editor of the *Bulletin*, presented Cauchy's paper on polyhedra to the society. In a report written by Biot and himself, Poisson made a recommendation favoring Cauchy's nomination to the Société Philomatique (28). But, Cauchy aimed far higher. His object was the Institut itself. After all, had not Legendre himself declared, following the research on polyhedra, that Cauchy should get a seat in the First Class of the Institut (29)? To reach such a goal, however, he needed another spectacular result, something like his proof of Euclid's theorem. According to a letter Louis-François wrote to his son, the young man should undertake investigations into Fermat's theorem on polygonal numbers. Addressing his son, Louis-François wrote:

> Your last paper on polyhedra made a deep impression on the Académie. If you prove one of Fermat's theorems, the way will be wide open for you. The moment is favorable for you. Do not let it slip by (30).

So, Cauchy began work in this area, work which would last for several more years.

The third research paper Cauchy sent from Cherbourg dealt with a quite different area of mathematics, the theory of the directrixes of conic sections. Unfortunately, this work was a failure in the sense that it was neither published nor read to a learned body. The reason for this seems to have been that this paper contained no new results at all; and we only know of its existence by way of a letter written by Augustin-Louis. Cited by Valson, this letter is undated, but was no doubt written sometime during the summer of 1812:

> It would seem that all my theorems have the good fortune of being found, on the first reading, to be either false or already known. But, might it not be with the result on the directrix of an ellipse as it was with the theorem on polyhedra? Though I lost my argument the first time, is it not possible that if another look be given to the matter, the theorem will indeed be found to be a new result? In any event, do not think that I attach great importance to that theorem. However, as I have gone through all the books on second-degree curves and have found nothing in them, I at least ask that you should find out from M. Poisson the title of the work where the theorem I discussed can be found (31).

Augustin-Louis was already contemplating his return to Paris when he wrote this letter, for he added:

> Moreover, I have some other, more interesting, theorems to work on; and I already have the material for one or several papers that I will write when I get back to Paris.

He had to return to Paris if he was to continue his research under favorable conditions and become situated in a position in the educational establishment as a professor, examiner, or academician. Such a position would leave him enough time for his mathematical pursuits. Besides, everything was happening in Paris; Paris was not only the political capital of the Empire, it was the scientific and intellectual capital of Europe.

From his father's connections with leading people in the capital and his growing reputation as a mathematician, Cauchy could well expect to get a position as an engineer—until something better came along—or as a teacher somewhere. But, matters did not wait for the changes he was contemplating. Badly overworked, he fell ill. He was no doubt physically exhausted by the harshness of the climate and the working conditions at the construction sites. But, even more than that, he had developed an acute case of strained nerves. The intensive intellectual activity, the long evenings at his desk piled high with books and papers, and the hours he spent tutoring youths who wanted to go on to the various schools of special study (32) served to undermine his morale, as well as his physical health. That he should have become exhausted at this time is all the less surprising when we consider the importance of the research work that he undertook while at Cherbourg, a place where conditions approached isolation, and, in the meanwhile, still managed to conduct his engineering work at the highest level. His research, of course, included not only those investigations that had been presented before the various learned bodies of the time, but also studies that he had undertaken on permutations and, no doubt, some of the material that, in the years to come, would go to make up his treatises on algebra and analysis.

We have no information about the exact nature of Cauchy's illness, and Valson's biography is unusually vague on the subject, mentioning only that 'Cauchy's health had never been particularly robust', and when his mother came to Cherbourg after he had taken ill, she 'found him in a weakened and depressed condition, which threatened to cut short this very promising career' (33). One point, however, is a letter dated January 12, 1814, in which Louis-François declared to Baron Costaz, the Directeur Général of the Ponts et Chaussées, that his son's health was 'profoundly altered by the heavy burdens he had to bear during the three years he spent at Cherbourg' (34). Both remarks are somewhat misleading, for the illness seems to have been as much emotional as organic, with psychological overtones that were as pronounced as the physical ones. All available information on Cauchy's personality tends to support this hypothesis. His cold, aloof, and inflexible personality was poor cover for his deep sensitivity. At once easily hurt and hotheaded, he was a loner and a person who pushed himself excessively. In his personal life, as in his work, he was both volatile and stubborn. Moreover, the view that this illness had psychological underpinnings is also supported by certain features of the malady itself. An ill-defined sickness, it lasted, off and on, for about a year. This long period of convalescence was periodically interrupted by a few weeks during which Cauchy was able to work. Numerous other facts point to the conclusion that throughout his life Augustin-Louis Cauchy suffered from a nervous condition.

During the summer of 1812, his condition gradually grew worse. His parents were greatly disturbed by his health. Towards the middle of September, they received a letter from Cauchy's father confessor, informing them that their son was planning to be married. Madame Cauchy, by now very anxious, immediately left Paris with her second son, Alexandre. She arrived in Cherbourg on September 24th and, finding her son in very poor health, took him back to the rue de Tournon (35).

Altogether, Cauchy had stayed in Cherbourg for close to three years. During his time there, he had learned to work as an engineer and had performed his assigned tasks excellently. But, his particular disposition, interests, and abilities had, by now, inclined him in another direction. Burdened with poor health and always more attracted by abstract questions than by concrete problems, he seems to have become increasingly disinterested in the goings-on at the construction site. At the same time, however, he was putting more and more of himself into his mathematical studies and had won his first laurels. He used his spare time to broaden his knowledge of pure science. A good deal of fruitful thoughts and ideas probably took root at this time, although it would still be a number of years before they would be fully developed and written in treatises and papers for posterity. During the years in Cherbourg, he also made his first scientific discoveries. They were sufficiently important to attract the attention of learned society back in Paris. Henceforward, the die was cast: his whole life, to use one of his favorite expressions, would be devoted to 'the search for truth'.

Chapter 3

The Waiting Room of the Academy

At home in Paris, back with his family, Augustin-Louis recovered his health. What he seemed to have needed most was the affection, closeness, and understanding that were typical of the Cauchy family. During this period, he worked contentedly on his new papers. First, he perfected his two-part memoir on symmetric functions, which he presented to the First Class of the Institut on November 30, 1812; the memoir was published in two articles in the beginning of 1815 (1).

In this important study, Cauchy created the calculus of substitutions, a topic that he would pursue in much more detail during 1845–1846 and would, in the years to come, play a major role in the development of group theory. Interestingly enough, in 1815, Cauchy seems to have had no real interest in the problem of the solvability of equations by radicals, although this avenue of investigation was the mathematical framework in which mathematicians had regarded permutations. Cauchy began the investigations that were to become the two-part study while he was in Cherbourg. At that time, he was working on number theory, especially on Fermat's theorem on polygonal numbers; and this research was to continue until 1815. In connection with his work in number theory, he made a careful study of Gauss' *Disquisitiones Arithmeticae*, which had been published in 1801 and translated into French in 1807. He seems to have mastered very quickly the methodology and sense of Gauss; and grasping the importance of the theory of forms that Gauss had used in proving Fermat's theorem on triangular numbers, he managed to simplify it and to generalize some of the results, particularly the ones on discriminants.

These investigations on number theory, he declared in the first article, led him to work on the 'theory of combinations' and from there to prove a theorem that was more general than the results obtained by Lagrange and by Ruffini on the number of values of a function of n given quantities: namely, if the number of distinct values assumed by a function of n quantities is less than the largest prime factor p of n, it is less than or equal to 2.

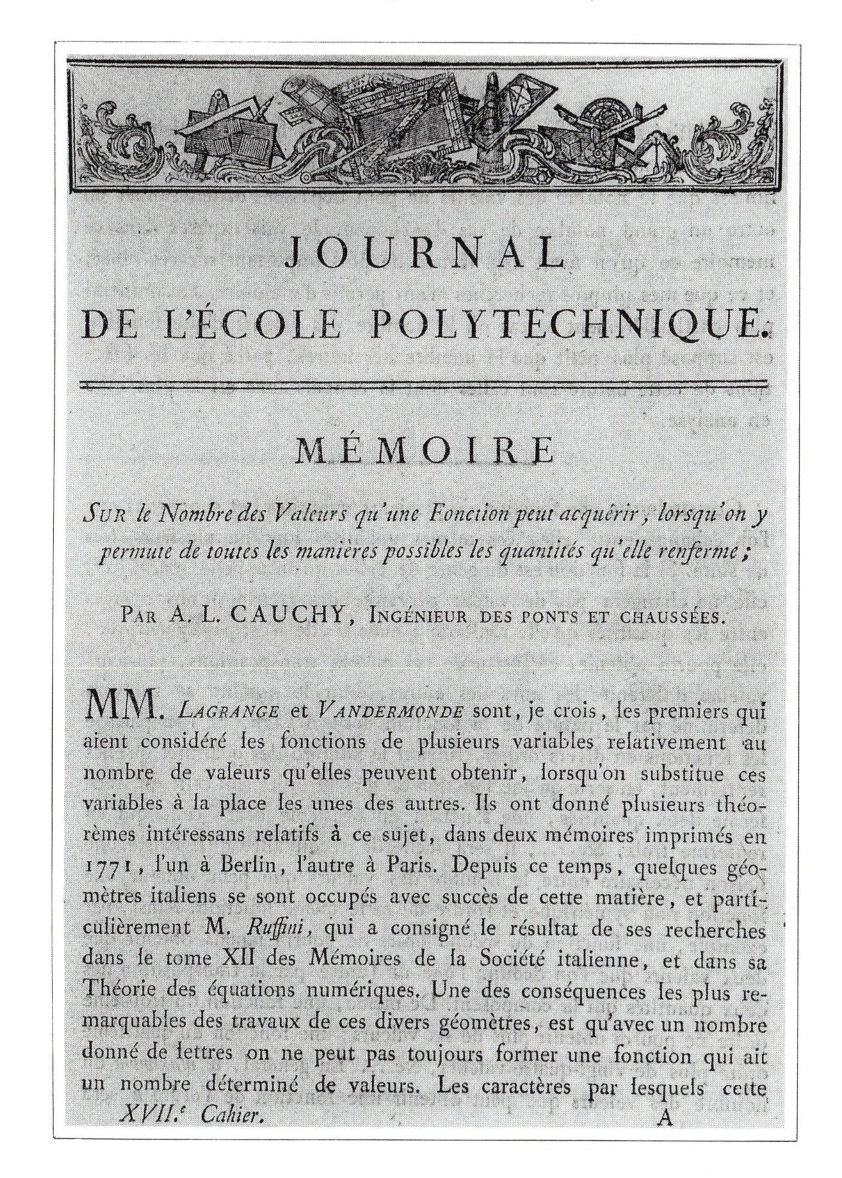
JOURNAL

DE L'ÉCOLE POLYTECHNIQUE.

MÉMOIRE

Sur le Nombre des Valeurs qu'une Fonction peut acquérir, lorsqu'on y permute de toutes les manières possibles les quantités qu'elle renferme;

Par A. L. CAUCHY, Ingénieur des ponts et chaussées.

MM. *Lagrange* et *Vandermonde* sont, je crois, les premiers qui aient considéré les fonctions de plusieurs variables relativement au nombre de valeurs qu'elles peuvent obtenir, lorsqu'on substitue ces variables à la place les unes des autres. Ils ont donné plusieurs théorèmes intéressans relatifs à ce sujet, dans deux mémoires imprimés en 1771, l'un à Berlin, l'autre à Paris. Depuis ce temps, quelques géomètres italiens se sont occupés avec succès de cette matière, et particulièrement M. *Ruffini*, qui a consigné le résultat de ses recherches dans le tome XII des Mémoires de la Société italienne, et dans sa Théorie des équations numériques. Une des conséquences les plus remarquables des travaux de ces divers géomètres, est qu'avec un nombre donné de lettres on ne peut pas toujours former une fonction qui ait un nombre déterminé de valeurs. Les caractères par lesquels cette

XVII.ᵉ Cahier. A

The calculus of substitutions, created by Cauchy in 1812. 'Mémoire sur le nombre des valeurs qu'une fonction peut acquérir, lorsqu'on y permute de toutes les manières possibles les quantités qu'elle renferme', *Journal de l'École Polytechnique*, **10**, 1815, pp. 1–28. Published by permission of the École Polytechnique.

In the proof of this result he introduced new concepts, notations, and methods, directly inspired by Gauss' *Disquisitiones*. He denoted

$$\binom{A_r}{A_s}$$

as the substitution transforming the permutation A_r of n letters into the permutation A_s of the same letters and

$$(\alpha, \beta)$$

as the transposition of two letters α and β in a given permutation. He defined the product

$$\binom{A_r}{A_s}\binom{A_s}{A_t}$$

to be the substitution

$$\binom{A_r}{A_t}$$

and used a method of proof that amounts to decomposing the group S_n into disjoint classes according to one of its subgroups. In the first article, he also gave a simple proof of Lagrange's theorem—the order of any subgroup of S_n is a divisor of $n!$—and determined all the subgroups of a cyclic group (2).

In the second article, Cauchy developed the abstract theory of determinants. First, he gave another result on substitutions: the decomposition of a substitution into disjoint cycles and the distinction between even and odd substitutions according to the number of disjoint cycles and, consequently, to the number of transpositions the given substitution contains. He also proved the existence of the alternating subgroup A_n of S_n. Then Cauchy introduced the notion of the 'symmetric alternating function $S(\pm K)$'. This function, deduced from a given function $K(a_1, \dots, a_n)$ of n given quantities, can be written in modern notation as

$$\sum_{\sigma \in G_n} \varepsilon_\sigma K(a_1, \dots, a_n)$$

(G_n is the substitution group on $\{1, \dots n\}$).
Cauchy showed that

$$S(\pm a_1^0 a_2^1 \cdots a_n^{n-1}) = \prod_{i<j} (a_i - a_j)$$

[$S(\pm a_1^0 a_2^1, \dots, a_n^{n-1})$ is a Vandermonde determinant] and that

$$S(\pm a_1^1 a_2^2, \dots, a_n^n) = a_1 a_2, \dots, a_n \prod_{i<j} (a_i - a_j). \qquad (3.1)$$

He used expression (3.1) to define the determinants as follows: replacing each exponent by a second subscript of the same value in the expansion of $S(\pm a_1^1 a_2^2, \ldots, a_n^n)$, he obtained the function $S(\pm a_{1,1}, a_{2,2}, \ldots, a_{n,n})$, which is the determinant (of nth order), denoted D_n, of the 'symmetric system' (i.e., the square matrix):

$$\left\{\begin{matrix} a_{1,1} & a_{1,2} & a_{1,n} \\ a_{2,1} & a_{2,2} & a_{2,n} \\ \vdots & \vdots & \vdots \\ a_{n,1} & a_{n,2} & a_{n,n} \end{matrix}\right. .$$

This symmetric system can be written more simply $(a_{1,n})$. For instance, since $S(\pm a_1^1 a_2^2) = a_1 a_2 (a_2 - a_1)$, we have $D_2 = a_{1,1}a_{2,2} - a_{1,2}a_{2,1}$ by replacing a_1 by $a_{1,1}$, a_2 by $a_{2,1}$, $a_1{}^2$ by $a_{1,2}$, and $a_2{}^2$ by $a_{2,1}$.

Thereafter, Cauchy gave the principal properties of determinants: D_n is null if two rows or columns of its symmetric system are equal; D_n changes by a sign if two rows or columns are interchanged; the determinant D_n of a symmetric system $(a_{1,n})$ is also the determinant of its conjugate (i.e., its transpose) $(a_{n,1})$. Cauchy proved the rule for the expansion of a determinant according to rows and columns. The most important result, however, concerns the product of the determinants: if $(m_{1,n})$ is the resulting system of two component systems $(a_{1,n})$ and $(\alpha_{1,n})$ [i.e., if $(m_{1,n})$ is the product of the two matrices $(a_{1,n})$ and $(\alpha_{1,n})$], then the determinant M_n of $(m_{1,n})$ is equal to the product of the determinants D_n of $(a_{1,n})$ and δ_n of $(\alpha_{1,n})$ (3).

The level of rigor and abstraction in his two-part memoir on symmetric functions of November 30, 1812, is certainly the first full-fledged example of Cauchy's mathematical style.

On February 1, 1813, the sick leave that had been granted to him came to an end. However, Cauchy had no desire to return to Cherbourg and resume his duties as an engineer. Desperate, he asked Prony, the director of the École des Ponts et Chaussées, for a position of associate professor. On February 3, the École's committee denied his request. Meanwhile, expressing a desire to see him [Cauchy] devote himself more completely to scientific research, the board 'urged' the central administration to assign him to an engineering position in Paris (4). Given this supportive recommendation and, no doubt, some other well-placed pressure from influential quarters, Cauchy was reassigned from the Ministry of the Marine to the Ministry of the Interior on March 17, 1813, and was quickly appointed to an engineering position at the Ourcq Canal (5).

He was thus returning to the construction project where he had worked so diligently and so successfully as a student engineer. Pierre-Simon Girard, who was still director of the Ourcq Canal and Paris Waterworks and who had a deep appreciation for young Cauchy's abilities, might well have been responsible for having his former protégé assigned once more to the Ourcq Canal project. In any event, Augustin-Louis was to replace the engineer Charles-Jean Lehot, who, because of engineering malfunctionings of the inner aqueduct, now found himself in Girard's bad graces. In spite of all his protestations and requests, Lehot was to be punished, the administration having decided to assign him far away from the capital, in the department of the Puy-de-Dôme (6). Before he could leave for Clermont-Ferrand, however, Lehot was to remain in Paris for a while. During this time, he was to complete a report on the works that had been under his supervision. In the meantime, Augustin-Louis had resumed his work as an engineer and was able to help Lehot in writing the required report (7).

So, Cauchy was once more working on an engineering project. But, even though he was still in Paris, his thoughts were elsewhere, far removed from the engineering routines of the Ourcq Canal project. Cauchy now contemplated resigning from the engineering service. Because of his father and his own abilities, he remained with the project, hopeful and expectant. On April 10, 1813, Lagrange died. The death of this eminent mathematician left two places vacant, one on the geometry staff at the Bureau des Longitudes and the other on the geometry staff at the Institut.

At the Bureau des Longitudes, Cauchy aspired to the position of librarian, which should have been vacant as a consequence of staff changes: Prony, supernumerary geometer at the Bureau since 1802, should have filled the vacancy created by Lagrange's passing; Prony's appointment to the Lagrange vacancy would mean that a supernumerary position would have to be filled, and the Bureau unanimously chose his librarian, Claude-Louise Mathieu, for that position; thus, a new librarian would have to be appointed. Cauchy won the majority of the votes for this position. On May 4, 1813, Lehot, still in disfavor because of his past performance, wrote to Count Molé:

> I have just learned that Monsieur Cauchy is about to be appointed to an important post at the Bureau des Longitudes; and, should that position materialize, he has decided to quit his job on the Ourcq Canal project. Sir, I dare hope that you would be so kind as to have me placed in that position (8).

Unfortunately, for both Lehot and Cauchy, the Minister of the Interior chose to disregard the Bureau's elections; the Bureau could only submit nominations for the Lagrange vacancy; its decisions were not binding. Accordingly, three candidates were nominated: Legendre, Poisson, and Prony. On May 26, 1813, Legendre was appointed to the position at the Bureau, and Cauchy lost all hopes of getting appointed to the librarianship since Mathieu would now remain in that position (9).

Cauchy was no longer contented with the First Class of the Institut. Nevertheless, with Poisson's help, Cauchy carefully laid the groundwork for his own candidacy. On April 12, 1813, Poisson submitted a report to the commission that was responsible for evaluating Cauchy's two-part study of November 30, 1812, on symmetric functions. Although Poisson's report was superficial, it was flattering to Cauchy. After examining the historical development of symmetric functions and giving a short account of the results contained in Cauchy's two studies, Poisson's report gave the grounds on which the commission's approval was based, stating:

> The questions that are examined in his [Cauchy's] latest studies are, without doubt, of less importance than his studies on polyhedra. Nevertheless, the present work gives new proof of the insights and abilities that he demonstrated in the study just mentioned (10).

Cauchy was able to insert this favorable report in the brochure which he had published for the elections and which contained the reports on all the studies he had submitted to the Institut (11). Moreover, in May 1813, the *Journal de l'École Polytechnique* published a brief article in which Cauchy gave several theorems on congruences he had taken from his studies on polygonal numbers. In particular, there was a theorem that had already been proved by Lagrange in 1770 (12).

Finally, a week before the geometry section was to submit its list of candidates, Cauchy presented a new study on the theory of equations to the First Class. Thus he artfully caught the eyes of the members of the Institut (13). Cauchy expected to give verbally a concise account of the contents of this paper at the First Class' meeting of May 17. But other readings filled up the meeting. In this account, which he published a few days later, Cauchy intended to pay his respects to Lagrange, 'the great geometer whose passing has saddened learned circles everywhere in Europe as much as it has saddened us', and he presented his own work as a response to a problem that Lagrange had posed in his *Traité de la Résolution des Équations Numériques de tous les Degrés.*

Lagrange had published that work in 1798, and in it he examined Descartes' rule on the number of positive roots of an algebraic equation. This rule had been proved by De Gua in 1748, and Lagrange's work proceeded to pose—but not solve—the more general problem of the a priori determination of the number of real roots, whether negative or positive, of an equation. While it may no doubt be that Lagrange's research was the source of Cauchy's inquiries on the subject, as Cauchy himself declared in his account, it is at best only indirectly so, for it was Poisson who had suggested this research problem to Cauchy. Moreover, nowhere else in the memoir Cauchy published on the subject two years later is Lagrange mentioned (14). The homage Cauchy paid to Lagrange thus appears to have been rendered with an eye on the coming election to the seat that Lagrange had left vacant.

Only eight days after the meeting, on May 24, 1813, Poisson, in a move that was unusual at the Institut but which can be explained in light of the coming election to the geometry section, gave his flattering report on Cauchy's papers. While emphasizing the young engineer's strengths, he pointed out certain methodological shortcomings and called on Cauchy 'to be more attentive [to them] if he would improve his work'. On the same day, the geometry section presented its list of candidates to the Institut: Cauchy was in second place, behind his young friend Jacques Binet, but ahead of Duvillard, Poinsot,

Joseph-Louis Lagrange (1736–1813), the best mathematician at the end of the eighteenth century.

Puissant, Ampère, and Parseval. Nevertheless, on May 31, 1813, Poinsot won by 23 votes over the obscure Duvillard on the third ballot. Cauchy had been eliminated on the first ballot, having received only 2 votes as opposed to 19 for Poinsot, 17 for Duvillard, and 12 for Binet (15).

There was now no chance that Cauchy would be able to give up his position as an engineer, at least for the moment. He was no doubt deeply hurt by these sharp checks on his ambitions, and being only partially recovered from his illnesses, he requested a six-months' leave of absence without pay to regain his health. This sick leave was granted on June 5, 1813 (16). However, he had to wait until mid-July to be replaced by the engineer Denoël before enjoying his leave (17).

Cauchy took advantage of this new sick leave to resume his research on the problem of the a priori determination of the nature and number of roots of an algebraic equation. The method he had propounded in May 1813 had, in effect, turned out to be difficult and applicable only to equations with simple roots. Thus, on October 18, 1813, he submitted a new paper to the First Class of the Institut, stating:

> Two months ago, I had the honor of submitting to the Class a method by which I could determine a priori the real roots of a given equation. The Class decided to honor it with their approval. However, although that method is capable of yielding a solution to the problem when applied to equations of any given degree, one must take account of the number of equal roots and of the multitude of special situations that are prejudicial to the results obtained. The method I propose today does not have these drawbacks. It allows us to determine the number or roots, positive and negative, in a general way and that is independent of any considerations of equal roots and of other special cases that might arise.

Cauchy first developed his ideas in geometric terms with no 'mathematical formulation', and then proceeded to develop an 'application to analysis of the preceding discussion'. Again, Poisson, assisted by Arago, evaluated this research. As the penciled-in marginal notes confirm, Poisson examined the manuscript very carefully. It seems that he was quite dissatisfied with the study, for in his comments, he demanded clarifications and concrete details. Cauchy wrote several supplementary notes, probably during the following months; and, at last, on November 22, 1813, he submitted a new study on this subject to the First Class of the Institut (18).

Cauchy had now spent many months working on this problem, and although Poisson's report was favorable, the undertaking was only partially successful. To be sure, he had shown for the first time that it is always possible—at least in theory—to determine a priori the nature and number of roots of an algebraic equation. In practice, however, this method led to long, complicated calculations and was accordingly almost unusable.

Thus, the year 1813, which Cauchy had spent researching equations, clearly seems, on balance, to have been less fruitful than 1812 had been or 1814 would be. Overall, this decline in Cauchy's output can perhaps be explained in terms of his poor health. In any case, over the course of his life, Cauchy would come back several times to tackle the problem of the a priori determination of the nature and number of roots of an equation. But, by then, as we will see, he would have the resources of analysis at his command, singular integrals and, later on, the calculus of residues.

In January 1814, when Augustin-Louis' sick-leave was due to expire, Louis-François wrote to Baron Costaz asking that an additional three months' leave be granted his son whose health had been 'profoundly undermined by his work load' during the three years of service at Cherbourg (19). Following this supplementary leave, it seems that Augustin-Louis resumed his duties at the Ponts et Chaussées in March 1814, although the relevant records are silent on this point (20). In any case, the events of 1814 and 1815 brought work on the Ourcq Canal project to a stop, and Cauchy thus had the time he needed to devote himself to research. After an unfruitful year in 1813, the next two years would be intensely creative.

On March 7, 1814, as the Emperor was battling the allies, who were now only a few kilometers outside Paris, Cauchy submitted another paper to the First Class of the Institut. Dealing with the theory of errors (21), this study was, as he put it, a 'work undertaken on command'. Obviously mindful of placing himself under the patronage of a man such as Laplace, who had great influence not only with the Institut but with civil officials also, the young engineer acknowledged that he had been 'guided to this research topic by M. Laplace' and was 'duty-bound to comply with Laplace's requests'. Again, a commission was charged with the responsibility of evaluating the paper. This commission, which was composed of Laplace and Poisson, made no report on the study.

But, Cauchy was already preparing another study, suggested by the reading of Laplace's works, which would have quite a different scope. This long study, entitled 'Sur les intégrales définies', set forth some entirely new methods for calculating numerous definite integrals. Truly, it was the starting point of all of Cauchy's work in analysis, especially in complex analysis. The study was submitted to the First Class of the Institut on August 11, 1814. A commission composed of Lacroix and Legendre, the latter acting as reporter, was charged with the responsibility of evaluating this study, and a favorable report was issued. Cauchy was now established as one of the most gifted young mathematicians of his day.

Still, two years had now passed since his return from Cherbourg, and he had not been able to quit his position at the Ponts et Chaussées. New opportunities for entering the Institut occurred by the end of 1814: the deaths of Charles Bossut on January 14 and of Pierre Lévêque on 16 October 16, 1814, created two vacancies in the First Class of the Institut, as well as one vacancy as Examiner of Naval Students. According to a letter Laplace wrote to Charles-Louis Huguet de Semonville, an important official in the chamber of Peers (22), Cauchy was canvasing for the latter position. In this letter, Laplace declared that:

> M. Cauchy [Louis-François] has just informed me of the death of M. Lévêque, my colleague at the Institut and an examiner of naval students. He would like his son to fill this position as examiner. I can assure you that no one is more qualified to fill this important position than that young man.

Alluding to Augustin-Louis' candidacy for the vacancy that Bossut's death had created on the geometry section of the First Class, Laplace added:

> The younger M. Cauchy is a very distinguished geometer whose many excellent papers dealing with different topics in science prove to the

> Institut that he possesses an unusual ability in mathematics. Of all those who aspire to the present vacancy at the Institut, he is the one who seems to me to have the greatest right to it. I believe that he would be a worthy replacement for the geometers in this illustrious body. I take great pleasure in recommending him (23).

The elections were to be held on November 24, 1814, and Augustin-Louis carefully prepared for it, writing up a printed prospectus on his works (24). Above all, his father, taking advantage of his position in the Chamber of Peers, used his influence to 'swing the election' in his son's favor. Following are portions of a letter from Cuvier to Charles-Henri Dambray, President of the Chamber of Peers and Keeper of the Seal, which, like the letter from Laplace cited above, provides clear proof of the unwarranted interference that now took place:

> Sir, young M. Cauchy, about whom your excellency has requested me to write, is really one of our most distinguished geometers. All the great masters of science have attested to his ability, and it cannot be doubted that the section will find that it has an excellent addition in him.

Cuvier, noticeably sensitive to pressure from higher authority, thought it wise to add that:

> The interest that your excellency shows in this adds immensely to my own desire to see him [take his place] among us, a desire that is unanimously shared by my colleagues, even those who might wish to see some of his rivals advanced ahead of him in the present competition (25).

Ampère, Cauchy's principal rival this time, thus had good ground for worrying and wrote:

> I have a very well supported competitor in the person of one of my [former] students from the École Polytechnique, now an engineer from the École des Ponts et Chaussées. He is the son of the Secretary of the Chamber of Peers, etc... Unless I submit papers each month from now until the nominations are over, I could be easily defeated (26).

Nevertheless, on November 28, 1814, Ampère was easily elected on the first ballot. Seven candidates had competed: Ampère, Binet, Cauchy, Duvillard, Francoeur, Parseval, and Puissant. Of the 52 votes, Ampère won 28, Cauchy 10, Binet 7, and Duvillard 6. Compared with the results of the May 1813 election, Cauchy had improved his score noticeably (27).

A few days later, on December 10, 1814, a vacancy was announced at the Société Philomatique de Paris, with which Cauchy had been a correspondent since 1812. On December 17 of that year, a committee composed of A. H. de Bonnard, Bosc, Collet-Descotils, and Poisson selected four candidates for the vacancy, Cauchy, Lepelletier, Thillaye, and Henri. On December 31, Cauchy was elected a member of the Society (28). This was surely a consolation because, instead of having become a member of the Institut, where his age was a handicap in the view of many members, Cauchy would now take his place on a learned society that was then regarded as a 'waiting room' for future members of the First Class of the Institut. Cauchy made a new and final attempt to get into the First Class when Napoleon Bonaparte, having escaped from Elba and reentered Paris (March 20, 1815) gave up his place in the mechanics section. On May 1, 1815, the section submitted a list of seven candidates for this vacancy to the First Class: Bréguet, Hachette, Gengembre, Cagniard-Latour, Molard, Cauchy, and Binet. The names of three other candidates were added to this initial list: Girard, Lenoir, and Janvier. The election took place on May 8, 1815, and of the 53 votes cast, Molard, the winning candidate, received 28; Gengembre was next with 9; Bréguet followed with 8; Hachette received 5; Janvier ranked next with 2, and Cagniard-Latour was last with only 1. Cauchy received not a single vote in this competition (29).

At this time, Augustin-Louis was only 26 years old. He was now looked upon as a gifted, young mathematician with a bright future ahead of him, a scientist following in the footsteps of his brilliant mentors, Lagrange, Laplace, and Poisson. Yet, his job situation was bleak indeed; in spite of his efforts and the strong support that he had received, he still had not been able to secure a position in the learned community. Indeed, such positions were now becoming increasingly difficult to come by, in education, at the Bureau des Longitudes and the Institut. All doors remained closed. In all fairness, however, the Ponts et Chaussées director had been very cooperative in giving him an assignment on the Ourcq Canal project and granting him many leaves of absence, which allowed him to work on his research in mathematics. But, even at that, had Cauchy not had an influential, well-to-do, understanding father who helped him materially and morally, it is doubtful that after his return from Cherbourg, he could have undertaken the research that opened (albeit with the aid of some politicking) the way to a brilliant scientific career.

It is difficult to use a particular case to generalize about the whole learned community in France during the last years of the First Empire. Yet, it does seem that after the great flowering of scientific institutions that had taken place between 1795 and 1800, a blossoming that allowed many young men, such as Poisson, Arago, Biot, and Ampère, to advance very rapidly to important positions, the revival slackened off in proportion to the number of very young people who happened to have 'made it'. For the next generation, the generation of Cauchy, Fresnel, and Navier, success was more difficult to come by because the avenues leading to positions were narrower and more congested. It is perhaps possible to identify this as one of the causes for the relative decline of science in France a few years later.

Chapter 4
A Man of Science

On July 18, 1815, after the defeat at Waterloo and the Emperor's second abdication on June 22, and thanks largely to the intrigues of the regicide Fouché, Louis XVIII reentered Paris on July 18, 1815, amidst the rejoicing of the royalists. A new era had now dawned in France. It would be a new era for Augustin-Louis Cauchy also; an age more to his liking, more compatible with his interest and ambitions. Since 1812, he had been nurturing a deep desire to embark on a career in science. He had produced studies of merit and had the support of influential people, but he had not been able to realize his hopes.

Things would be different now. His father, Louis-François—ever the loyal supporter of whoever happened to be holding the reins of power—had managed to preserve his position in the Chamber of Peers, the new name of the Senate after the Hundred Days. Thus, Louis-François had also salvaged his political influence. Augustin-Louis could count on him using this influence to advance his career (1). Moreover, in this period of reaction, which followed Napoleon's fall, Augustin-Louis could count on the goodwill and support of powerful royalists whose friendship he had been nurturing since 1808 when he joined the Congrégation. Henceforth, Augustin-Louis' political views and leanings would be ultra.

For Augustin-Louis, the Restoration Era would be a time of rebirth after the somber years of the Great Revolution and First Empire. Thus, in a letter dated September 3, 1815, the young man expressed his views on the current situation:

> You have no doubt heard that some 40 hours of prayers have been offered here, prayers in repentance for the excesses and crimes committed during the Revolution. This is as it should be: nobody should stop praying, there is so much to set right in France. Besides, it is rather a consolation to see that the prayers offered in the matter of the elections appear to have borne good fruit; it seems one can generally be satisfied with the choice of deputies (2).

In such a frame of mind, he could identify his own interests with those of France and promptly accept political interventions and influence peddling, backstairs maneuvers that were solicited by him or by those close to him.

At the Institut, no less than at the École Polytechnique, there was now a move afoot to replace those scientists who had become politically compromised by their overly close association with the now discredited Revolution and Bonapartist regime. For Cauchy, then, it was a question of taking advantage of what his fellow royalists saw as offerings of atonement.

It was at the École Polytechnique that Cauchy was first able to assess the effect that the new order would have on his career. On November 2, 1815, when he first presented himself as a candidate seeking the chair in mechanics that Poisson had vacated, the Conseil de Perfectionnement (Improvements Committee) at the École had preferred Jacques Binet to him. Binet, of course, was already a répétiteur (tutor) in analysis and, like Cauchy, a member of the Congrégation. But, the matter did not rest here; for, a few days later, the Governor of the École appointed Cauchy assistant professor of analysis, responsible for the first division (that is, the second year) course.

The two courses in analysis at the École were taught by Ampère and Poinsot, the latter having occupied the chair in analysis since 1809 when he replaced Labey. Holding the rank of associate professor, Poinsot was lax about his duties. He continually used his poor health as an excuse for not teaching his courses himself. During 1812–1813 and 1813–1814, the Governor of the École had appointed Antoine Reynaud, a répétiteur, as Poinsot's substitute. Reynaud was such a mediocre teacher that at the beginning of the November 1815 term the students sent a delegation to Poinsot 'imploring him to please be kind enough not to get sick anymore' (3). Poinsot, however, was unwilling to resume his teaching duties until he had wrung from the administration a change in the schedule when his class sessions would be given.

Poinsot's hostility to teaching his assigned classes in the normal way sprang from his displeasure with his professorial rank. Being only an associate professor, he was required to split his salary with Labey, the full professor whom he had replaced and who was still incumbent.

This was the early 19th century, and of course, there were no retirement funds at the École. The oldest professors, men who were frequently ill and who could no longer teach, preferred to let themselves be 'substituted for', rather than resign outright. By this device, the incumbent full professor took part of the salary while his replacement assumed the full teaching load. Poinsot, of course, was not the only member of the faculty at the École who was caught in this salary-splitting trap: Francois Arago in applied analysis and Alexis Petit in physics were in the same situation. The Conseil de Perfectionnement had made several unsuccessful attempts to come to grips with this problem. On November 2, on the advice of a special commission that was headed by Laplace, the Conseil decided to retire those professors who no longer taught, naming them professors emeriti. The chairs were affected to the extent that the actual teachers would obtain the full professorships. This measure would not affect pay, of course: salary splitting would continue. However, it did offer Poinsot, Ampère, and Petit titles of full professor along with the hopes and chances of getting their chair's full salary once the respective emeritus was dead.

On November 15, 1815 several days after this measure was effected, Poinsot declared to the Governor that his health would not permit him to teach at the scheduled time; and, having this time been unable to get a change of schedule (as he had done before), he asked to be replaced. Instead of asking that Reynaud (whose work had been unsatisfactory) be named as the replacement, Poinsot proposed the répétiteur Lefébure de Fourcy. Upon receiving Poinsot's letter, the Governor of the École sought the advice of Durivau, the Directeur des Études (Dean of Studies), who considered Cauchy a better choice (4). Durivau's advice was accepted. Thus, the Governor took the early opportunity of fulfilling a secret promise to admit Cauchy into the faculty of the École. The position carried a salary of only 1500 francs, a mere fourth of that received by a full professor and the same amount répétiteurs were paid. Yet, Cauchy accepted the offer—no doubt because the Governor, now exasperated by Poinsot's whims, gave him grounds to expect that a definitive appointment would be soon forthcoming (5). On December 2, 1815, the Governor informed the Conseil de Perfectionnement of his decision and gave notice that Cauchy would teach the first division (second year) course for the entire year. Moreover, the Conseil decided, much to Poinsot's displeasure, to suspend his (Poinsot's) nomination to full professor, which had been advanced a month earlier. Thus, the way was now open for Cauchy to displace Poinsot once and for all.

By appointing a young engineer to the faculty of the École (6)—and to the most prestigious of all the professorial chairs at that—the Governor had broken a well-established tradition, which had also worked against Cauchy, and in favor of Jacques Binet, on November 2. The rule was to appoint professors from among the répétiteurs on staff at the École. To be sure, the importance of the works Cauchy had written could easily have justified such an exception to the tradition. But, the recommendations of several of the most illustrious men on the faculty—particularly of Laplace, Cauchy's protector—would doubtlessly have not sufficed to gain him the appointment had not political events worked in his favor. The forced resignations of Monge, Guyton-Morveau, Hassenfratz, and Lacroix had provided an elegant opportunity to staff the École with politically acceptable scientists. The replacement of Poinsot by Cauchy was obviously part of this great purging enterprise, an undertaking that would culminate several months later in the reform of the École Polytechnique (7).

So, Cauchy was able to satisfy his academic ambitions during the period of reaction that followed the Second Restoration. He had already been presented to the Institut three times—in 1813, 1814, and 1815—and each time his ambitions had been checked. Still, he could hope that in 1816 he would be elected on his own merit. Thus, it was during the course of the second semester of 1815 when he presented two outstanding studies, works that undisputably made him equal to the greatest.

Dealing with the theory of waves, the first work was presented anonymously to the Institut on October 2, 1815. This study, which competed for the Grand Prix in mathematics, is one of Cauchy's most carefully written works. On December 26, 1815, it took the prize of 3000 francs (8). But it was another study, presented to the Institut on November 13, 1815, that made him famous (9). In this work, he gave the general proof of Fermat's conjecture on polygonal numbers. 'Every number', wrote Fermat in a letter of 1636 to Father Mersenne, 'is the sum of three cubes, of four squares, and so on, indefinitely'. Lagrange had proved the conjecture for the squares in 1770 and Legendre for the cubes in 1798. Gauss gave a new proof of the theorem for cubes and squares in his *Disquisitiones Arithmeticae.* His demonstration was based on the theory of binary forms. Cauchy, who had read *Disquisitiones,* had been wrestling with the general proof of the conjecture since 1812. Cauchy's proof used the general proposition that it is always possible to decompose an integer into four squares, the positive roots of which add up to a sum that is equal to a given whole number of the same parity (10). The proof of Fermat's theorem on polygonal numbers made him a veritable sensation in the world of mathematics. The announcement of his proof may have supported his appointment to the École Polytechnique a few days later. Following Cauchy's discovery, the First Class of the Institut proposed a new research topic for its Grand Prix de Mathématiques: Fermat's last theorem (11). In spite of the efforts of the contestants, among them Gabriel Lamé and Sophie Germain, nobody took the prize.

Cauchy's studies on the theory of waves and on Fermat's theorem opened the Institut's doors to him. He needed only to wait for the next election, but even this was not necessary, because, as in the situation at the École Polytechnique, Cauchy took advantage of the purge that had been undertaken by the new regime. On March 21, 1816, a royal ordinance reorganized the Institut. The Académie des Sciences was reestablished under its former name. At the same time, several members of the First Class were replaced by newcomers; in mechanics, Carnot and Monge were removed for political reasons, and Louis XVIII appointed Cauchy and Bréguet to their places.

These purges were regarded by the Académie, and by learned society in general, as contemptible affronts. Carnot and especially Monge were respected scientists, first-rate minds. Cauchy nevertheless accepted his appointment without hesitating. Judgment has been harsh on him for his insensitive attitude. 'In the corridors and in the drawing rooms', wrote Bertrand, 'invectives and slanders are accepted, no doubt in good faith, by savants worthy of respect as well as by other important persons'. This politically imposed appointment to the Académie caused considerable harm to young Augustin-Louis, who promptly proceeded to make enemies for himself: 'Cauchy', Bertrand added, 'found few defenders. He has seen more than one

Gaspard Monge (1746–1818). Cauchy replaced him at the Académie des Sciences in 1816.

friend who, though naturally tolerant and decent, turned away and refused to call him "brother"' (12).

A few days after this nomination, Cauchy's first year of teaching at the École Polytechnique came to a premature end, with the disbanding of the student body (13). The royalists regarded the Polytechnicians with dislike and mistrust, for these students had given proof of their liberal and Bonapartist way of thinking. Since the Hundred Days, their lack of discipline had been on the increase. An incident within the École brought the royalists' anger to a head: the répétiteur Lefébure de Fourcy, a dyed-in-the-wool conservative, was subjected to a boisterous display of disrespect by the students in the second division, and the students in the first division joined their comrades. Among the disruptive youths was Auguste Comte, who had an intense dislike for the arrogant répétiteur. On April 12, 1815, the student assembly demonstrated against the administration for having punished the students for their actions. Here the matter might have rested with no further ado. But the authorities seized on the incident as a pretext for attacking the republican, antireligious mood that prevailed at the École. On April 13, the student body was disbanded by royal ordinance, and on the next day they were sent home. The teaching faculty was now put on half-pay until further notice and a five-member commission, headed by Laplace, was appointed to prepare a reorganization plan for the institution (14).

On September 4, 1816, a new ordinance that profoundly modified the bylaws of the École was issued. The principal step taken was the demilitarization of the institution's structure. This reorganization however, was merely a pretext for getting rid of a certain number of professors. The purge that had begun in November 1815 was now to be completed. A campaign led by the reactionaries got under way. One of the participants in this movement was Lamennais, who published an anonymous pamphlet entitled *Quelques Réflexions sur l'École Polytechnique* (15) Published in June 1816, this little brochure was designed to put pressure on Laplace's commission, and its most obvious result was the dismissal of Andrieux, the professor of fine arts. Moreover, this pressure may not have been unrelated to Cauchy's appointment as full professor in analysis and mechanics. Thus, the liberal Poinsot was set aside to the advantage of the extremely conservative Cauchy, who, it will be recalled, had already been rewarded six months earlier with a political appointment to the Academie des Sciences.

In the climate of reaction that prevailed in France during those days, the effects of the political atmosphere cannot be overlooked. Yet, there are nonpolitical reasons that might have justified such an appointment. As we have seen, Poinsot had not taught his classes since 1812; moreover, the commission decided to combine the analysis and mechanics courses under a single teacher. This would mean considerably more work for the full professors responsible for these courses. It would require teachers who were completely available—something that Poinsot certainly was not. In selecting Ampère and Cauchy, the commission had certainly chosen two scholars of undisputed intellectual qualities (even though both would turn out to be rather poor teachers), who consistently taught their assigned courses during all the years they were on the faculty of the École. The situations of the two professors, Cauchy and Ampère, were quite different: Ampère had taught analysis since 1809, when he replaced Lacroix; Cauchy, on the other hand, had only taught analysis for one year—and then only as a replacement. Thus, it is clear that Cauchy benefited from the special favor of the authorities.

Cauchy was now 27 years old and a recognized scientist. He was also a teacher, and preparation for his courses at the École absorbed his attention almost completely. In subsequent years, he also took on teaching jobs at the Collège de France and at the Faculté des Sciences. He could not become a full professor at these institutions since there were no vacancies and had to content himself with being a substitute or a replacement for professors who were on leave.

The first opportunity had to do with the chair in mathematical physics at the Collège de France. The incumbent professor in that chair, J.-B. Biot, was preparing for a geodesy expedition to Scotland and the Shetland Islands. On

The Montagne Sainte-Geneviève in 1840. Lithograph. In the foreground, the École Polytechnique, in the middle ground, left, the Collège Royal Henri IV, formerly École Centrale du Panthéon, which the young Cauchy attended in 1802–1804, and, in the background, the Panthéon. Published by permission of the École Polytechnique.

November 10, 1816, he proposed to the faculty council that he should be replaced by Cauchy if his planned expedition materialized (16). In May 1817, Biot left Paris, but it is not known exactly when Cauchy began teaching at the Collège. In his 1817 course at the Collège de France, Cauchy presented for the very first time the integration methods that he had discovered in 1814, but that to date had remained unpublished (17).

Biot returned to France and shortly afterward seems to have resumed teaching. Following his defeat in the election to the position of Secretary of the Académie des Sciences, he temporarily withdrew from active participation in scientific life. In 1824, he left with his son on a year-long trip to Italy and Sicily. Before leaving, he asked that Cauchy teach his courses during his absence (18). A year later, when he returned to France, he wrote a letter to the administrator of the Collège proposing that Cauchy become his regular substitute at 2000 francs, to be deducted from his annual salary starting January 1, 1826 (19). Thus, Cauchy taught courses at the Collège de France, first as a replacement and then as a regular substitute, from 1824 to 1830. There he developed mathematical methods that were applicable to the physical sciences. In particular, he presented methods and techniques that could be used to solve certain classes of linear differential equations with constant coefficients, both ordinary and partial differential equations. Moreover, in 1830, he delivered lectures on the theory of light (20).

Aside from this, Cauchy was appointed substitute professor, replacing Poisson on the Faculté des Sciences de Paris (21). Poisson had given up his teaching position on the Faculté des Sciences once he became a member of the Royal Committee on Public Instruction in 1820. At that time, Ampère had agreed to teach the course in mechanics in Poisson's stead (22). But, the task proved to be too much of a burden, and during the academic year 1820–1821, he interrupted his teaching for a while. Cauchy was quite willing to act temporarily as Ampère's replacement (23). At the beginning of the October 1821 term, Ampère made up his mind to retire and suggested to Poisson that he be replaced by Cauchy. This appointment to the Université marked an important point in Cauchy's scientific career.

Since 1816, his time had been monopolized by preparation for the courses he taught at the École Polytechnique. Now that the first part of his course on analysis, entitled *Analyse Algébrique*, had been published, the course he taught at the Faculté presented him with a chance to focus attention once again on the mechanics of deformable bodies, a subject that he had not worked on since 1815. At the Faculté, as at the École Polytechnique, a large number of his lectures were devoted to the mechanics of solids, approaching the topic from the standpoint of the theory of linear moments. However, in other lectures, he dealt with fluid mechanics and the methods of solving equations [for sound as well as the wave equation and the heat equation]. It is in this context that Cauchy gave a full exposition of his general theory of elasticity in 1821 and 1822 (24). On December 1823, he was made an associate professor at the Faculté and remained on staff until 1830 (25).

Established in 1815 and 1816 by virtue of his political connections, Cauchy's reputation in scientific circles continued to develop along the lines that it started out on. At the outset of his career, he had the firm support of Laplace and Poisson and considered himself as being their disciple. But very slowly, as the years went by, he began to become more independent. His own ability was now confirmed, and the positions in scientific institutions that he had obtained between 1815 and 1816 secured for him a greater measure of independence from those who had been his patrons. The ever-increasing degree of mathematical rigor that Cauchy advocated led him to criticize certain works of Laplace and Poisson.

In this regard, the introduction to *Analyse Algébrique* is significant. Even though he mentions in 'recognition' the names of Laplace and Poisson, as regards the principle of 'the generality of algebra', which they (Laplace and Poisson) subscribed to, he wrote:

> As to the methods, I have sought to endow them with all the rigor that is required in geometry, doing so in such a way that no recourse to reasons based on the generality of algebra is needed. Such reasons, although quite commonly admitted, especially in going from convergent series and from real expressions to imaginary ones, can be only regarded, it seems to me, as inductions that can sometimes be suitable for guessing

> the truth. But, they agree very poorly with the vaunted exactitude claimed by the mathematical sciences. It should even be noted that they [reasons based on the generality principle] attribute an indefinite (i.e., infinite) realm of applicability to algebraic formulas when, in truth, most of these formulas hold only under certain conditions and for certain quantities satisfying those conditions.

These remarks were aimed directly at Laplace and Poisson, the former having based his theory of generating functions on the consideration of generally divergent series, while the latter had advanced a method of computing definite integrals by intuitively passing from the real line to the complex domain (26). Elsewhere in this same introduction, Cauchy let it be known that he was opposed to the philosophical conclusions Laplace had drawn from his theory on the probability of testimonies (27).

Relations between Cauchy and Laplace and Poisson deteriorated as the years passed. During these years, of course, Laplace had all but retired from any active role in science, and his influence, though still considerable, had declined accordingly. But, the unkind remarks that he made around 1825 before the Conseil de Perfectionnement of the École about Cauchy raise serious questions about his former protégé as a teacher, if not as a mathematician. More decidedly, these remarks reveal the lack of sympathy that now separated the two men.

For his own part, Poisson was still very much active in the world of science and continued to play an important role in the different scientific institutions of Paris. Moreover, competition was very sharp between him and Cauchy. They tended to do research in the same broad areas: theory of definite integrals, theory of elasticity, and linear partial differential equations. Indeed, there was sometimes a veritable race between the two. But their competition was always kept confined to scientific matters, so that, on the whole, the two men managed to get along. An examination of their scholarly activities tends to confirm this impression: Poisson was a member of 19 of the 42 evaluation commissions with Cauchy as reporter between 1818 and 1823. No other member of the Académie collaborated so closely and so frequently with Cauchy during this period as did Poisson (28).

Relations between Cauchy and Poisson deteriorated during 1824 to 1825. This was triggered by the evaluation report on a study dealing with the solutions of differential equations that Brisson presented to the Académie on November 17, 1823 (29). The evaluation commission, consisting of Cauchy, Laplace, Fourier, and Poisson, with the first acting as reporter, was unable to reach an understanding on the report that Cauchy drew up: Fourier and Cauchy being very favorably disposed to Brisson's work, while Poisson was adamantly opposed to it. Very prudently, old Laplace kept his distance from all the brouhaha; and, when in a letter of March 10, 1825, Brisson demanded that Cauchy's report be read, Laplace decided to withdraw

Siméon-Denis Poisson (1781–1840), the French rival of Cauchy in analysis and physical mathematics. Photograph by J. L. Charmet, permission of the Académie des Sciences.

from the commission, claiming the pressing need to attend to other business.

Up until this time, the dispute was confined to the commission, but it became public on March 21, 1825, when Cauchy read his report to the Académie, and under Poisson's instigation, the Académie refused to adopt it. It was a stunning humiliation for Cauchy insofar as it was highly unusual for the Académie des Sciences to reject a report's conclusions. Thanks to Fourier, however, Cauchy was able to get his revenge two weeks later. Following a letter from Brisson, the Académie decided at its meeting of April 4, 1824, that a new commission should be appointed to evaluate Brisson's study—and Poisson was to be excluded from it, although Cauchy would still be reporter. The second report was adopted by the Académie on June 13, 1825.

In the wake of the Brisson affair, Cauchy and Poisson seem to have avoided being on the same commissions. Nevertheless, in 1827 and 1829 similar situations arose. However, in these cases no report was submitted to the Académie. Scholarly collaboration between the two suddenly burst forth again in 1830. Over a period of eight months, they worked together on six commissions, something that had not been seen since 1823. No reports were presented, however, for Cauchy left France in September 1830. Although nothing is known about the reasons why the two scholars so suddenly changed their attitudes, it is doubtful that there was a real reconciliation.

The scientific rivalry between Cauchy and Poisson had, in fact, been very much a reality for the three years preceding the former's departure from France in the fall of 1830. The issue on which it centered was the molecular theory of elasticity, a research area that was then in full bloom. On several points, Cauchy sallied forth to express criticisms or issue an objection to something that Poisson would have communicated to the Académie. Thus, a veritable argument broke out between them in April 1829 about some points on the theory of fluid flow. Again, in June 1830, Cauchy made some biting remarks to the Académie on a question of priority between himself and Poisson relative to fluid flow and the theory of light (30).

It is difficult to determine the precise nature of Cauchy's relations with other members of the Académie des Sciences. He kept away from all the cabals and shifting groups and maintained a distant coolness but perfect politeness toward his colleagues. In general, he avoided getting involved in the debates and squabbles that ripple across the quiet of academia from time to time. Nevertheless, the nature of the relationships he maintained with the other academicians seems to have been determined, in large part, by political and religious considerations. Besides, no one had forgotten the conditions under which he had come to be on the Académie in the first place. Cauchy stayed on good terms with those of his fellows who were conservative in their political learnings and staunchly Catholic in their religious life.

He was especially close to Ampère, who had been his répétiteur in analysis at the École Polytechnique during 1805–1806, and they almost always stood together in the face of criticisms from the various committees. They obviously saw a lot of each other at the École, where, since 1816, they occupied the twin chairs of analysis and mechanics. Ampère, being less interested than Cauchy in analysis and continuing his research on physics, tended to let his colleague take the initiative in preparing the courses they taught. The two men collaborated quite regularly at the Académie. From 1816 to 1830, Cauchy was the reporter on 96 evaluation commissions at the Académie; Ampère was on 26 of these, more than any other member of that body. Ampère thought highly of Cauchy. According to Valson, he used to attend his colleague's classes, sitting right alongside the ordinary students. As for Cauchy, he publicly thanked Ampère in the introductions and forewords to his textbooks for his wise advice and observations (34). Later, on several occasions, he pointed to Ampère as the exemplar of a genial, deeply Christian scholar and scientist. It should be noted, however, that at no time did Cauchy ever show any great interest in electrodynamics, Ampère's field of research.

On the other hand, Cauchy was interested in optics during the decade of the 1820's, and that was Fresnel's research area. Unfortunately, we know almost nothing about his relationship with the founder of the wave theory of light. They were the same age and had the same intellectual background, as well as the same political and religious learnings. But, Fresnel died prematurely in 1827, just at a time when Cauchy began to work on his own theory of light.

Cauchy's relationship with colleagues of a liberal persuasion seems to have been determined by political considerations. He could hardly have been held in high regard by Arago and Prony. Cauchy and his old teacher, Prony, had a mutual dislike dating from the École Polytechnique. Of course, Cauchy was frankly detested by Poinsot, whom he ousted from a chair at the École (32) and who reproached him for having plagiarized his theory of couples and renamed it the theory of linear moments (33).

In spite of all this, however, it would be incorrect to regard the personal relationships between the academicians as consisting only of opinionated quarrels. For example, we know nothing of how Fourier and Cauchy got along. In spite of the differences in their political, philosophical, and religious learnings, nothing indicates that relations between these two men were bad.

Outside the Académie, Cauchy was bound to a certain number of fellow scholar–scientists whose work he esteemed. In this respect, particular mention should be made of Jacques Binet, with whom he had become close during the time they were both students at the École Polytechnique. Binet shared Cauchy's extreme conservative notions and, also like Cauchy, was a member of the Congrégation. In 1816, Binet became Inspecteur des Études at the École Polytechnique and in that capacity was able to spend time regularly with his old friend (34).

Relations between Cauchy and the engineer Barnabé Brisson were altogether different between 1823 and 1828. Cauchy attached considerable importance to the works that Brisson submitted to the Académie des Sciences, although the engineer was totally opposed to him at first, as Brisson had been one of Monge's favorite disciples and was also Monge's nephew by marriage. But, Brisson was also the brother-in-law of Biot, with whom Cauchy was on good terms.

One of the most difficult challenges facing an established scientist is that of finding just the right tone in dealing with young, as yet unknown, scholars who ask for advice and protection. The real challenge lies in criticizing without discouraging, protecting without smothering, listening without taking over another's work. In his youth, Cauchy had the wise, strong support of established mathematicians, such as Lagrange, Laplace, and Poisson. While it is true that he was never head of a school, it is also true that he lacked the moral authority of Laplace, just as he lacked the intellectual generosity and openness of Fourier. But, his reputation as a mean-spirited, callous scholar (35), a reputation that was eagerly hawked about in liberal publications and confirmed by several unfortunate slights and misdeeds that have since been shown to have been pinned on him, was largely unjustified. As a matter of fact, throughout his long career, particularly during the 1820's, he showed a sincere, but admittedly clumsy, concern for the aspiring young who sought him out.

At the Académie Cauchy shouldered a good deal of the work on the commissions that were responsible for evaluating submitted studies. Indeed, between 1816 and 1830, he evaluated 32 papers outside the commissions. Most of these were of no importance, and he merely gave the Académie verbal reports on them (36). Acting for the commissions, he actually gave the Académie reports on 43 papers, one verbal and 42 written. Another 18, which were submitted in 1829 and 1830, probably could not be evaluated until after Cauchy left the country in 1830 (37).

Among the young mathematicians whose studies Cauchy examined were Poncelet, Libri, Lamé, Clapeyron, Rouche, Woisard, Abel, Ostrogradski, Sturm, Galois, Liouville, and Duhamel. Several of these youths were disappointed with Cauchy's attitude.

Poncelet's work on projective geometry was criticized for its lack of rigor. Years later, Poncelet could still recall with anger and bitterness how one day in June 1820 Cauchy had literally 'sent him packing'.

> Fearing, and with good reasons, that twelve years of work and ceaseless meditation would not clarify a deceptive problem and perhaps even make me the subject of ridicule in the eyes of my superiors, of my friends, and of everybody interested in geometry, though indifferent and a bit indulgent, I managed to approach my too-rigid judge at his residence at No. 7 rue Serpente. I caught him just as he was leaving for Saint-Sulpice. During this very short and very rapid walk, I quickly perceived that I had in no way earned his regards or his respect as a scientist, and that it might even be impossible to get him to understand me. Humble petitioner that I was, I thus restricted myself to respectfully informing him that the objectionable points and difficulties that he believed he saw in the adaptation of the principle of continuity to geometry were essentially results of the insufficient attention that had heretofore been accorded to the law of signs, a law that had absorbed my attention since 1813, when I was in Russia, and especially since my return to France in 1814. I explained that the mathematical discussion of this law could have preceded my communication with the Académie, had the esteemed men on that body not dissuaded me from doing so. However, without allowing me to say anything else, he abruptly walked off, referring me to the forthcoming publication of his *Leçons à l'École Polytechnique*, where, according to him, 'the question would be very properly explored'. (38)

A short while later, Abel also complained about Cauchy's behavior in this regard. He had sent his study 'Propriété générale d'une classe très étendue de fonctions transcendantes' to Cauchy several days before submitting it to the Académie on October 30, 1826. But, Cauchy refused to pay any attention to it.

Jean Victor Poncelet (1788–1867), the inventor of projective geometry. Photograph by J. L. Charmet, permission of the Académie des Sciences.

> Cauchy is mad and there is nothing that can be done about him, although, right now, he is the only one who knows how mathematics should be done.

Abel wrote to his friend Holmboe on October 24. A little further on in this letter, Abel elaborated:

> I have completed a big paper on a certain class of functions to present to the Institut. That will take place on Monday. I showed it to Cauchy, but he hardly took the time to glance at it. Without boasting, I dare say it is good, and I am curious to see what judgment will be given at the Institut (39).

Niels Henrik Abel (1802–1829).

Unfortunately, Abel was not to have that satisfaction. He wrote:

> I waited every day for the decision on the paper I handed in at the Institut. But, the sluggish men there did not finish with it. Legendre and Cauchy were judges, Cauchy being 'reporter', Legendre merely going along (40).

When Abel's untimely death occurred on April 6, 1829, Cauchy still had not given a report on the 1826 paper, in spite of several protests from Legendre. The report he finally did give, on June 29, 1829, was hasty, nasty, and superficial, unworthy of both his own brilliance and the real importance of the study he had judged (41).

Galois received the very first time around the kind of attention that Cauchy ordinarily did not give. In fact, on May 25, 1829, Cauchy agreed to present Galois' first research papers to the Académie himself, even though Galois was a very young man and completely unknown. It was a step he had taken only once before, on June 27, 1825, when he presented Frizon's study 'Sommation des puissances semblables des racines d'une équation et calcul des fractions continues'. On January 18, 1830, Cauchy remained at home indisposed.

Evariste Galois (1811–1832).

However, he sent a letter to the Académie declaring his intentions to present an evaluation of Galois' work the following week. Intentions aside, Cauchy did not make the report on January 25 nor at any time later. Moreover, he made no further reference to Galois, either in his own works or in his subsequent statements to the Académie.

We can assume, as does R. Taton (42), that the decision not to present the report was made in agreement with Galois in light of Galois' correction of the paper and his participation in the Grand Prix de Mathématiques. However, the fact remains that in December 1831, Galois complained bitterly that 'parts [of his paper of February 1820] had been sent in 1829, but no report on it has followed, and I have found it impossible to retrieve the manuscripts'. These remarks were obviously aimed at Cauchy, and they raise the possibility that he never returned to Galois the paper that had been submitted to him (43).

Not every young mathematician who sought Cauchy's help suffered because of his indifference or lack of understanding as did Poncelet, Abel, and

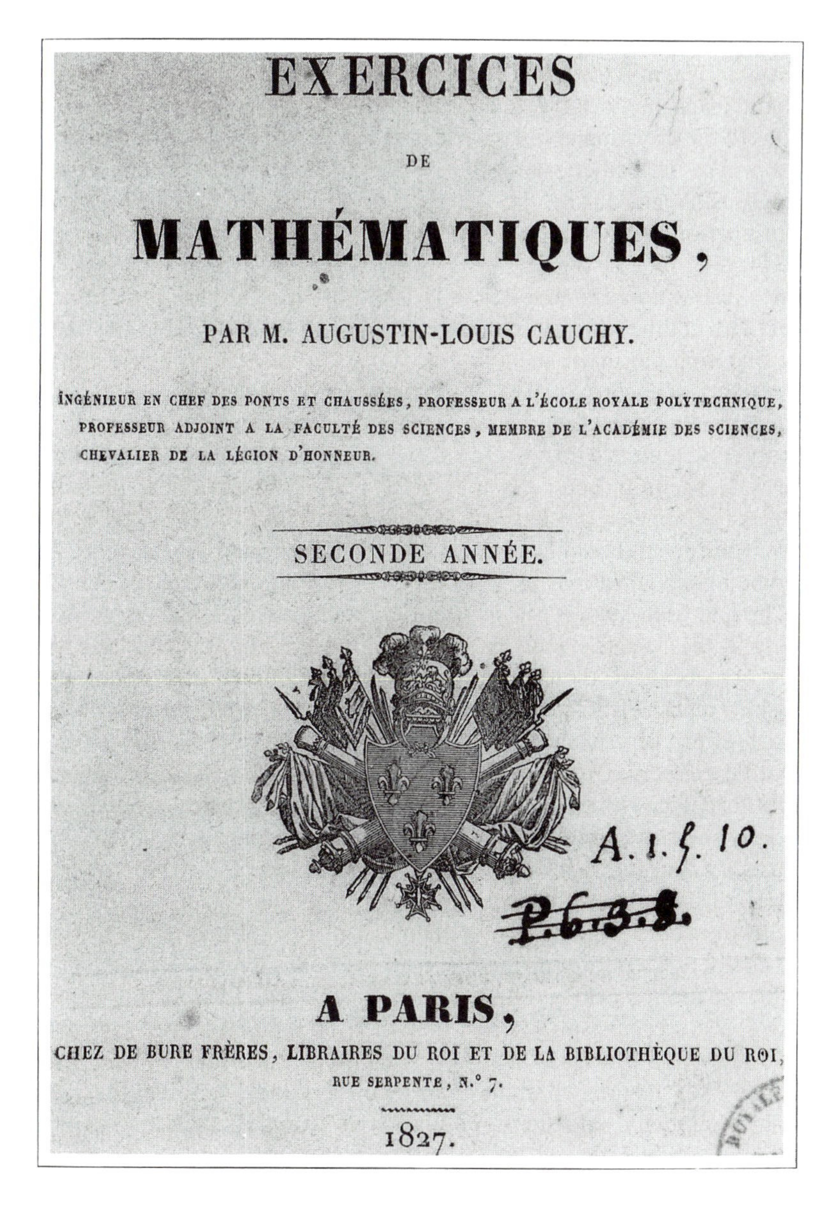

EXERCICES

DE

MATHÉMATIQUES,

PAR M. AUGUSTIN-LOUIS CAUCHY.

INGÉNIEUR EN CHEF DES PONTS ET CHAUSSÉES, PROFESSEUR A L'ÉCOLE ROYALE POLYTECHNIQUE, PROFESSEUR ADJOINT A LA FACULTÉ DES SCIENCES, MEMBRE DE L'ACADÉMIE DES SCIENCES, CHEVALIER DE LA LÉGION D'HONNEUR.

SECONDE ANNÉE.

A PARIS,

CHEZ DE BURE FRÈRES, LIBRAIRES DU ROI ET DE LA BIBLIOTHÈQUE DU ROI, RUE SERPENTE, N.° 7.

1827.

Exercices de Mathématiques, **2**, 1827. Collection of papers by Cauchy, published by his father-in-law, de Bure, between 1826 and 1830. Published by permission of the École Polytechnique.

Galois. Ostrogradski, for example, had been taking courses under Cauchy since his arrival in Paris in 1824 and in 1825 had fondly referred to him as his 'brilliant teacher'. As for Cauchy, he praised Ostrogradski for his contributions to the development of the calculus of residues and cited him in several of his works. It further seems that on several occasions Cauchy came to Ostrogradski's aid during the latter's stay in Paris by getting him out of debtor's prison, where the young Russian had been thrown for not paying his rent (44).

The Restoration era was certainly the most fruitful period of Cauchy's career. The number of research papers that he presented during this period, some one hundred in all, including textbooks, articles in scientific journals, and extracts, was considerable (45). Starting in 1826, he published a review entitled *Exercices de Mathématiques*, which he himself edited. In fact, he was the only contributor to this periodical, which appeared in regular installments until 1830. By editing and publishing his own work, he was able to present the results of his immense productivity more quickly and more thoroughly to the public. The appearance of *Exercices* seems to be connected with his retirement, for some unexplained reason, from the Société Philomatique in 1825. Until then, he had been able to get his summaries, extracts, and papers published quickly in the Société's *Bulletin*. The fact that Cauchy was connected to a family of publishers, the de Bure family, provided him with the means of publishing a periodical that was devoted solely to his own research studies, an unheard-of privilege in the history of mathematics.

During this period, Cauchy's creative work was dominated by three crucial themes: teaching, with emphasis on the foundations of classical mechanics and especially on analysis; mathematical physics, with special interest in the theory of elasticity and its application to the theory of light; and finally, higher analysis, with emphasis on the development of the theory of functions and the calculus of residues. Each of these themes will be explored in the following three chapters. Here, we see that, as with his dual career as a professor and member of the Académie, different research interests took shape, each one acting as a source of inspiration for the other. Thus, in his immensely creative mind, there was an interplay of problems, methods, and results, which between 1821 and 1825 culminated in his great textbooks, the creation of the general theory of elasticity, and the development of complex integration and the calculus of residues.

Chapter 5
Teaching at the École Polytechnique

In the preceding chapter, we saw how Cauchy became a member of the faculty of the École Polytechnique, first, in November 1815, by replacing Poinsot as professor of analysis and then, in September 1816, by obtaining an appointment as a full professor of analysis and mechanics. After the closing of the École in May, classes did not resume until January 1817. However, the Conseil d'Instruction (Curriculum Committee) of the École began to hold a series of sessions on November 15, 1816, in order to organize the instructional program for the academic year 1816–1817. The courses in analysis and mechanics, which would henceforth be taught by the same professor, were of particular concern.

The first piece of business facing the Conseil d'Instruction had to do with the selection of répétiteurs who would work under Cauchy and Ampère. Cauchy proposed that Coriolis, a young engineer from the Ponts et Chaussées, serve as tutor for his courses. Born in 1792, Coriolis had entered the École Polytechnique in 1808 and was only a few years younger than Cauchy. Cauchy and Coriolis had never worked together, either at the École or in the engineering services, but they shared the same political and religious views. Endorsing Cauchy's choice, the Conseil d'Instruction selected Coriolis on November 28, 1816, preferring him to Paul Binet, brother of the Inspecteur des Études (Dean of Studies) and to Destainville. Coriolis served in this capacity until 1830, when he temporarily replaced Cauchy. It is beyond doubt that he exercised a certain amount of influence on Cauchy's teaching (1).

Meanwhile, the basic issue facing the Conseil d'Instruction at its meeting of November 15 was that of defining and restructuring the École's academic programs. On his own authority, Cauchy requested a change in the organization of the course in analysis and mechanics:

> M. Cauchy asks for a change in the structure of the analysis course and the mechanics course. He describes the inconvenience of not being able to present the differential and integral calculus completely and

> thoroughly during the first year, but rather to be forced to veer off and devote time to statics and dynamics. As a result, it is necessary to go back and teach differential and integral calculus again during the second year, as well as dynamics. He notes that changes in the structure of the curriculum had been requested several times; but such changes could not be effected because the courses in analysis and mechanics were divided between four professors—a situation that does not exist today. This fact should allow a better arrangement.

His position was supported by his colleague Ampère and by the Inspecteur des Études, Jacques Binet (2). Two meetings later, on December 11, 1816, Ampère submitted an instructional plan for the first year analysis course, a plan that had been developed and written by Cauchy. This plan can be found in Ampère's papers (3).

In line with what he had proposed at the November 15 session, Cauchy planned that the entire course in analysis would be given during the first year and that mechanics would be restricted to the second year.

He also proposed some important changes in the École's analysis program. As was customary, the course would begin with a section on algebraic analysis (4). This initial section introduced three innovations. The first was an instructional unit entitled 'Imaginary Expressions'. It was to be taught before DeMoivre's theorem and the imaginary exponential were introduced. This unit was followed by one on the difference between continuous and discontinuous functions, a topic that was totally neglected in the traditional program. Finally, there was a unit devoted to the rules governing the convergence of series. The plan gives few details about these innovations, so it is impossible to state exactly which specific topics would have been covered in these instructional units. However, it can be assumed that at this stage Cauchy already had some of the important results that were to appear in 1821 in *Analyse Algébrique*. Moreover, examination of his study Sur les Intégrales Définies, which had been presented to the Académie on August 22, 1814, suggests that by this time he had begun developing the concepts of limit and continuity in the form that they would have in 1821 (5).

The second section of the plan was a course on the calculus of finite differences. Cauchy meant to define the finite differences and integrals of first and higher orders, to develop the analogy between powers and differences, first brought out by Leibniz, and to introduce 'the simplest notions about integration of some finite difference equations'. The interpolation formula would end this section. The addition of a complete course on the calculus of finite differences was an innovation at the École. That was probably useful for introducing infinitesimal calculus, as well as being of interest in its own right. The third section, entitled 'Differential and Integral Calculus', was modeled on the second section: Cauchy drew parallels between the two calculi, contrary to custom. The fourth section of the plan dealt with 'The Application of Integral Calculus to Geometry', a subject treated in the second year.

Cauchy had also written a report in which he gave the reasons for the proposed changes. However, in spite of the fact that this position was supported by Ampère, who said that the plan left 'nothing to be desired', Cauchy's program was not adopted by the commission that was charged with the task of proposing new programs to the Conseil de Perfectionnement (Improvements Committee). Statics was reintroduced into the first-year curriculum and only a few insignificant changes were made in the analysis course. Although the commission gave no reasons to justify its rejection of Cauchy's and Ampère's proposals, it is not hard to figure out what they were. The views of the commission and of Cauchy and Ampère were diametrically opposed. In the latters' opinion, understanding, assimilating, and using the principles of mechanics required such a thorough knowledge of analysis that it was necessary to devote the entire first year to analysis and to restrict the mechanics to the second year. On the other hand, in the commission's view, the analysis was only a tool—albeit an indispensable one—for mastering concrete problems in construction, ballistics, engineering, design, etc. The École had been founded not for the sake of mathematics and mathematicians, but for the training of engineers and the development of the engineering sciences. The professors should introduce analysis in as quick and convenient a way as possible and present instruction in mechanics and its application parallel to instruction in analysis during the first year.

The École Polytechnique reopened on January 17, 1817, in the presence of the Duc d'Angoulême, the King's nephew. Cauchy, in agreement with Ampère, taught the second-division (i.e., first-year) course. Later, Cauchy and Ampère took turns teaching the second division. The one who taught the second division also taught the same students in first division (i.e., the second year) the next academic year. Thus, Cauchy taught the second-division course in 1817, 1818–1819, 1820–1821, 1822–1823, 1824–1825, 1826–1827, and 1828–1829, and the first-division course in 1817–1818, 1819–1820, 1821–1822, 1823–1824, 1825–1826, 1827–1828, and 1829–1830. At first, he was busy developing and refining his lectures. As a result, by 1821 he had considerably slowed the pace at which he submitted papers to the Académie des Sciences. Eventually, he devoted himself more and more to his personal research. The curriculum's registers (registres d'instruction), kept by Jacques Binet, Inspecteur des Études, give valuable information about the evolution of his teaching (6); from the beginning, the originality of his lectures was evident. Cauchy did not follow the instructional program and gave free play to his inspiration instead, changing his teaching year after year, at least until 1823.

In his second division courses of 1817 and, especially, 1818–1819, he devoted a great deal of attention to algebraic analysis: the principal theorems about means of several quantities; the explanation of the method of limits and the definition of continuous function [as far back as 1817, he stated the intermediate-value theorem for such functions, which he used in his first proof of the fundamental theorem of algebra (7)]; a thorough study of ordinary functions with imaginary values of the variable; rules of convergence applied to the binomial series expansion and to the expansions of e^x, $\cos x$, $\sin x$, and $\log(1+x)$. All these matters are summarized without any change in the *Analyse Algébrique* of 1821. The section on differential and integral calculus was much more succinct. The examination of the curriculum's registers does not allow us to specify its standard of rigor. In his lectures, Cauchy probably used the method of limits (the derivative defined as the limit of a difference quotient and the integral as the limit of finite sums). He probably avoided series expansions whose convergence was not demonstrated (as far back as 1817, Cauchy put off the study of Taylor-series expansions to the end of the second-division course, after the lectures on integral calculus).

In his first division course of 1817–1818, he explained his method of integration by passing from real to imaginary variables and dwelt on Euler's method of polygonal approximation for solving differential equations. There is no indication that he had deduced an existence theorem for solutions of a differential equation (Cauchy's problem) from Euler's method by this time. Cauchy gave also the first general method for solving first-order partial differential equations, now called Cauchy's method of characteristics. He presented a paper on this matter to the Académie on December 21, 1818 (8).

His mechanics teaching was less innovative. Cauchy expounded the principles of mechanics in the tradition of Euler, insisting upon the notions of force and torque and relegating the Lagrangian principle of virtual velocities to the end of the course. In the second-division course of 1820–1821, he developed the general theory of resultants and linear moments, which would constitute the framework of his further research in mechanics.

Incontestably, Cauchy taught with zeal at the École Polytechnique. His best students took advantage of his teaching. Charles Combes, who had been the major (first passed) of the class of 1818, wrote about Cauchy in 1857:

> We all found that this professor was extremely energetic, good natured, and tireless. I often heard him repeat and review, for several hours on end, whole lessons that we had not understood clearly; we would then become impressed by the elegant clarity of his analysis, an analysis dry and tedious. Indeed, M. Cauchy had the genius of Euler, Lagrange, Laplace, Gauss and Jacobi, and his love for teaching, which bordered on pure zeal, brought with it a kindness, a simplicity, and warmth of heart that he retained until the end of his life (9).

However, the originality of his lectures with respect to the official program soon provoked unfavorable reaction within the École. The first evidence consists of some 'Remarks on the lack of progress in the 2nd division analysis course', extracted from the minutes of the March 4, 1819, meeting of the Conseil d'Instruction (10). In these 'Remarks', Arago, the professor of applied analysis, complained about the poor training (in analysis) of some of the students in the second division. He attributed this lack of preparedness to the fact that 'the analysis course is behind schedule' and 'does not keep pace with the course in applied analysis'. In particular, he criticized the fact that the lectures on the applications of differential calculus to geometry had not yet been presented. Cauchy's position was that the delays (that Arago spoke of) were a result of the enormous amount of material that had to be covered, and he proposed that the material on statics be abridged, so that the material on analysis could be completed. That, he alleged, was what had been done in previous years. However, neither of these explanations nor the support he received from Jacques Binet, the Inspecteur de Études, sufficed to calm down Cauchy's critics at this meeting, and the assembly proceeded to examine Cauchy's teaching methods. Adopting Arago's position, Petit, the professor of physics, expressed concern about the delays in starting the material on differential calculus. Specifically, the record shows that:

> He [Petit] asked that this material [on differential calculus] be presented without certain notions from algebra, which mainly had to do with series and which, he alleged, the students would never have occasion to use in the [engineering] services. Moreover, he insisted that the method of infinitesimals be used. This method, he claimed, was one that, in his opinion, the students seemed to be largely unfamiliar with and, at the same time, was so useful that it should be thoroughly known (11).

Finally, the director of the École repeated the criticisms that had been leveled and warned Cauchy that:

> It is the opinion of many persons that instruction in pure mathematics is being carried too far at the École and that such an uncalled-for extravagance is prejudicial to the other branches (12).

He also asked Cauchy to adhere strictly to the official syllabus and 'to devote time to questions at the end of class so as to introduce the students to numerical applications'. This warning did nothing to produce a fundamental change.

On June 15, 1820, the Conseil d'Instruction directed Cauchy and Ampère to revise their courses (13). The students would benefit by these revisions, particularly in those areas that the professors did not cover in class. In this way, it would be possible to control, at least a posteriori, the content of the courses in question. Cauchy took this opportunity to inform the Conseil d'Instruction that 'its desires relative to the analysis course would soon be fulfilled by the publication of a work now being printed, the first volume of this work soon to appear'. During a discussion of the principle of continuity, around this same time, he referred Poncelet to the forthcoming publication of his lectures at the École Polytechnique (14). These two facts lead us to believe that by the spring of 1820 Cauchy had already written what was to become known as his *Analyse Algébrique*. Printing delays prevented the work's release in time for the opening of school in the fall of 1820, and perhaps Cauchy took advantage of this delay to add a number of new notes at the end of the work during the 1820–1821 academic year. Ampère eagerly awaited the publication of the material that Cauchy had prepared 'in order to help with any work that needed to be done on it [the course material], as well as to be sure that their instructional approaches followed the same plant'.

The first part of the *Cours d'Analyse*, entitled *Analyse Algébrique*, was finally published by de Bure Publications in June 1821 (15). Certainly, the publication of this work constitutes a landmark in the history of analysis. The task the author had set before himself was indeed ambitious. His aim, as he explained in the introduction, was to endow proof in analysis with precisely the same level of rigor that was used in geometry since Euclid's time (16). Such a goal, of course, demanded that he undertake a considerable investigation of the very foundations of analysis.

In the opening discussions, Cauchy quickly introduced the concepts of number (i.e., the absolute measure of magnitude), real quantities (i.e., numbers preceded by a $+$ or $-$ sign), and algebraic operations ($+$, $-$, $\times$, and $\div$). But, most important of all, he laid the cornerstones for the new structure by giving the definition of the limit of a variable quantity—that is, of a quantity that 'assumes values successively, each value differing from the other'. He said:

> When the successively attributed values of one variable indefinitely approach a fixed value in such a way that they finally differ from it by as little as desired, then that fixed value is called the limit of all the others (17).

The limit concept allowed him to define infinitesimals:

> As the successive numerical values of the same variable decrease indefinitely, so as to become less than any preassigned given number, this variable becomes what is called an infinitesimal or an infinitely small quantity. A variable of this type has zero as limit (18).

Likewise, an infinitely large quantity is taken as a variable whose numerical values continually increase in such a way as to become greater than any preassigned number.

In the preliminaries, Cauchy also introduced the arithmetical notion of mean of several quantities: the quantity x is a mean of the quantities $a, a' \, a'', \ldots$ if it satisfies the inequalities

$$\inf(a, a', a'', \ldots) \leqslant x \leqslant \sup(a, a', a'', \ldots).$$

A mean of a, a', a'',... is denoted by Cauchy $M(a, a', a'', \ldots)$. Cauchy showed that

$$\frac{a + a' + a'' + \cdots}{b + b' + b'' + \cdots} = M\left(\frac{a}{b}, \frac{a'}{b'}, \frac{a''}{b''}, \ldots\right)$$

if the quantities b, b', b'',... all have the same sign;

$$\sqrt[(B+B'+B''+\cdots)]{A + A' + A''} + \cdots = M(\sqrt[B]{A}, \sqrt[B']{A'}, \sqrt[B']{A''}, \ldots)$$

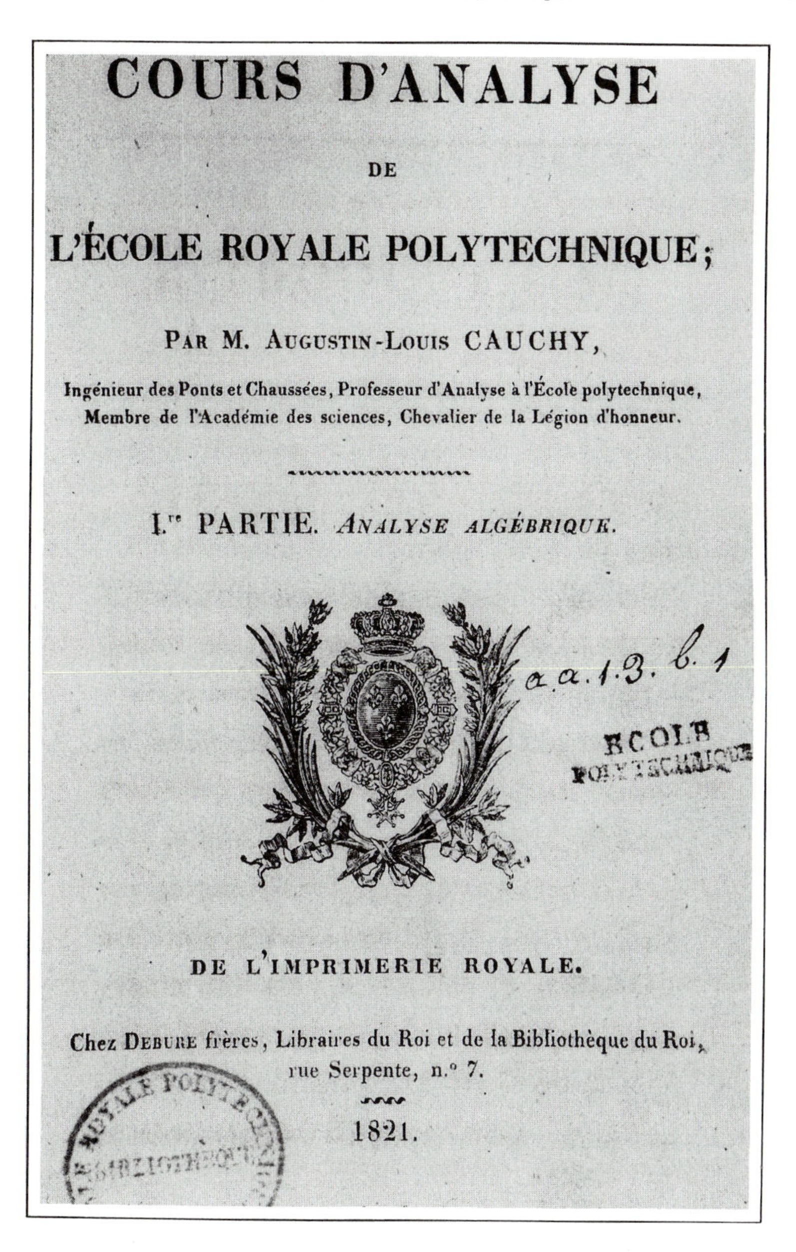

COURS D'ANALYSE

DE

L'ÉCOLE ROYALE POLYTECHNIQUE;

PAR M. AUGUSTIN-LOUIS CAUCHY,

Ingénieur des Ponts et Chaussées, Professeur d'Analyse à l'École polytechnique, Membre de l'Académie des sciences, Chevalier de la Légion d'honneur.

I.re PARTIE. *ANALYSE ALGÉBRIQUE.*

DE L'IMPRIMERIE ROYALE.

Chez DEBURE frères, Libraires du Roi et de la Bibliothèque du Roi, rue Serpente, n.° 7.

1821.

Title page of the *Analyse Algébrique*, 1821. Published by permission of the École Polytechnique.

if A, A', A'', B, B', B'' all are (positive) numbers;

$$\alpha a + \alpha' a' + \alpha'' a'' + \ldots = (\alpha + \alpha' + \alpha'' + \ldots) M(a, a', a'', \ldots)$$

if the quantities $\alpha, \alpha', \alpha'', \ldots$ all have the same sign.

Cauchy skillfully used these means, instead of inequalities, as powerful tools to prove many theorems of his analysis course: the mean-value theorem, theorems on the convergence of series, the proof of the existence of the definite integral of a continuous function and of a solution of a differential equation.

Chapter I was devoted to real functions. Cauchy defined a function in a rather general way as a variable that can be expressed by means of one or several other variables, which he called 'independent variables'. If the function is multivalued, Cauchy denoted it $f((x))$. He then examined several different types of functions: explicit, implicit, simple, algebraic, and trigonometric.

In Chapter II, he reexamined infinitesimals. Using the notion of limit, he compared infinitesimal quantities in terms of orders of magnitude and then introduced the important notion of continuity on an interval. This concept, he asserted, 'should be ranked among the matters that are closely connected with the investigation of infinitesimals'.

Cauchy gave the following definition of the continuity of a real function of one variable:

> Let $f(x)$ be a function of the variable x and suppose that, for each value of x between two given bounds, this function constantly takes one finite value. If, from a value of x between those bounds, one attributes to the variable x an infinitely small increment α, the function itself will receive as an increment the difference
>
> $$f(x + \alpha) - f(x),$$
>
> which will depend at the same time on the new variable and on the value of x. This being granted, the function $f(x)$ will be a continuous function of the variable x between the two assigned bounds if, for each value of x between those bounds, the numerical value of the difference $f(x+\alpha) - f(x)$ decreases indefinitely with α. In other words, the function $f(x)$ remains continuous with respect to x between the given bounds, if, between these bounds, an infinitely small increment in the variable always produces an infinitely small increment in the function itself (19).

The continuity of the function was established at each point of the interval. Cauchy used the same infinitesimal α, however, and did not distinguish between uniform continuity and simple continuity. This confusion is one of the most serious flaws in Cauchy's development of the course. Thus, when he extended the notion of continuity to functions of several variables, he gave a theorem that is only true for a function uniformly continuous with respect to each variable:

34 COURS D'ANALYSE.

numérique de cette variable, le polynome finit par être constamment de même signe que son premier terme.

§. 2.e *De la continuité des Fonctions.*

Parmi les objets qui se rattachent à la considération des infiniment petits, on doit placer les notions relatives à la continuité ou à la discontinuité des fonctions. Examinons d'abord sous ce point de vue les fonctions d'une seule variable.

Soit $f(x)$ une fonction de la variable x, et supposons que, pour chaque valeur de x intermédiaire entre deux limites données, cette fonction admette constamment une valeur unique et finie. Si, en partant d'une valeur de x comprise entre ces limites, on attribue à la variable x un accroissement infiniment petit α, la fonction elle-même recevra pour accroissement la différence

$$f(x+\alpha)-f(x),$$

qui dépendra en même temps de la nouvelle variable α et de la valeur de x. Cela posé, la fonction $f(x)$ sera, entre les deux limites assignées à la variable x, fonction *continue* de cette variable, si, pour chaque valeur de x intermédiaire entre ces limites, la valeur numérique de la différence

$$f(x+\alpha)-f(x)$$

décroît indéfiniment avec celle de α. En d'autres termes, *la fonction $f(x)$ restera continue par rap-*

I.re PARTIE. CHAP. II. 35

port à x entre les limites données, si, entre ces limites, un accroissement infiniment petit de la variable produit toujours un accroissement infiniment petit de la fonction elle-même.

On dit encore que la fonction $f(x)$ est, dans le voisinage d'une valeur particulière attribuée à la variable x, fonction continue de cette variable, toutes les fois qu'elle est continue entre deux limites de x, même très-rapprochées, qui renferment la valeur dont il s'agit.

Enfin, lorsqu'une fonction $f(x)$ cesse d'être continue dans le voisinage d'une valeur particulière de la variable x, on dit qu'elle devient alors *discontinue*, et qu'il y a pour cette valeur particulière *solution de continuité*.

D'après ces explications, il sera facile de reconnaître entre quelles limites une fonction donnée de la variable x est continue par rapport à cette variable. Ainsi, par exemple, la fonction $\sin. x$, admettant pour chaque valeur particulière de la variable x une valeur unique et finie, sera continue entre deux limites quelconques de cette variable, attendu que la valeur numérique de $\sin.(\frac{1}{2}\alpha)$, et par suite celle de la différence

$$\sin.(x+\alpha)-\sin.x=2\sin.(\tfrac{1}{2}\alpha)\cos.(x+\tfrac{1}{2}\alpha),$$

décroissent indéfiniment avec celle de α, quelle que soit d'ailleurs la valeur finie que l'on attribue à x. En général, si l'on envisage sous le rapport de la continuité les onze fonctions simples que nous avons

C*

Definition of a continuous function, *Analyse Algébrique*, 1821, pp. 34–35. Published by permission of the École Polytechnique.

> If the variables $x, y, z, \ldots$ have the fixed and determined quantities X, Y, $Z, \ldots$ as limits and if the function $f(x, y, z, \ldots)$ is continuous with respect to each variable $x, y, z, \ldots$ in the neighborhood of the system of values X, Y, $Z, \ldots$, $f(x, y, z, \ldots)$ will have the limit $f(X, Y, Z, \ldots)$ (20).

Finally, he stated the intermediate-value theorem (21) and sought to determine the value of certain functions at singular points by continuous extension.

The notions Cauchy presented in the first part of *Analyse Algébrique* were not new (22). Following D'Alembert, several mathematicians had attempted to base analysis on the limit concept. Some, notable are Lhuillier and Carnot, had obtained results that anticipated Cauchy's. Furthermore, the method of limits was the approach used in teaching calculus at the École when Cauchy was a student there. In the meanwhile, in 1811, the Conseil d'Instruction had renounced this method, which 'to be truthful, lacked rigor and was not amenable to applications' and reinstated the former method of infinitesimals as the approach for teaching calculus. Mathematicians did not ignore the properties of continuity of the functions commonly studied; but, in spite of the works of Arbogast, they continued to use Euler's definition according to which a continuous function is a function that is defined by a single analytic expression (23).

While *Analyse Algébrique* made use of then current notions on limits, continuity, and infinitesimals, it broke with the traditional texts treating such concepts. On the one hand, the definitions given by Cauchy in *Analyse Algébrique* are much clearer, more precise, and more general than those that had been used up to then. On the other hand, the relations between the basic concepts are more neatly exhibited. By defining infinitesimals as variables that converge to zero, Cauchy, as he explained in the foreword to *Calcul Infinitésimal*, reconciled the rigor of taking limits (which was given there for the first time in terms of inequalities) with the simplicity of the results obtained by considering infinitesimals directly (which he generally used in his course because of the requirements of the official program of the École) (24).

The set of basic notions thus given formed a logical, coherent basis on which it would be possible to erect the structure of the course in analysis.

Chapter V was devoted to five functional equations:

$$\phi(x+y)=\phi(x)+\phi(y)$$
$$\phi(x+y)=\phi(x)\times\phi(y)$$
$$\phi(xy)=\phi(x)+\phi(y)$$
$$\phi(xy)=\phi(x)\times\phi(y)$$
$$\phi(y+x)+\phi(y-x)=2\phi(x)\,\phi(y)\,(\phi \text{ continuous}).$$

Cauchy used the second of these for studying the binomial series in the next chapter.

In this chapter (Chapter VI), Cauchy took up the theory of series. He defined a series as a sequence of partial sums,

$$s_n = u_0 + u_1 + \cdots + u_{n-1}.$$

The sum of the series is equal to

$$\lim_{n\to\infty} s_n,$$

when this limit exists. In that case, the series is said to converge; otherwise, it diverges. If the series diverges, then it has no sum. This is the reason that Cauchy devoted the main part of the chapter to establishing certain tests for the convergence of series. The most general of these tests, known today as 'Cauchy's test', was presented in the following terms:

> In order that the series u_0, u_1, u_2, $u_3, \ldots u_n$, $u_{n+1}, \ldots$, etc... shall be convergent, (...) it is necessary that, for increasing values of n, (...) the sums of the quantities u_n, u_{n+1}, u_{n+2}, etc.... taken from the first one, in any desired number, end up by constantly assuming numerical values that are less than any assignable limit. Conversely, whenever these different conditions are fulfilled, the convergence of the series is guaranteed (25).

Cauchy stated, but did not prove, the converse. In the same section, Cauchy stated that the sum of a series of continuous functions is continuous. In 1826, Abel was the first to note the insufficiency of this statement by giving the famous counterexample

$$f(x) = \sum_{n=0}^{\infty} \frac{\sin(2n+1)x}{2n+1},$$

which is discontinuous at π (26).

Cauchy proved the rules governing absolutely convergent series, 'Cauchy's rule' and 'D'Alembert's rule', and the ones regarding sums and products, etc. He first gave a rule for determining the radius of convergence of power series, and he showed the uniqueness of the power series expansion of a continuous function. Then, he investigated the series expansions of some important functions in their circle of convergence: Newton's binomial series, the exponential, the logarithm of $(1+x)$. But, it was not until 1831, with the theorem of Turin, that he succeeded in determining the context of the complex function theory, the conditions under which a function shall have an infinite series expansion.

Chapters VII through X were devoted to the theory of imaginary expressions. Cauchy took an imaginary expression as a symbolic expression that has no meaning in itself but is equivalent to two real quantities. In the same way, he regarded an imaginary equation as "the representation of two equations connecting real quantities." In order to obtain the product of two imaginary expressions, he used the ordinary rules of multiplication in algebra by treating $\sqrt{-1}$ as a real quantity with square -1. Cauchy examined the trigonometric form of imaginary expressions, as well as the integer powers of such expressions. He also investigated fractional and irrational powers of imaginary expressions, paying particular attention to the nth roots of unity. Finally, Cauchy defined imaginary functions of the form $\phi(x) + \chi(x)\sqrt{-1}$, where $\phi(x)$ and $\chi(x)$ are real (complex functions of a real variable). He also extended the definition of e^z, $\sin z$, and $\cos z$ to complex z by means of convergent 'imaginary series' with the general term $p_n + q_n\sqrt{-1}$ and defined the multivalued functions $\log((z))$, $\arcsin((z))$, $\arccos((z))$, etc. (27).

Analyse Algébrique had a considerable influence on young mathematicians. 'It should be read,' wrote Abel in 1826, 'by any analyst who likes rigor in mathematical investigations' (28). Nevertheless, it was not well-thought of at the École and was never used by the students there for the following reason. A few months before its publication, a serious incident took place in Cauchy's second-division course. On April 12, 1821, at the end of a lecture that had gone into overtime, the incident occurred just at the moment the class monitor turned his back. Cauchy was booed and hissed by five or six students whose identities were never determined by the administration.

Baron Bouchu, the Governor of the École, took the matter seriously. He submitted a report on it to the Minister of the Interior (29), and he ordered the class leaders to apologize for the class' misconduct to Cauchy, who refused to continue teaching the class following this affront. In his report, Baron Bouchu blamed the students and the professor equally, raising questions about Cauchy's method of teaching. Cauchy, he felt, respected neither the official program nor the official schedule:

> It cannot be denied that M. Cauchy, by his own stubborness and by extravagantly extending his class sessions [beyond the allotted time], unwisely pushed his students to the thoughtless insult that took place...
>
> I can no longer hide the fact that for the past five years, he has been given many warnings to simplify his teaching methods so as to bring them into line with the official program.

The Minister of the Interior, however, did not find these explanations convincing. The whistling and booing, he believed, was a manifestation of the students' hostility toward a teacher who did not bother to hide his extremely royalist views. Following the assassination of the Duc de Berry (the Dauphin) in February 1820, the government had once again curtailed civil liberties. The liberal opposition, always very popular among the students, as well as in the army where the Charbonnerie movement had taken roots, became radicalized. Cauchy, of course, was not highly regarded by the students at the École Polytechnique, because they had not forgotten that at the Académie he had replaced Monge, the scholar who had founded the École and whose memory was still held sacred by the students. The affair was discussed by the Conseil d'Instruction on April 17:

> The Inspecteur [des Études] recalls that at the meeting before last, M. Cauchy stated that, after 53 [teaching] lessons, he still needed 7 additional ones to complete the analysis course. The former number was not enough for him, and a total of 66 was required instead of the 50 contemplated by the schedule. This situation meant 'slow-down' of some five weeks in the combined progress of the course in analysis and the course in mechanics... M. Cauchy stated that he would be able to give his reasons for an increase in the number of lessons in the analysis course at the end of the year. He stated that the amount of material for the first-year course had been increased and that he had not gone beyond what was required by the curriculum. He recalled that for five years it has not been possible, either for him or for M. Ampère, his colleague, to keep within the prescribed number of lessons for the first-year course; while, for the second-year course, the limits outlined were always adhered to (30).

No doubt, Baron Bouchu was anxious to protect the École from the government's wrath, and accordingly, on April 21, he submitted a second, much more severe report concerning Cauchy (31). In this document, the Baron explained that the students had complained to their parents about the analysis course, but had not dared to register their grievances with the authorities, for fear of being labeled agitators and troublemakers.

> These considerations explain, on one hand, why the students were displeased; and, on the other, it is easy to see that they should have felt discontent. In fact, whatever their attitude towards the course, they cannot lose sight of the fact that the material [in the course] is essential, if they are to pass the examinations, which are exclusively based on the syllabus. They are given copies of the syllabus, and it outlines the required rules of instruction, just as it outlines the duties of the professor.
>
> These course syllabi are the fruit of many years of experience, and they prescribe, for M. Cauchy's course, a total of 85 lectures, 50 for analysis and 35 for mechanics. The neglect of the material on mechanics as a result of an excess of lectures on analysis must not be allowed, given that the students must still answer all the questions dealing with the other part of the course even if it has not been entirely taught. All of the students willingly acknowledged M. Cauchy's enthusiasm and appreciated his outstanding abilities. However, though he [Cauchy] is not the only professor here who possesses superior abilities and excellent principles, he is the only one who perseveres in disregarding the official syllabus...

The incident of April 1821 had some important consequences. It clearly showed that even if Cauchy was able to interest and inspire the most gifted students, he was unable to adapt his course to the level of the majority of the students who came to the École with only the most rudimentary knowledge of mathematics and who were required to master a good number of topics and concepts in two years of study. Testimonies of contemporaries corroborate this impression. In his speech at Cauchy's funeral, Charles Dupin observed that:

> Once our illustrious confrere had been appointed to the École Polytechnique, he was not content to follow the programs and lectures that the eminent professors who preceded him had devised. He restructured, so to speak, the material on algebraic and infinitesimal analysis along lines and by methods that he deemed appropriate. He was found to be—and why should it not be said here? For it is the only reproach that seems possible to make, and it is a reproach which is, as it were, a praise. He was found to be too learned, too brilliant, for the students who came to him in large numbers expecting to learn the practical material required for the public services and not his thoughts and erudition at the Institut. These practical applications demanded above all else that his methods and means be clear, simple, common, and quick first, as military and civil requirements often are (32).

This is confirmed by a comparison that J. Bertrand drew between Poinsot and Cauchy:

> When Cauchy replaced Poinsot, the students were divided [in their opinion]. 'Poinsot did not teach us anything', remarked the students

> who liked the new course; 'Cauchy will disenchant them with science forever', said Poinsot who never bothered to hide his views... Poinsot, it is true, did not teach many things in any given lecture; but what he did present was presented very well indeed! Cauchy, on the other hand, was forever going beyond bounds, and only a few, very gifted students could understand him. This elite indeed found him praiseworthy (33).

Cauchy's lectures were too ambitious to be presented in the time allotted for his courses in the official schedule. Consequently, he was forced to sacrifice the applications and exercises that the Conseil contemplated in its official program. The writing and editing of *Analyse Algébrique*, which took an inordinate amount of time, had further upset the balance between algebraic analysis and the remainder of the course. Applications was the area suffering the most. This lack of balance created a situation that was unacceptable to the students and to the Conseils at the École.

Thus, in November 1821, when he resumed teaching the second-division course, Cauchy had to reduce the number of lectures that preceded differential calculus (34). Abandoning publication of the remainder of the *Cours d'Analyse*, only the first part (*Analyse Algébrique*) of which had appeared in 1821, Cauchy began to write up summaries of his lectures. He had been ordered to do so by the Conseil d'Instruction:

> In order to satisfy the repeated requests of the Conseil d'Instruction, the professors of analysis and mechanics [Cauchy and Ampère] are now busy writing summaries of the most difficult points in their courses. We regret that it took so long for the work in the first-division classes to get started and that it has not been carried out quickly. However, these summaries should be very useful to the students and will help them a great deal (35).

These summaries (in a volume entitled *Résumé des Leçons Données à l'École Royale Polytechnique sur le Calcul Infinitésimal*) were published together on August 11, 1823, by Bure Publications (36).

The first part of *Calcul Infinitésimal* of 1823 was devoted to differential calculus. After a brief exposition in the first two lectures of the basic concepts of analysis, which he had developed in the initial sections of *Analyse Algébrique*, Cauchy introduced the notion of the derivative of a function of one variable $y = f(x)$ in the third lecture. The term derivative and the accompanying notation $f'(x)$ were borrowed from Lagrange (37). Lagrange, however, developed a theory of differentiable functions that was always assumed to be analytic, while Cauchy used the method of limits: if $f(x)$ is continuous, then its derivative is the limit of the difference quotient:

$$\frac{\Delta y}{\Delta x} = \frac{f(x+i) - f(x)}{i}$$

as i tends to 0.

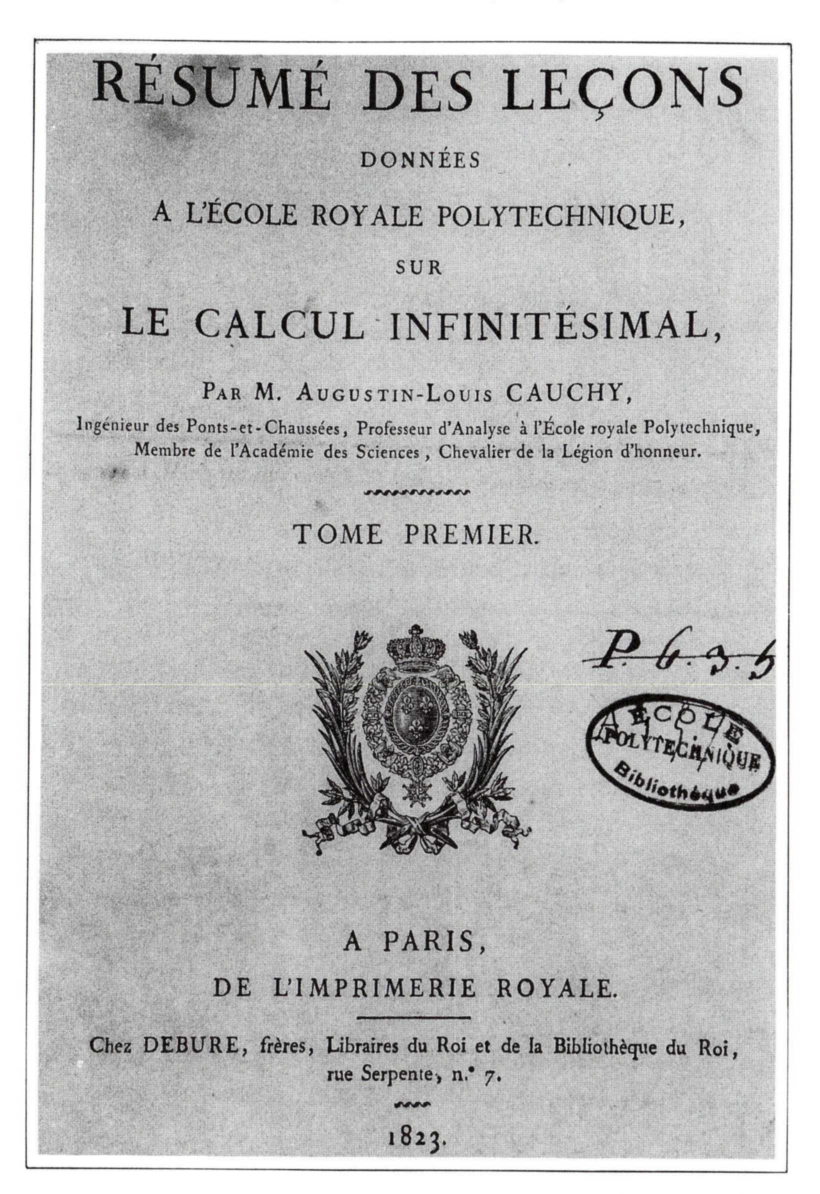

RÉSUMÉ DES LEÇONS

DONNÉES

A L'ÉCOLE ROYALE POLYTECHNIQUE,

SUR

LE CALCUL INFINITÉSIMAL,

PAR M. AUGUSTIN-LOUIS CAUCHY,

Ingénieur des Ponts-et-Chaussées, Professeur d'Analyse à l'École royale Polytechnique, Membre de l'Académie des Sciences, Chevalier de la Légion d'honneur.

TOME PREMIER.

A PARIS,

DE L'IMPRIMERIE ROYALE.

Chez DEBURE, frères, Libraires du Roi et de la Bibliothèque du Roi, rue Serpente, n.° 7.

1823.

Title page of the *Calcul Infinitésimal*, 1823. Only the Tome Premier has been published. Published by permission of the École Polytechnique.

COURS D'ANALYSE. 9

TROISIÈME LEÇON.

Dérivées des Fonctions d'une seule Variable.

LORSQUE la fonction $y = f(x)$ reste continue entre deux limites données de la variable x, et que l'on assigne à cette variable une valeur comprise entre les deux limites dont il s'agit, un accroissement infiniment petit, attribué à la variable, produit un accroissement infiniment petit de la fonction elle-même. Par conséquent, si l'on pose alors $\Delta x = i$, les deux termes du *rapport aux différences*

$$(1) \qquad \frac{\Delta y}{\Delta x} = \frac{f(x+i) - f(x)}{i}$$

seront des quantités infiniment petites. Mais, tandis que ces deux termes s'approcheront indéfiniment et simultanément de la limite zéro, le rapport lui-même pourra converger vers une autre limite, soit positive, soit négative. Cette limite, lorsqu'elle existe, a une valeur déterminée, pour chaque valeur particulière de x; mais elle varie avec x. Ainsi, par exemple, si l'on prend $f(x) = x^m$, m désignant un nombre entier, le rapport entre les différences infiniment petites sera

$$\frac{(x+i)^m - x^m}{i} = mx^{m-1} + \frac{m(m-1)}{1.2} x^{m-2} i + \ldots + i^{m-1}$$

et il aura pour limite la quantité mx^{m-1}, c'est-à-dire, une nouvelle fonction de la variable x. Il en sera de même en général; seulement, la forme de la fonction nouvelle qui servira de limite au rapport $\frac{f(x+i) - f(x)}{i}$ dépendra de la forme de la fonction proposée $y = f(x)$. Pour indiquer cette dépendance, on donne à la nouvelle fonction le nom de *fonction dérivée*, et on la désigne, à l'aide d'un accent, par la notation

$$y' \quad \text{ou} \quad f'(x),$$

Dans la recherche des dérivées des fonctions d'une seule variable x, il est utile de distinguer les fonctions que l'on nomme *simples*, et que

Leçons de M. Cauchy. B

Third lesson of the *Calcul Infinitésimal*, 1823. Definition of the derivative of a continuous function. Published by permission of the École Polytechnique.

Cauchy had no doubts about a continuous function being differentiable. The notion of the differential of a function of a single variable was linked in a natural way to the notion of the derivative: the differential dy of $y = f(x)$ is not an infinitesimal but the finite expression $f'(x)dx$, wherein dx is an arbitrary constant (38).

In the fifth lecture, Cauchy defined the differential of a complex function of a real variable $u + v\sqrt{-1}$ by the formula $du + dv\sqrt{-1}$.

After having applied the notion of the derivative to maxima and minima problems and to the determination of expressions involving indeterminate forms, he presented the mean-value theorem in the seventh lecture: if f and f' are continuous between x and $x+i$, then

$$f(x+i) - f(x) = if'(x+\theta i) \qquad (0 \leqslant \theta < 1). \tag{5.1}$$

He constantly used this theorem in the remainder of the lectures. But, although he made use of the very first 'epsilonizations' of the concept of the limit in this proof, Cauchy lacked the basic notion of uniform continuity, a notion that was absolutely necessary if his argument was to be completely rigorous (f' must be uniformly continuous between x and $x+i$). In the subsequent lectures, Cauchy generalized his definitions of the derivative and differential to the case of functions of several variables. But, the proofs for these generalizations were far from rigorous. Moreover, his generalizations did not question whether a function with partial derivatives with respect to each of its variables is indeed differentiable. He simply took this for granted.

The second part of the *Calcul Infinitésimal* was devoted to the integral calculus. Cauchy began by giving a new definition of the integral of a continuous function $f(x)$ in the twenty-first lecture: by transforming Euler's method of approximation into an existence theorem, he defined the integral

$$\int_{x_0}^{X} f(x)dx$$

$$\left[\text{the} \int_{x_0}^{X} \text{sign had been proposed by Fourier}\right]$$

as the limit, when the interval $[x_0, X]$ is indefinitely subdivided, of the sums

$$(x_1 - x_0)f(x_1) + (x_2 - x_1)f(x_2) + \cdots + (x_{n-1} - x_{n-2})f(x_{n-1}) + (X - x_{n-1})f(X)$$

or

$$(x_1 - x_0)f(x_0) + (x_2 - x_1)f(x_1) + \cdots + (X - x_{n-1})f(x_{n-1}),$$

where

$$x_1, x_2, \ldots, x_{n-1} \text{ lie between } x_0 \text{ and } X.$$

But his proof of existence, like his proof of the mean-value theorem, fell short because of a lack of the notion of uniform continuity. In the following lecture, Cauchy derived from the definition of the definite integral a proof of the integral mean-value theorem:

$$\int_{x_0}^{X} \phi(x)\chi(x)\,dx = \phi(\xi)\int_{x_0}^{X} \chi(x)\,dx \qquad (x_0 \leqslant \xi < X, \phi \geqslant 0). \tag{5.2}$$

Cauchy generalized his definition of the definite integral. First, he introduced the notion of imaginary integral:

$$\int_{x_0}^{X} u(x) + v(x)\sqrt{-1}\,dx = \int_{x_0}^{X} u(x)\,dx + \int_{x_0}^{X} v(x)\,dx\,\sqrt{-1}. \tag{5.3}$$

In the twenty-fourth lesson, he defined improper definite integrals and investigated the case where the function to be integrated is infinite at some points of the interval of integration by using the notions of principal value of an integral and of a singular integral (39).

Later, in the twenty-fifth lecture, Cauchy proved the fundamental theorem of the integral calculus, which reduces the problem of evaluating a definite integral to that of calculating the primitive of the function to be integrated. In order to prove this theorem, he showed that the indefinite integral

$$y(x) = \int_{x_0}^{x} f(\xi)d\xi$$

is the solution of the differential equation $\frac{dy}{dx} = f(x)$, satisfying the initial condition $y(x_0) = 0$. This procedure initiated the study of the ordinary differential equations, which was restricted to the first-division course.

The concluding material on the integral calculus dealt with MacLaurin's and Taylor's formulas. By using the equation

$$\frac{d}{dx}\int_{x_0}^{x} (x-z)^m f(z)dz = m\int_{x_0}^{x} (x-z)^{m-1} f(z)\,dz$$

Cauchy proved, in the thirty-fifth and thirty-sixth lectures, the MacLaurin's and Taylor's formulas with integral remainder

$$F(x) = F(0) + \frac{x}{1}F'(0) + \frac{x^2}{2!}F''(0) + \cdots + \frac{x^{n-1}}{(n-1)!}F^{(n-1)}(0)$$
$$+ \int_0^x \frac{(x-z)^{n-1}}{(n-1)!}F^{(n)}(z)\,dz$$

and

$$f(x+h) = f(x) + \frac{h}{1}f'(x) + \frac{h^2}{2!}f''(x) + \cdots + \frac{h^{n-1}}{(n-1)!}f^{(n-1)}(u)$$
$$+ \int_0^h \frac{(h-z)^{n-1}}{(n-1)!}f^{(n)}(x+z)\,dz,$$

where F (resp. f) and all of its derivatives up to order n were implicitly supposed to be continuous in the neighborhood of 0 (resp. x).

From the MacLaurin's and Taylor's formulas with an integral remainder, Cauchy derived the MacLaurin's and Taylor's formula's with the Lagrange remainders:

$$\frac{x^n}{n!}F^{(n)}(\theta x) \quad \text{and} \quad \frac{h^n}{n!}f^{(n)}(x+\theta h) \qquad (0 \leqslant \theta < 1)$$

and with the Cauchy's remainders:

$$\frac{x(x-\theta x)^{n-1}}{(n-1)!}F^{(n)}(\theta x)$$

and

$$\frac{(h-\theta h)^{n-1}}{(n-1)!}hf^{(n)}(x+\theta h) \quad (0 \leqslant \theta < 1).$$

Afterward, in the thirty-seventh lecture, Cauchy obtained the MacLaurin and Taylor series expansion of indefinitely differential functions F in the neighborhood of 0 and f in the neighborhood of x:

$$F(x) = F(0) + \frac{x}{1}F'(0) + \frac{x^2}{2!}F''(0) + \cdots + \frac{x^n}{n!}F^{(n)}(0) + \cdots$$

and

$$f(x+h) = f(x) + \frac{h}{1}f'(x) + \frac{h^2}{2!}f''(x) + \cdots + \frac{h^n}{n!}f^{(n)}(x) + \cdots$$

by assuming that the remainder of the MacLaurin's and Taylor's formulas tends to 0 as n tends to infinity. This condition of convergence is obviously necessary, but it is not sufficient. In the thirty-eighth lecture, Cauchy gave tests of convergence for the MacLaurin and Taylor series expansions. Moreover, he observed that a convergent MacLaurin series can have a sum different from the function it represents.

In fact, Cauchy had discovered a counterexample, the function e^{-1/x^2}, continuously extended to 0, whose MacLaurin series expansion yields the null series. He first gave this counterexample in the note 'Sur le développement en séries et sur l'intégration des équations différentielles', which he presented to the Académie on January 22, 1822. The context was a critique of the method of the undetermined coefficients used in the theory of differential equations: Cauchy deduced from his counterexample that the MacLaurin series expansion of a function $f(x)$ is the series expansion of many other functions, for instance, $f(x) + e^{-1/x^2}$ and, consequently, that the method of the undetermined coefficients could not guarantee the generality of the solution of a differential equation. The mathematicians of the era, such as Lagrange, Laplace, and Lacroix, regarded the Taylor series expansion as the basis of differential calculus. That Cauchy should have relegated his treatment of this topic to the end of the discussion on integral calculus is closely connected with his adoption of the method of limits as the cornerstone of analysis.

Until Cauchy, in fact, it was thought that a function continuous in the sense of Euler could always be expanded in a power series whose coefficients were determined by use of Taylor's formula. Lagrange made use of this Taylor series expansion in his *Theorie des Fonctions Analytiques* of 1797 to define the derivative: if $f(x)$ is a given function, called the primitive function, then the derivative $f'(x)$ is the coefficient of the second term in the infinite series expansion of $f(x)$. The attention that Cauchy gave to problems relating to the convergence of series, from 1815, and later, his discovery of the counter-example e^{-1/x^2}, convinced him of the insufficiency of Lagrange's point of view as the basis for analysis (40). However, in light of the importance of the MacLaurin's and Taylor's formulas in analysis, Cauchy decided to include a new proof of them that did not use the integral calculus. This proof directly led to formulas with Lagrange remainder. Hence the MacLaurin and Taylor formulas could be presented in the course on differential calculus (41).

Far from pleasing the Conseils at the École, Cauchy's *Calcul Infinitésimal* provoked the same criticisms that his *Analyse Algébrique* had. On December 29, 1823, the matter came before the Conseil de Perfectionnement (42). A discussion ensued, and the outcome was the appointment of a commission by the Minister of the Interior. Laplace, Poisson, and Prony were its members. Its purpose was to work 'in concert with the professors of analysis and mechanics' (i.e., with Cauchy and Ampère). This commission was authorized 'to make such modifications in the course brochures (that Cauchy was preparing) that it might deem necessary, so as 'to increase their usefulness', and to present its results at the next meeting of the Conseil de Perfectionnement, in order 'that revised brochures could be distributed to the students the next year'.

To expedite the commission's work, Cauchy and Ampère were to hand over their brochures to the commission, which functioned as a veritable board of censorship, during the first semester of 1824. Ampère furnished part of the material on the second-division course in analysis (43). Cauchy, meanwhile, stalled for time. He was late in submitting the material for the first-division course. On May 6, 1824, he explained to the Conseil d'Instruction that 'he hopes to be able to devote more time to it [the rewriting of the material for the second-year course], since the revision of the study that he has been working on is now finished' (44). 'But', he went on, 'there might be another delay because the royal printing service has all of its presses busy printing material on the laws of finance'. In effect, then, between May and July 1824, 13 lectures for the first-division course were printed. (As a matter of fact, only part of the thirteenth lecture was included.) This material made up the first part of the second volume of the *Résumé des Leçons Données à L'École Polytechnique sur le Calcul Infinitésimal*. However, the printing of this material was suddenly interrupted during the summer of 1824, so that Cauchy's course material, like Ampère's, remained incomplete (45).

These 13 lectures dealt with ordinary differential equations. Just as in the course on integral calculus, where Cauchy had transformed a method of approximation into a definition of the integral, he gave here the first rigorous proof of the local existence and uniqueness of the solution of a first-order differential equation $y' = f(x, y)$, satisfying a given initial condition $y(x_0) = y_0$ by the method that would later be called the Cauchy–Lipschitz method. This method, derived from Euler's method of polygonal approximation, was based on a thorough study of the differences equations:

$$\left.\begin{aligned} y_1 - y_0 &= (x_1 - x_0)f(x_0) \\ y_2 - y_1 &= (x_2 - x_1)f(x_1) \\ &\vdots \\ y - y_{n-1} &= (x - x_{n-1})f(x_{n-1}), \end{aligned}\right. \qquad (x_0 \leqslant x_1 \leqslant x_2 \leqslant \cdots \leqslant x_{n-1} \leqslant x)$$

when x tends to x_0 (46).

Nothing in the minutes of the meetings of the Conseil d'Instruction or the Conseil de Perfectionnement explains the sudden interruption of the printing of the material from Cauchy's and Ampère's lectures. Thus, lacking evidence to the contrary, one suspects that the commission, highly displeased with the material that Cauchy and Ampère had already handed in, was responsible for the interruption of the printing. Laplace even asserted on one occasion that 'some of the material handed in (to the commission) by one of the professors of analysis [Cauchy] was so unintelligible to him that he could only make sense of it after a third reading' (47). In light of the disagreement between Cauchy and Ampère and the commission, the printing would not be resumed during the following year.

During the academic year 1824–1825, Cauchy's performance as teacher of the second-division course was continually the subject of criticism in the Conseil d'Instruction. The harshest critic was Arago, the professor of applied analysis. Arago was in a good position to judge the level of training in mathematics of the students coming from the École, since he was an admissions examiner at the École du Génie et de l'Artillerie in Metz (48). The Inspecteur des Études, J. Binet, also reproached Cauchy with digressions from the official program, as well as omissions, particularly where geometry was concerned. Cauchy replied to his critics, asserting:

> It should be noted that the lectures that were devoted to this material were not a loss for the students insofar as the discussion of these topics would serve to shorten the treatment needed for other topics later on. This will not result, therefore, in the need for additional lectures. However, it will not be possible to cover the entire course in the number of lectures alloted by the official program. The impossibility of doing so is a point that this teacher has continually raised over a ten-year period and that he has again striven to demonstrate by detailing the various parts of the course and by indicating the number of lectures that he devoted to each part, a number that he deemed to be too small. Over the preceding years, he [Cauchy] was granted a quarter-hour extension on the half-hour that is devoted to the question–answer period that precedes each lecture. This year, he used the entire half-hour for the question–answer period; and this, accordingly, diminished the length of the lecture by one fifth. Thus, it is impossible to cover the entire course within the time limits set by the official schedule (49).

When the school opened in 1825, the Conseil d'Instruction decided to make significant changes in the analysis program. Real changes in the program were what Cauchy and Ampère had been requesting. They had complained for a long time that the program was unwieldy and that it was impossible to cover all the material in the allotted time. Now there would be changes, but only at a price! A very high price it would be, too, because, 'with these changes, there will no longer be any preliminary algebraic material before differential calculus, which will begin the course immediately' (50)! Thus, the changes conceived by the Conseil d'Instruction struck a blow at Cauchy's very method of teaching, for the avowed aim of the official scheme was nothing less than a return to the old way of studying calculus by means of infinitesimals, a method that the Conseil regarded as not only simpler but more efficient as well.

Astonished and offended to see the heart of his instructional approach reduced to nothing, Cauchy presented his arguments to the Conseil d'Instruction. The minutes of the meeting include verbatim his reply to those who had criticized him and his work:

> To be sure, there are individuals who believe that certain portions of the analysis and mechanics course, particularly the first-year course, demand too much of the students. Whether that belief is well founded or not, it has nothing to do with the professor's method. It has rather to do with the large number of topics that have been added to the course since the reorganization of the École, to the program for the first-year course, and to the level of rigor that the professors require in their proofs and arguments. By simply comparing the new methods [i.e., Cauchy's methods] to those formerly used, anyone can see, without too much trouble, that the new methods are simpler [than the old ones], when they are not more rigorous. As for the remainder, I suggest that if some rigor be sacrificed, as one of the analysis professors proposed at the meeting last November 24th (51), then experience will soon show that the new methods, far from leaving the students in the dark, will enable them to learn in less time and with less effort all that they might have learned by the former methods. That much can easily be assured. Thus, for example, the professor of analysis in the first-division course was able to explain and develop the second part of the infinitesimal calculus in fewer lectures than the (official) program allots to it (52).

After this last speech in his own defense, Cauchy was obliged to submit to the Conseil's demands and to give up the publication of *Calcul Infinitésimal* (53). In response to the criticism that he was sacrificing geometric applications to pure analysis, Cauchy undertook the publication of a three-volume work entitled *Leçons sur les Applications du Calcul Infinitésimal à la Géométrie* (54), in which he presented the mathematical framework of his work in mechanics. The first volume, devoted to differential calculus, appeared in July 1826. In this elementary treatment of differential geometry, Cauchy developed, among other things, the theory of the radii and centers of curvature of an arbitrary curve and the theory of the orders of contact of curves and of curved surfaces. In the fifteenth lecture, he examined problems relating to the centers, diameters, and axes of curved surfaces, paying special attention to the quadrics. This study led him to the problem of determining the three eigenvalues of a linear symmetric mapping of a three-dimensional space. These values, he showed, are always real (55). In addition, from May 1826 to April 1827, he published numerous studies in *Exercices de Mathématiques* that were based on his mechanics teaching at the École and at the Faculté des Sciences (56).

Starting in November 1826, the courses in analysis and mechanics—particularly Cauchy's courses—were closely supervised by the Conseil de Perfectionnement. Each year, the examiners of the graduates made a report to the Conseil on the courses taught by Cauchy and Ampère. Prony was responsible for 'overseeing' Cauchy's teaching. He was generally critical in the series of annual reports that he sent in for 1825–1826, 1826–1827, 1827–1828, 1828–1829, and 1829–1830. In his report on the academic year 1825–1826, Prony criticized the excessive sophistication of Cauchy's teaching:

> I will finish my observations on the course in pure analysis by manifesting the desire to see the use of the algorithm of imaginaries [i.e., complex numbers] reduced to what is strictly necessary. I have been astonished, for instance, to see the expression of the element of a curve, given in polar coordinates, derived from an analysis using this algorithm; it follows much more quickly and with greater ease from a consideration of infinitesimals. It is quite true that the introduction of the imaginaries into analytic calculations is often very useful. However, the fact remains that by using them unnecessarily in a mathematics course at the École Polytechnique one deviates from a very important goal: the goal of learning to exercise thought and develop powers of judgment (57).

November 17, 1826, the Conseil de Perfectionnement appointed a new commission in analysis, mechanics, social arithmetic, and geodesy. This new commission, made up of Laplace, Biot, Binet, Poisson, Prony, and Puissant, was to function within the framework of a general reorganization of the academic programs, because the pamphlets and brochures that professors Cauchy and Ampère were required to submit to the Conseil de Perfectionnement were not always published. On this occasion, the minutes of the Conseil's meeting state that:

> The Marquis de Laplace called the Conseil's attention to the attempts made by the Conseil over several years to improve education and training in mathematics by a simplification of instructional methods. He also referred to the fact that the commission that had been appointed by the Minister of the Interior to implement this goal, of which he [Laplace] was president, had tried in vain for three years to get the professors to submit written material on their courses that would satisfy the

> commission. He added that, as a consequence of the requests of the Conseil de Perfectionnement at its last meeting, and as a result of the Conseil's urging, the Minister of the Interior had directed the professors in question to present printed material on what they intended to do in their courses... but the other professor concerned, Monsieur Cauchy, has only submitted some pamphlets that the commission found unsatisfactory. As a result, it has been impossible, at least up to now, to get him to comply with the Conseil's instructions and the Minister's decision (58).

Following this declaration from Laplace, the Gouverneur of the École declared that 'in case other attempts are made, and are unsuccessful, then he [the Gouverneur] would consider himself as obliged to propose to the government that it take harsh measures, which he would find very distasteful'.

The grave warnings from the administration of the École notwithstanding, the development of the materials on the analysis and mechanics course made no progress in the following years. Prony's report on the academic year 1826–1827 insinuated that the students did not make use of Cauchy's teaching in the examinations for passing from the second to first division:

> According to the unfavorable opinion expressed by the analysis professor on the performance of the students during the 1827 [i.e., 1826–1827] year, one would have expected some fairly unsatisfactory results on the examinations. However, these results were better than anyone had grounds for expecting. Of the 143 students tested, 100 to 110 showed a more or less sustained mastery. It is to be believed that those students who neglected their studies during the course worked hard to catch up the last time during the 5 or 6 weeks between the end of the course and the examination period. But this kind of improvised instruction has drawbacks that could become serious, and it is necessary to take definite steps to prevent the students from being again in the situation of having to substitute learning by cramming. Learning, indeed, will leave no traces in memory, it will not interest and profit the students if it does not follow a methodically gradual course. This graduation is a condition that is incompatible with great haste. Moreover, it should be considered that such cramming has a bad influence on the health of the students, and one can cite examples of this (59).

In the same report, Prony censured Cauchy's mechanics teaching, which seemed to him too abstract.

On the basis of a report made by Poisson, in the name of the commission appointed on November 17, 1826, the Conseil de Perfectionnement decided on February 15, 1828 to take charge of the printing of the brochures. The plan was for stenographers to make exact records of the lectures given in the course on analysis and mechanics, as well as in other courses. These records were then to be revised and corrected by the professors and then printed. In addition, the professors of analysis and mechanics were to hand over to the commission complete topical outlines of their courses. These outlines would serve as basic guides for the construction of new syllabi and programs. Once this material had been approved by the Inspecteur des Études and by the examiners, it would be printed. 'The material for both divisions', reads the instructions, 'should not exceed two volumes of 400 pages in quarto format, one for analysis and the other for mechanics. The main purpose of these volumes will be to guide the students in the direction of those applications that are most necessary in the public services'. The commission set very strict timetables for the course stenographers to complete their tasks, as well as for the professors to have the topical outlines for the new course material written and submitted.

These arrangements prompted protests of indignation from the professors of analysis and mechanics. Cauchy sent a letter (which, unfortunately, cannot be found in the archives of the École Polytechnique) to the Minister of the Interior. Ampère resigned from the École in May 1828, after declaring that it was impossible to complete the required two volumes of 400 pages in quarto in one year. Finally, the Conseil d'Instruction decided against making stenographic records, because their corrections would have required too much work from the professors (60).

As with all the other decisions made by the various conseils at the École, the resolutions of February 15 were thus dead letters, completely ineffectual. Cauchy was content with the June 1828 publication of the second volume of the *Leçons sur les applications du Calcul Infinitésimal à la Géométrie*, which was devoted to the applications of the integral calculus. In this work, Cauchy investigated the classical methods used for the rectification of curves, the quadrature of surfaces, and the computation of volumes.

Prony was still unsatisfied with Cauchy's teaching. He wrote in the report on the academic year 1827–1828 (first-division course):

> The professor considered several objections that were made last year concerning the instructional system. Nevertheless, there remains a vagueness in the exposition and the use of general theories. Sometimes a student who can discuss generalities quite well comes to a complete stop or is embarrassed when it comes to applications to particular cases. It is to be feared that abstract theories, isolated and on their own, are not remembered. He [Prony] continues to insist on the suppression of considerations that have to do with the method of limits and that extend proofs without making them concrete. This observation is of particular importance when it has to do with a course for youths who will become engineers (61).

In May 1829, Cauchy published his *Leçons sur le Calcul Différentiel*, which replaced the first part of *Calcul Infinitésimal* of 1823, which, by then, had gone out of print (62). This work was more logically developed than the work of 1823. In this new study, Cauchy introduced the theory of the order of infinitesimals, which he used to prove Taylor's formula (with Lagrange and

Cauchy's remainders), now included once again in the treatment of differential calculus. Moreover, in the eleventh, twelfth, and thirteenth lectures, he developed a theory of functions of a complex variable. This theory had not been presented at all in *Calcul Infinitésimal* of 1823. In developing it, he extended the definition of the derivative of a real function to functions of a complex variable, and this extension allowed him to apply Taylor's formula to complex functions. Unfortunately, however, the crucial point, the difference between differentiability in the real and complex cases, eluded him altogether. Thus, he continued to believe that, just as in the case of functions of a real variable, a continuous function of a complex variable is differentiable.

Obviously, *Leçons sur le Calcul Différentiel* did not meet the terms that had been laid down by the Conseil, and Prony was obliged to declare to the Conseil de Perfectionnement that:

> The professor of analysis and mechanics has now published the portion of his course material on differential calculus. It can be seen that he intended to comply—up to a certain point, at least—with the conditions imposed by the Conseil. However, those conditions have been only imperfectly fulfilled in this new work; and in general, the methods used in this publication are not the same as those that have been taught (in his classes). That being the case, the present students, like those in the past, have no way of reviewing and studying the material they have covered in class except by way of their in-class notes (63).

By the end of the academic year 1829–1830, Cauchy had finally written and gradually submitted the complete topical outlines for his first- and second-year courses in analysis (64). But, no work was done on the mechanics course. Here, matters rested. During the summer of 1830, following the July Revolution, Cauchy left France and did not return to the École when the November 1830 term began. Neither the third volume of his *Leçons sur les Applications du Calcul Infinitésimal à la Géométrie* nor *Leçons sur le Calcul Intégral* were ever published (65).

Chapter 6

From the Theory of Waves to the Theory of Light

Mathematical physics, like teaching, was an extremely fruitful source of inspiration for Cauchy. The study of various physical phenomena, of course, provided a natural setting for many difficult mathematical problems. The challenge facing Cauchy was to develop and articulate new mechanical models consistent with the experimental facts and to solve the equations of equilibrium and motion derived from them. In these areas, Cauchy undertook fundamental research work, which required a great deal of time and effort. His main contribution to mathematical physics remains today the continuum theory of elasticity; however, his longest and most ambitious research efforts centered on the molecular theory of light. In fact, the purely mathematical theories, such as the theory of elliptic functions and the theory of algebraic equations, having few physical applications, never particularly interested Cauchy. So, apart from his arithmetical works, most of his studies dealt with mechanics or mathematical physics in one way or another, especially those on linear differential equations, complex function theory, and linear algebra. In this chapter, we will adopt the current tensorial notation with the summation convention.[1] This unimportant anachronism simplifies the involved mathematical writing that Cauchy used in his papers on mechanics.

Cauchy undertook his first work on mathematical physics for the Grand Prix de Mathématiques at the Académie des Sciences in 1815. The subject treated in this competition was the theory of wave propagation on the surface of a liquid (1). Laplace had already examined the question in the case of a liquid of constant depth. He concluded from his investigation that waves on the surface of a liquid are propagated with a constant velocity. Not long after that, solving the problem for the case where the depth of the liquid is very small and

---

[1] The symbol of derivation relative to time is ∂_0; ∂_i, ∂_j, and ∂_k are the symbols of derivation relative to the space coordinates. According to the summation convention, whenever a lowercase italic subscript appears twice in the same monomial, this monomial stands for the sum of the three terms obtained by successively giving to this index the values 1, 2, and 3.

constant, Lagrange determined that the wave velocity was proportional to the square root of the depth. By assuming that the wave propagation was perceptible only at a very small depth, he generalized his formulas to the case of a liquid of indefinite depth in his *Mécanique Analytique* (2). On January 3, 1814, the commission of the Grand Prix, which had been appointed on November 15, 1813, a few months after Lagrange's death, proposed a problem dealing with the matter for the competition:

> A heavy mass of fluid, originally at rest, has been set in motion by the action of a given cause. Determine the shape of the boundary surface at the end of a specified time and the velocity of each of the molecules belonging to this surface.

In principle, the competitors were to work in secret, without discussing their results with others. However, Cauchy kept Laplace informed from the very beginning, of the conclusion he reached in his research, namely, that wave propagation within an incompressible fluid, that is, a liquid, is not uniform, but is uniformly accelerated. A discussion with Poisson, who was also investigating this problem (and who had also reached the same conclusions, albeit by a heuristic approach), convinced Cauchy that his findings were correct. Accordingly, on July 24, 1815, he presented a note entitled 'Sur le problème des ondes' to the Académie. In this unpublished note, he gave the four laws of wave propagation within a liquid that were later stated in the third part of the award-winning study (3).

Cauchy submitted his paper anonymously on October 2, 1815. He generalized the problem that the Académie had originally posed and investigated the motion of an arbitrary molecule, that is, particle of the liquid in the case of a potential flow caused by a perturbation that he assumed to be initially weak and confined to a small area of the boundary surface. The initial conditions are the ordinates $\Xi_3 = F(\xi_1, \xi_2)$ of the boundary surface and the initial impulse $Q^{(0)}$ on it.

In the first part of this study, Cauchy examined the state of the liquid at the initial time. He pointed out that if the initial impulse $q(\xi_i, 0)$ is $q^{(0)}$ then the initial velocity $v_i^{(0)} = v_i(\xi_j, 0)$ of any particle with the Lagrangian coordinates (ξ_i, t) can be expressed in the form

$$v_i^{(0)} = -\frac{1}{\rho}\partial_i q^{(0)}$$

where ρ is the density of the liquid. Thus, in modern terms, the initial velocity field derives from the velocity potential $\dfrac{q^{(0)}}{\rho}$. By using the theory of determinants he had developed in 1812, Cauchy then established the equation of continuity for a liquid in Lagrangian variables (ξ_i, t):

$$\det(\partial_i x_j) = 1,$$

where $x_i(\xi_j, t)$ are the coordinates of the particle in a given rectangular Cartesian coordinate system. In particular, for $t=0$,

$$\det(\partial_i \xi_j) = 1.$$

From this Lagrangian expression, he deduced the equation of continuity for a liquid in Eulerian variables $v_i(x_j, t)$ at the initial time

$$\partial_i v_i^{(0)} = 0$$

and obtained the Laplace equation

$$\partial_{ii} q^{(0)} = 0$$

for the initial impulse $q^{(0)}$ within the liquid, since the flow is potential. He applied these equations to the boundary surface and expressed the solutions satisfying the boundary conditions by means of Fourier integrals. Thus, he obtained, in integral form, the values of $q^{(0)}$ and $v_i^{(0)}$ at all points of the liquid.

In the second part, Cauchy investigated the state of the liquid at an arbitrary time. Using the so-called fundamental property of the fluids, that is, the equality of the pressures around any point, he derived Euler's dynamic equation

$$a_i = F_i + \frac{1}{\rho}\partial_i p, \tag{6.1}$$

where a_i is the acceleration of a particle, F_i is a given specific body force, and p is the pressure on the particle. He expressed the acceleration a_i first in Lagrangian variables and then in Eulerian variables. Then, he proved that there is potential flow at any time:

$$v_i(x_j, t) = -\frac{1}{\rho}\partial_i q(x_j, t), \tag{6.2}$$

and he deduced from the equation of continuity $\partial_i v_i = 0$ combined with Eq. (6.2) a Laplace equation:

$$\partial_{jj} q = 0.$$

Then, Cauchy assumed that F_i is derived from a potential ϕ and transformed Euler's equation Eq. (6.1), into the Cauchy–Lagrange equation:

$$\frac{p - \partial_0 q}{\rho} + \frac{v_i v_i}{2} - \phi = 0.$$

If ϕ is the gravity and v_i is infinitesimal, this equation yields the value of the pressure p at an arbitrary point in the liquid:

$$p = \partial_0 q - \rho g x_3.$$

Cauchy applied these results to the boundary surface where p was assumed to vanish. Thus, knowing the impulsion Q on the surface, he had the ordinates X_3

of the surface:

$$X_3 = \frac{1}{\rho g} \partial_0 Q$$

and, after differentiating, its vertical velocity:

$$V_3 = \frac{1}{\rho g} \partial_0^2 Q \tag{6.3}$$

From Eqs. (6.2) and (6.3), it follows that

$$g \, \partial_3 Q = - \partial_0^2 Q.$$

This equation substituted in the Laplace equation yields the partial differential equation

$$\partial_0^4 Q + g^2(\partial_1^2 Q + \partial_2^2 Q) = 0 \tag{6.4}$$

for the impulse at any time on the boundary surface. Cauchy solved Eqs. (6.2), (6.3), and (6.4) once again by Fourier transforms. The solutions have to satisfy the initial conditions. In this way, he obtained the values of the unknowns q, Q, V_i, x_i, $p(x_i)$, etc., depending on the initial conditions, in integral form. He discussed the results in the final part of his study. He showed that this discussion amounted to investigating nonconvergent integrals of the type

$$K = \int_0^{+\infty} \cos \sqrt{2k\mu} \cos \mu \, \mathrm{d}\mu \qquad (k > 0)$$

and to approximating them by means of asymptotic expansions. So, he obtained the laws of the wave propagation within a liquid and the shape of the boundary surface at any time.

During the time of his research, Cauchy made additional investigations: he studied the case in which the depth of the liquid is finite, and he discussed the equation of continuity and the equations of motion for an elastic fluid, in order to determine the wave motion on the boundary surface between air and water. He never did publish these supplements (4).

Cauchy's paper easily won the prize in the competition. Still, it did not attract much attention, because it was eclipsed by two studies that Poisson presented when the competition was over. Whereas Cauchy had to wait until 1827 to see his paper published in the *Savants Étrangers*, Poisson published his own studies in the *Mémoires* of the Académie as far back as 1818. Poisson used methods similar to Cauchy's. However, by giving a clearer and more thorough discussion of the integral formulas, Poisson established the existence of a wave propagation with a constant velocity in addition to the uniformly accelerated wave propagation. Cauchy took advantage of the delay in the publication of his paper of 1815 by adding, in 1821, 2 new notes to the 13 initial notes. He later added 5 other notes to this paper before giving the final manuscript to the secretary of the Académie on May 27, 1824 (5).

Computations relative to the wave propagation on the surface of an incompressible fluid. Manuscript by Cauchy, 1815, Sorbonne Library, ms 2057. Published by permission of the École Polytechnique.

Over the following years, Cauchy did not present any further studies on mathematical physics or mechanics to the Académie, even though he was a member of the mechanics section. Nevertheless, he did not completely abandon these disciplines; he continued to teach them at the École Polytechnique and, in 1817, at the Collège de France. This teaching gave him the opportunity to master the mechanical concepts and the mathematical tools he needed to create continuum mechanics. Cauchy stated the fundamental equations of hydrostatics and hydrodynamics by Euler's method in his course on mechanics at the École Polytechnique. Unlike Euler, however, he based the theory of perfect fluids on the property that the hydrostatic pressure is normal to the surface on which it acts, and he argued that the property of the hydrostatic pressure expected to be equal in all directions can be derived from it.

In October 1821, Cauchy began teaching a course on mechanics at the Faculté des Sciences, and this course no doubt provided the inspiration for further research in mathematical physics: on the one hand, he carried on the investigation of the linear partial differential equations with constant coefficients, using Fourier transforms, and on the other hand, he began to develop his continuum theory of elasticity. Thus, he submitted a major paper to the Académie on continuum mechanics on September 30, 1822 (6). In this paper, entitled 'Recherches sur l'équilibre et le mouvement intérieur des corps solides ou fluides, élastiques ou non élastiques', he set forth the basis of his continuum theory of elasticity, which he thought to be applicable to nonmaterial media, such as luminiferous ether or caloric fluid, as well as to solids and fluids.

By this time, the problem of the elasticity had begun to be thoroughly investigated from three different points of view: an empirical point of view in the research on the strength of the materials, a physical one in Fresnel's research on the wave propagation of light, and a mathematical one in a series of studies on the small deformations of curves and surfaces (7). As an engineer, Cauchy knew the empirical approach of the study of the strength of materials. Inspired by the works of Coulomb, he had undertaken some investigations on the strength of the bridges and the theory of the arches at the École des Ponts et Chaussées in 1809 and 1810 (8). He had also worked on the site of the Ourcq Canal under Girard, who had published *Théorie Analytique de la Résistance des Matériaux*, a work of considerable renown, in 1798. He remained interested in these practical problems, and on August 9, 1819, he even presented a detailed report to the Académie on a paper of Duleau's on the strength of wrought iron (9).

On the other hand, the theory of elastic surfaces had been of interest to the Parisian scientific community since 1810, when, following a series of spectacular experiments by Chladni on the nodes of a vibrating plate, the Académie des Sciences announced a competition dealing with this matter for the Grand Prix de Mathématiques: the Académie demanded that the competitors 'give a mathematical theory of the vibrations of elastic surfaces' and 'compare this theory with experimental results'. In spite of the difficulty of the problem, Sophie Germain successfully determined the solution and was awarded the prize in the 1816 competition, after an unsuccessful showing in the 1812 competition and an honorable mention in that of 1814. In the meanwhile, Poisson presented the paper 'Sur les surfaces élastiques' to the Académie on August 1, 1814 (10). In this work, he used a molecular hypothesis to obtain the equation of a flexible elastic surface equally stretched in all directions, which Sophie Germain had already established with help from Lagrange in 1813. Sophie Germain published her results in 1821 in a work that she promptly sent to Cauchy (11).

In fact, a few months before, on August 14, 1820, Navier had submitted to the Académie a paper on the same theory of elastic plates (12). He had also distributed lithographic copies of his paper to a few members in the scientific community, Cauchy especially. Motivated by an engineer's concern, namely, the application of rational mechanics to the study of the strength of materials, Navier calculated the small deformations of a weighted plate equally stretched in all directions. His paper began with a derivation of the equation that Sophie Germain and Poisson had obtained for an elastic surface. However, his derivation took the thickness of the plate into account. In his analysis of the elastic forces, Navier considered separately the forces produced by elongations and contractions of the plate and those produced by the bending. In order to obtain mathematical expressions for the latter forces, Navier examined an arbitrary small element of the plate that, before the bending deformation, has the shape of a right circular cylinder whose elevation is precisely equal to the thickness of the plate and that, after the bending, is assumed to have the shape of a truncated cone, one base of the cylinder dilating and the other contracting. Navier assumed, in general incorrectly, that during the bending the elastic forces act normally to the faces of the cone (13).

A reading of Navier's study inspired Cauchy's paper of September 29, 1822. As has been noted, Navier had separated the forces produced by the contractions and dilations of the plate and the bending forces. In contrast, Cauchy combined the elastic forces transmitted onto an arbitrary isolated region inside the solid. In order to describe these internal forces, which are known today as stresses, Cauchy took the very natural route of generalizing the well-known definition of hydrostatic pressure (14). Like hydrostatic pressure, the stresses at a point inside a solid body are contact forces that act on each surface element passing through the point. However, there is a great difference between fluids and solid bodies. In a fluid, the pressure is always normal to the surface element on which it acts. We have seen that Cauchy, at the École Polytechnique, based his lectures on hydrostatics on this principle (15). In a solid, however, the stresses generally are not normal to the surface on which they act, contrary to Navier's assumption.

The application of the classical laws of equilibrium to conveniently chosen elements of volume allowed Cauchy to express the stresses at a given point in terms of nine stress-components T_{ij}, of which six are pairwise equal ($T_{ij} = T_{ji}$), with respect to a rectangular Cartesian coordinate system. In modern terms, the state of stress at a point of a continuum was mathematically defined by a symmetric tensor, namely, the stress tensor **T**. Cauchy's method of proof is today regarded as a classic. In the case of a fluid at rest, the nine stress components reduce to three nonnull components that are all equal to the hydrostatic pressure p. Cauchy developed these first results by 1821.

> I was at that point [Cauchy wrote in 1822] when M. Fresnel came to talk with me about some investigations on light that, as yet, he had only presented in part to the Institut. I learned that he had obtained a theorem analogous to my own, his result being based on certain laws according to which the elasticity emanating from a single given point varies in different directions (16).

In two supplements to his study 'Sur la double réfraction', one of which was presented in January 1822 and the other the following March, Fresnel had, in fact, determined the molecular forces acting upon a single, slightly displaced molecule of ether from other molecules around it; the results were thus similar, in this very special case, to the ones Cauchy had obtained in the general case of a continuum (17). In developing this analysis, Fresnel was attempting to construct a molecular ether across which transverse vibrations could be propagated. In this way, he would be able to substantiate his hypothesis that light waves were exclusively transverse.

Still, as Cauchy wrote in 1823, he was as yet a long way from being able to set forth the general equations of equilibrium and motion inside a solid. Accordingly, Fresnel's achievement was, in a sense, a source of inspiration for Cauchy and an encouragement to continue his own investigations. Finally, during the second half of 1822, he obtained the desired results. Let us see how he proceeded. Cauchy considered only the infinitesimal displacements u_i of any point x_i in the neighborhood of a referential point $x_i^{(0)}$. Since the u_i are infinitesimal, the velocity v_i of x_i is reduced to $\partial_0 u_i$ and its acceleration to $\partial_0 v_i$. By examining the stresses on two opposite faces of an arbitrary small parallelipiped of the continuum, subjected to a specific body force F_i, Cauchy first determined the equation of equilibrium. Under the same hypothesis, assuming in addition a specific inertial force $-a_i$, he easily deduced the equation of motion in a continuum, which generalized Euler's dynamic equation, Eq. (6.1). In tensorial notation, this equation, known today as Cauchy's dynamic equation, may be written as

$$a_i = F_i + \frac{1}{\rho}\,\partial_j T_{ij}. \tag{6.5}$$

However, Cauchy still had to express the six independent stress components at a point as functions of the displacements. In his study of 1822, he assumed that only the linear strains, i.e., linear condensations and dilations about a point, needed to be taken into account. The key idea, no doubt inspired by Fresnel's work, thus consisted of investigating the geometrical properties of the stresses

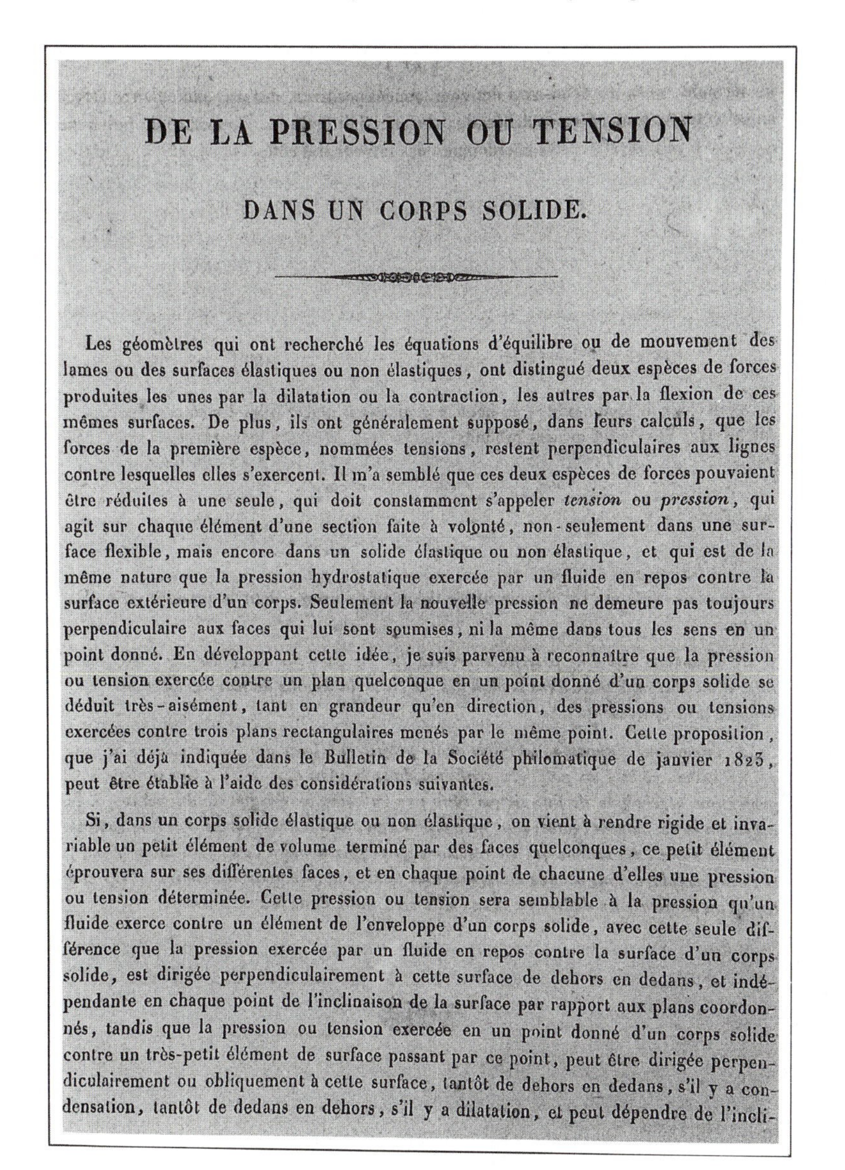

DE LA PRESSION OU TENSION

DANS UN CORPS SOLIDE.

Les géomètres qui ont recherché les équations d'équilibre ou de mouvement des lames ou des surfaces élastiques ou non élastiques, ont distingué deux espèces de forces produites les unes par la dilatation ou la contraction, les autres par la flexion de ces mêmes surfaces. De plus, ils ont généralement supposé, dans leurs calculs, que les forces de la première espèce, nommées tensions, restent perpendiculaires aux lignes contre lesquelles elles s'exercent. Il m'a semblé que ces deux espèces de forces pouvaient être réduites à une seule, qui doit constamment s'appeler *tension* ou *pression*, qui agit sur chaque élément d'une section faite à volonté, non-seulement dans une surface flexible, mais encore dans un solide élastique ou non élastique, et qui est de la même nature que la pression hydrostatique exercée par un fluide en repos contre la surface extérieure d'un corps. Seulement la nouvelle pression ne demeure pas toujours perpendiculaire aux faces qui lui sont soumises, ni la même dans tous les sens en un point donné. En développant cette idée, je suis parvenu à reconnaître que la pression ou tension exercée contre un plan quelconque en un point donné d'un corps solide se déduit très-aisément, tant en grandeur qu'en direction, des pressions ou tensions exercées contre trois plans rectangulaires menés par le même point. Cette proposition, que j'ai déjà indiquée dans le Bulletin de la Société philomatique de janvier 1823, peut être établie à l'aide des considérations suivantes.

Si, dans un corps solide élastique ou non élastique, on vient à rendre rigide et invariable un petit élément de volume terminé par des faces quelconques, ce petit élément éprouvera sur ses différentes faces, et en chaque point de chacune d'elles une pression ou tension déterminée. Cette pression ou tension sera semblable à la pression qu'un fluide exerce contre un élément de l'enveloppe d'un corps solide, avec cette seule différence que la pression exercée par un fluide en repos contre la surface d'un corps solide, est dirigée perpendiculairement à cette surface de dehors en dedans, et indépendante en chaque point de l'inclinaison de la surface par rapport aux plans coordonnés, tandis que la pression ou tension exercée en un point donné d'un corps solide contre un très-petit élément de surface passant par ce point, peut être dirigée perpendiculairement ou obliquement à cette surface, tantôt de dehors en dedans, s'il y a condensation, tantôt de dedans en dehors, s'il y a dilatation, et peut dépendre de l'incli-

Definition of the stress in a continuum. 'De la pression ou tension dans un corps solide', *Exercices de Mathématiques*, **2**, 1827, p. 42. Published by permission of the École Polytechnique.

and strains. As a matter of fact, Cauchy had already developed the methods of linear algebra he needed to calculate the moment of inertia of a solid and to solve the linear differential equations. Cauchy succeeded therefore in reducing all the stresses, pressures, or tensions at a point to three principal stresses acting along the axes of a suitable chosen quadric, namely, Cauchy's stress quadric, generally an ellipsoid. The principal stresses, showed Cauchy, are normal stresses.

The analysis of the state of strains at the same point yielded a similar geometrical representation: the linear contractions or dilatations $\varepsilon^{(\boldsymbol{\mu})}$ along all the directions $\boldsymbol{\mu}$ about the point (i.e., the rates of extension) can be expressed by means of three principal strains along the axes of a second quadric, namely, Cauchy's strain quadric. Formally, Cauchy obtained the symmetric strain tensor $\mathbf{U}$, whose components are $U_{ij} = \partial_{(i}u_{j)}[\partial_{(i}u_{j)} = \frac{1}{2}(\partial_i u_j + \partial_j u_i)]$; he showed that $\varepsilon^{(\boldsymbol{\mu})} = U_{ij}\mu_i\mu_j$ and that the tensorial invariant U_{ij} represents the cubical dilatation. However, Cauchy did not investigate the kinematical properties of strain and never used the fundamental concept of shear.

Cauchy was now able to obtain, in the case of very small deformations, the expressions of the stresses as functions of the strains by using a double hypothesis, which he articulated in the following terms:

> That given, it is clear that in an elastic solid, if the tensions or pressures depend only on the condensations or dilations, then the principal tensions or pressures are directed in the direction of the principal condensations or dilations. Moreover, it is natural to assume, when the displacements of the molecules are very small, that the principal tensions or pressures are respectively proportional to the principal condensations or dilations (18).

Cauchy generalized Hooke's law. He implicitly assumed the isotropy of the continuum, obtaining the equation

$$T_{ij} = kU_{ij}. \tag{6.6}$$

Cauchy considered not only homogeneous (k constant) but also heterogeneous (k function of $x_i^{(0)}$) isotropic media. In order to formulate the equations of motion in a continuum, a substitution had to be made in his dynamic equation, Eq. (6.5), replacing the stress components by their respective expressions as functions of the strain components given by the linear relation, Eq. (6.6). If the continuum is homogeneous, this substitution yields the equation

$$a_i = F_i + \frac{k}{2\rho}(\partial_{jj}u_i + \partial_i U_{jj}), \tag{6.7}$$

from which

$$\partial_0^2 U_{ii} = \partial_i F_i + \frac{k}{\rho}\partial_{ii}U_{jj}. \tag{6.8}$$

Cauchy applied his theory to the study of the propagation of sound in an isotropic elastic body and obtained an equation identical to that of the propagation of sound in air. He also examined the case of a nonelastic body in which the stresses depend only on the instantaneous strains. He thus established a significant analogy between the propagation of heat and the wave propagation in an entirely nonelastic body.

In this work, which marks a turning point in the history of continuum mechanics, Cauchy gave a stunning exposition, a scientific study resplendent with intellectual clarity. 'Never had Cauchy given the world a work as mature from the outset as this', remarked Hans Freudenthal (19). It was not surprising. Cauchy's study of hydrodynamics, for which he had derived the equations in 1815, was a reliable guide to his research in continuum mechanics. Following Euler, Cauchy systematically developed the analogy between fluids and solids in his investigation of the continua, using the classical methods he taught at the École Polytechnique (20). This remark in no way detracts from the considerable merit of Cauchy's achievement.

Strange to say, six years were to pass before Cauchy published the results contained in the paper of 1822. In fact, Cauchy informed the Académie of his investigation on September 30, 1822, without giving a reading or lecture on it; he did not even give the manuscript to the Académie. Rather, he published an abstract in the January issue of the *Bulletin* of the Société Philomatique and gave a reading of this résumé at the Société's meeting on February 22 (21).

The work did not appear until 1827 and 1828 in a series of articles in *Exercices de Mathématiques* (22). These articles contained a number of improvements. However, Cauchy never published his paper, 'Equations du mouvement des fluides élastiques ou incompressibles', which was presented to the Académie on January 27, 1823 (23). Why did Cauchy wait almost four years to publish the details of his investigations of 1822? We think that two possible reasons can explain this surprising decision.

The first of these reasons has to do with the conditions surrounding both the presentation of the paper of September 1822 and the publication of the résumé in the January 1823 issue of the *Bulletin* of the Société Philomatique. When he presented his paper to the Académie, Cauchy acted in a way that was certainly clumsy, if not downright tasteless. He acknowledged that the work he was now presenting was inspired by Navier's paper of August 14, 1820, which actually had not yet been evaluated by the commission responsible for examining it. On the other hand, Cauchy's study focused on the same topic as Navier's famous paper, 'Sur les lois de l'équilibre et du mouvement des corps solides élastiques' of May 14, 1821, which was also awaiting evaluation by an academic commission. On October 6, 1822, a week after Cauchy's paper was presented to the Académie, Navier wrote a letter to the president of the Académie demanding that both of his papers be evaluated promptly (24). In this letter, he noted (but made no official protest or charges on the matter) that Cauchy had been doing research that was very similar to his own investigations.

Up until this time, the affair had been confined to the Académie. But, by publishing the abstract that he had read at the Société Philomatique on February 22, Cauchy brought matters to a head. On March 1, 1823, Navier called the Société's attention to the existence of his two papers (25). Thereafter, the scientific community began to take sides with Navier. The Société Philomatique resolved that Navier's observations would be included in the minutes of its meetings and would also appear in the coming issue of its *Bulletin.* Moreover, Augustin Fresnel published a quite harsh note about Cauchy and the quarrel over priority in the same *Bulletin* (26). First, Fresnel called attention to the fact that Navier's paper of August 14, 1820, had not been *published* (Fresnel himself underlining the word) and remarked (incorrectly) that Cauchy was the reporter of the evaluating commission (27). After questioning Cauchy's interpretation of Navier's paper of August 14, 1820, Fresnel went on to call attention to the existence of Navier's second paper of May 14, 1821. He concluded his note with these words:

> The work that M. Cauchy has just published seems to have the greatest similarity to the paper that was discussed here [that is, Navier's second paper], and it is important that the date of this paper be recalled and certified.

Fresnel was personally concerned, as it were, in this matter and clearly suspected that Cauchy might have plagiarized Navier. But such was not the case with Fourier. In the *Analyse des Travaux de l'Académie* for 1822, after having mentioned the study Cauchy announced on September 30, 1822, Fourier nevertheless pointed out that Navier and Fresnel had investigated the same problem (28). Cauchy seems to have been aware of these remarks and rumors, mutterings that were so often repeated that they could not have been made in innocence. He thus gave up the idea of publishing his study right away and of presenting a second paper on the theory of thin plates that he had been working on (29). Finally, he kept his research works on continuum mechanics to himself until Navier published his own paper in 1827 (30). The postponement was unfortunate. Cauchy nevertheless took advantage of the long delay between the presentation of his study before the Académie in 1822 and its publication in *Exercices de Mathématiques* of 1827 and 1828 to make an important change in his theory of elasticity for isotropic bodies.

As has been noted, in 1822, he had assumed a linear relation between the principal stresses and the principal strains that was determined by a single elasticity coefficient k (see Eq. 6.6). This continuum theory with a single coefficient for isotropic media was incompatible with the theory that Navier had developed from a molecular model in 1822. In the study 'Sur les équations qui expriment les conditions d'équilibre ou les lois du mouvement intérieur d'un corps solide élastique ou inélastique', published in 1828, Cauchy gave a new hypothesis. Regarding each principal stress as being composed of two distinct parts, he assumed that one was proportional to the principal strain and the other to the cubical dilatation U_{ii}. Thus, in tensorial notation, the relations between stress components and strain components yield the equation

$$T_{ij} = kU_{ij} + KU_{kk}\delta_{ij} \tag{6.9}$$

instead of Eq. (6.6) and, if the continuum is homogeneous,

$$a_i = F_i + \frac{k}{2\rho}\partial_{jj}u_i + \frac{k+2K}{2\rho}\partial_i U_{jj} \tag{6.10}$$

from which

$$\partial_0^2 U_{ii} = \partial_i F_i + \frac{k+K}{\rho}\partial_{ii} U_{jj}. \tag{6.11}$$

In this way, Cauchy obtained the continuum theory with two elasticity coefficients K and k for isotropic media. Navier's theory, which was derived from the molecular hypothesis, was thus the particular case $k = 2K$ of the more general theoretical framework that Cauchy had now formulated.

Another reason probably prompted Cauchy to delay the publication of his 1822 study. He had based his theory of elasticity on the classical hydrostatic model and, accordingly, had regarded solids and fluids as continua. But, advances in molecular physics had made this concept out of date in the 1820's. The French physicomathematicians now tried to explain all sorts of phenomena, such as heat, sound, light, electricity, and magnetism, as well as the elastic properties of bodies, in terms of attractions and repulsions between the material molecules constituting the bodies and, in a wider sense, between the immaterial molecules constituting fluids like caloric, light and electricity. In France, the majority of the great figures who worked in mathematical physics were won over to this theoretical position: Laplace, of course, and his disciple Poisson, but also, to a certain extent, Ampère, Fresnel, and Navier. The molecular model was thus, in various forms, so much a part of the thinking of that era that many physicomathematicians, such as Fresnel in optics and Ampère in electrodynamics, interpreted physical theories they had conceived and developed by means of other schematic conceptualizations of the real world in terms of this model. Navier adopted the molecular model in his paper of May 14, 1821, regarding an elastic body as 'a collection of molecules that are located at extremely small distances from each other'. Moreover, in a résumé published in the *Bulletin* of the Société Philomatique, he (incorrectly) gave the view that in 1820 he had deduced the expression for the elastic moment of a plate from a molecular hypothesis. In reality, however, no such notion is found anywhere in the original study.

Cauchy, of course, could not be indifferent to the success that the molecular hypothesis was enjoying among physicists. The papers of Poisson, Navier, and Fresnel encouraged him to work on the basis of this hypothesis (31). But, as we saw earlier, Cauchy had waited until Navier published his paper of May 14, 1821 before he shared his research with the Académie. On October 1, 1827, Poisson, outstripping Cauchy, asserted in a note that he read at the Académie that he was 'presently engaged in a very far-reaching study of the laws governing the equilibrium and motion of elastic bodies' (32). In order to preserve his priority rights, Cauchy, on that very same day, announced (in a hurried fashion) a study in progress 'Sur l'équilibre et le mouvement intérieur d'un corps solide considéré comme un système de molécules distinctes les unes des autres' and registered a rough copy of it; the work, hastily written and edited, was deposited in a sealed envelope with the Académie (33). Accordingly, on April 21, 1828, when Poisson read the paper he had announced on October 1, 1827, Cauchy made an objection, probably in order to defend his priority rights and to recall the existence of his sealed envelope (34).

This incident underscores the unhealthy atmosphere in which the theory of elasticity developed after 1814. Irrelevant bickerings and quarrels and continual protests, claims, and objections on matters of priority ruffled the peace of the little world of the scientists who labored there. No doubt they were all responsible for the situation, and Cauchy was particularly blameworthy. But it is likely that Poisson should bear a larger share of the responsibility than the others. After all, did he not often use his position to take more than his due share of the glory? There were frequent enough complaints as regards his behavior in this respect; and Cauchy, who had been on bad terms with him since 1825, simply did not trust him. From that day on, until 1830, these two scholars, brilliant scientists with roughly concurrent research interests, worked in an atmosphere that was poisoned with suspicion and hostile polemics.

Shortly after the presentation of his paper to the Académie, Poisson published an abstract of the work in the April 1828 issue of the *Annales de Chimie et de Physique*. This publication was followed by endless polemics with Navier in the succeeding issues of the same periodical. Cauchy kept away from all the clamor. However, he was probably upset at seeing the theory evolve without his participation, particularly as Poisson was preparing a printed edition of his paper (35). So, during the summer of 1828, he actively worked on three new studies that were to be published in *Exercices de Mathématiques*. The first of these studies was a major investigation, 'Sur les équations qui expriment les conditions d'équilibre ou les lois du mouvement intérieur d'un corps solide, élastique ou non élastique'. As was noted earlier, in this study Cauchy presented a corrected version of the continuum theory of elasticity that he had first articulated in 1822. There is no going back on it. The other two works, 'Sur l'équilibre et le mouvement d'un système de points matériels sollicités par des forces d'attraction ou de répulsion mutuelle' and 'De la pression ou tension dans un système de points matériels,' were both based on a molecular hypothesis (36). No doubt fearing that he might be overtaken in a race to publish first, he once again deposited sealed envelopes containing the rough drafts of both of these works with the Académie on August 18, 1828 (37).

The study 'Sur l'équilibre...' was an improved version of the rough copy he had deposited in a sealed envelope with the Académie on October 1, 1827. This work rested on the analysis of the motion of a given molecule M under the action F_M of all the neighboring molecules M_p. Cauchy assumed that between two molecules M and M_{p} there exists a central force of attraction or repulsion which is proportional to their masses m and m_{p} and is a function of their respective distance r_{p}. Then, the action F_M is of the form

$$F_M = m \sum_{\mathrm{p}} m_{\mathrm{p}} f(r_{\mathrm{p}}).$$

Let us denote by u_i the displacement of the molecule M, by $u_i + \Delta_{\mathrm{p}} u_i$ the displacement of the molecule M_{p}, by $\boldsymbol{\mu}_{\mathrm{p}}$ the unit vector of direction MM_{p} and by $\varepsilon^{(\boldsymbol{\mu}_{\mathrm{p}})}$ the rate of extension in the direction MM_{p}. Thus Cauchy obtained the equation

$$\partial_0^2 u_i = \sum_{\mathrm{p}} m_{\mathrm{p}} \left(\mu_{\mathrm{p}i} + \frac{\Delta_{\mathrm{p}} u_i}{r_{\mathrm{p}}} \right) \frac{f(r_{\mathrm{p}}(1 + \varepsilon^{(\boldsymbol{\mu}_{\mathrm{p}})}))}{1 + \varepsilon^{(\boldsymbol{\mu}_{\mathrm{p}})}} \tag{6.12}$$

for the acceleration of the molecule M. By assuming that u_i is very small in comparison to r_{p}, he showed that

$$\varepsilon^{(\boldsymbol{\mu}_{\mathrm{p}})} = \mu_{\mathrm{p}i} \frac{\Delta_{\mathrm{p}} u_i}{r_{\mathrm{p}}}$$

and

$$\partial_0^2 u_i = \sum_{\mathrm{p}} m_{\mathrm{p}} \left\{ f(r_{\mathrm{p}}) \frac{\Delta_{\mathrm{p}} u_i}{r_{\mathrm{p}}} + [r_{\mathrm{p}} f'(r_{\mathrm{p}}) - f(r_{\mathrm{p}})] \mu_{\mathrm{p}i} \mu_{\mathrm{p}j} \frac{\Delta_{\mathrm{p}} u_i}{r_{\mathrm{p}}} \right\}. \tag{6.13}$$

The hypothesis that the forces existing between the molecules decrease very rapidly with the distance enabled him to neglect infinitesimals of the second order in the infinite series expansion of the finite differences $\Delta_{\mathrm{p}} u_i$. Afterward, Cauchy made some assumptions relative to the distribution of the molecules in the system. He first considered a crystal with three rectangular axes subject to an external body of force F_i. From Eq. (6.13), he deduced the equations of motion:

$$\begin{aligned}
\partial_0^2 u_1 &= F_1 + (L + G) \partial_1^2 u_1 + (R + H) \partial_2^2 u_1 \\
&\qquad + (Q + I) \partial_3^2 u_1 + 2R \, \partial_{12} u_2 + 2Q \, \partial_{31} u_3, \\
\partial_0^2 u_2 &= F_2 + (R + G) \partial_1^2 u_2 + (M + H) \partial_2^2 u_2 \\
&\qquad + (P + I) \partial_3^2 u_2 + 2P \, \partial_{23} u_3 + 2R \, \partial_{12} u_1, \\
\partial_0^2 u_3 &= F_3 + (Q + G) \partial_1^2 u_3 + (P + H) \partial_2^2 u_3 \\
&\qquad + (N + I) \partial_3^2 u_3 + 2Q \, \partial_{13} u_1 + 2P \, \partial_{32} u_2,
\end{aligned} \tag{6.14}$$

which depend on nine constants. If $G = H = I$, $L = M = N$, and $P = Q = R$ (three constants), the crystal has cubic symmetry, and if $G = H$, $L = M = 3R$ and $P = Q$ (five constants), the crystal is uniaxial with axis x_3.

This theory was thus more general than the one he had obtained in his investigation of the problem in 1822, in which the bodies were regarded as isotropic continua. Cauchy encountered difficulties in deriving the case of isotropic elastic bodies from the molecular model. After a first attempt in the rough draft of October 1, 1827, an attempt that was based on a debatable hypothesis, Cauchy developed a molecular theory with two constants for isotropic elastic bodies. For this theory, he used his molecular theory of the crystals with cubic symmetry as a starting point, considering the invariance of the elasticity in an arbitrary rotation about each molecule. Then, the equation of motion in an isotropic medium is

$$\partial_0^2 u_i = F_i + (R + G)\,\partial_{jj} u_i + 2R\,\partial_i U_{jj}, \tag{6.15}$$

where $U_{ij} = \partial_{(i} u_{j)}$.

Wrongly, Cauchy still thought that this molecular theory with two constants was equivalent to the one he deduced from the continuum hypothesis [compare Eq. (6.15) with Eq. (6.10)] and that it was more general than the molecular one-constant theory Navier and Poisson had developed (38). Cauchy took up this point again in the second paper, 'De la pression ou tension dans un système de points matériels'. In this study, he introduced the concept of stress (pressure or tension) into the molecular theory of elasticity. In case the original state of the system is stress free, the constants G, H, I vanish. Thus, for an isotropic body, Cauchy found the stress–strain linear relation

$$T_{ij} = R(2U_{ij} + U_{kk}\delta_{ij})\rho \tag{6.16}$$

[compare with Eq. (6.9)] and the equation of motion

$$\partial_0^2 u_i = F_i + R(\partial_{jj} u_i + 2\partial_i U_{jj}) \tag{6.17}$$

instead of Eq. (6.15). This equation with one constant is identical to Navier's.

During the following months, Cauchy was very occupied with the problem of applications of his molecular theory. In his study of April 21, 1828, Poisson had deduced the equations of equilibrium and motion of elastic strings, rods, membranes, and plates. Cauchy simply refused to be outdone. Accordingly, he presented to the Académie a number of studies on the problems of the motion of plates, laminas, and rods. His first research, like that of Poisson, dealt with isotropic bodies (39). Later on, he undertook the investigation of anisotropic plates and rods (40). At this juncture, he noted, for the first time, that the molecular theory of elasticity depends, in general, on 15 coefficients (41).

Cauchy also applied his molecular theory to the problem of the equilibrium and motion of perfect fluids. As we saw earlier, this problem had interested him since 1815 and was in a sense at the origin of his continuum theory of elasticity. He advanced the hypothesis that in a fluid the action that the neighbouring molecules exert on a given molecule is imperceptible; this assumption enabled him to replace the finite sums by integrals in the equations of equilibrium and motion. He showed that one could easily deduce the characteristic property of fluids, that is, the equality of the pressure in all directions. The proof of the characteristic property of fluids within the framework of molecular theory gave occasion for a sharp dispute between Cauchy and Poisson in April 1828 (42).

In the meanwhile, the theory of light became the newest area of application of the theory of elasticity (43). Between 1816 and 1822, Fresnel had developed a wave theory of light that allowed him to explain a number of optical phenomena: interference, diffraction, polarization, reflection, and refraction in refringent and birefringent bodies, with one or two optical axes. Fresnel's theory was based on careful experiments. Analytical formulas, such as, for example, the expression of the Fresnel surface that allows the calculation of the direction of refracted rays in a doubly refracting crystal with two optical axes, had been obtained by inductive methods. However, in 1821, Fresnel sought to construct a molecular model whose properties would account for the nature of light. According to this model, light waves would correspond to molecular vibrations in the ether. For light waves, Fresnel showed, the ether molecules—unlike the molecules of air for sound waves—vibrated transversely. This hypothesis presented some difficulties and was rejected by Poisson (44).

As we have seen, Fresnel's work inspired Cauchy, who, at the time, was working on his continuum theory of elasticity. However, Cauchy had not attempted to apply his continuum theory to the study of light waves, and it was only in 1828 that he first took up this problem. Could it have been that he wanted to leave to Fresnel, who died prematurely in 1827, the exclusive honor of making fundamental investigations in this area? It is more probable that Cauchy delayed embarking on any research in this area until he had overcome certain mathematical obstacles.

On the one hand, in order to examine the problem of double refraction, he needed to generalize his theory of elasticity to some anisotropic crystals. Accordingly, he developed his molecular theory, which, in turn, would serve as the basis for his theory of light. On another level, he needed to solve the differential equations representing the motion of the ether molecules. At any rate, in 1827, Cauchy published a study entitled 'Sur l'application des résidus aux questions de physique mathématique' in which he explored a general method of solving linear partial differential equations with constant coefficients.

In a short study published at the beginning of 1829, Cauchy presented the first conclusions he had reached on the basis of his research on the theory of light (45). He pointed out that a shock initially produced at an arbitrary point of an isotropic system of molecules is propagated in the form of two spherical waves, one of which vanishes with the initial cubical dilation, while the other corresponds to the molecular vibrations parallel to the plane containing the initial vibrations. In the case of a uniaxial crystal, he had recourse to Huygen's theorem on double refraction.

> We think [Cauchy wrote] that it should be concluded that the equations of motion of light are contained in those that express the motion of a

system of molecules in which the molecules deviate but a little from an equilibrium position.

Here, Cauchy referred to Eqs. (6.14) (with $F_i = 0$), which are the basic equations of his theory of light (without dispersion), and to those that can be derived from them, such as Eq. (6.15). Cauchy took more than a year to realize the program he had enunciated in his paper of 1829. This delay probably stems from theoretical difficulties for integrating Eqs. (6.14). During the following months Cauchy came to grips with these difficulties. In several papers, especially in a paper of April 12, he showed by using a change of variables that if a differential equation is homogeneous, the sextuple integral representing his general solution reduced to a quadruple integral (46). He further proved that in the case of an initial shock at a given point the motion of the molecules is propagated like a wave, the vibratory phenomena being perceptible only in the neighborhood of a wave surface that can be determined at any time. The theorem can be applied to a system of ether molecules. Once this latter point was established. Cauchy, in 1830, was able to use two distinct mathematical methods to derive his theory of light from the analysis of the propagation of a local disturbance in the system of ether molecules.

The first method, which Cauchy published in *Exercices de Mathématiques*, was based on Huygen's principle of superposition of little motions (47). It consisted of first finding only the simple solutions of Eq. (6.14) corresponding to a given plane wave. Let us denote by **n** the unit vector that is oriented in the direction of propagation of this wave front. Then, for a molecule whose distance from the plane wave at the initial time is h, Eq. (6.14) can be written

$$\partial_0^2 u_i = \partial_h^2 a_{ij}(\mathbf{n}) u_j, \tag{6.16}$$

where $a_{ij}(\mathbf{n})$ is symmetrical.

By using an orthogonal transformation, Cauchy reduced the problem to solving the equation

$$\partial_0^2 u_i = \partial_h^2 s_{(i)}^2(\mathbf{n})\, u_j,$$

where the $s_{(i)}^2(\mathbf{n})$ are the three real roots of the characteristic equation $\det(a_{ij}(\mathbf{n}) - s) = 0$. Cauchy assumed that these roots are positive. Thus, he obtained three plane waves, whose equations are given by

$$h - s_{(i)}(\mathbf{n})t = 0.$$

For each $s_{(i)}(\mathbf{n})$, the displacements of the molecules are parallel to the principal direction. In other words, the three plane waves are polarized. In order to construct the wave surface that is propagated from a local disturbance, Cauchy considered wave fronts that are slightly inclined to each other (the direction **n** varies continuously). The wave surface is the envelope of all these planes. In general, it has three sheets, each of them being the envelope of the system of wave fronts that corresponds to one of the three principal values $s_{(i)}(\mathbf{n})$.

The second method developed by Cauchy led to the same results. It consisted of directly investigating the general solution (given in the form of a quadruple integral) of Eq. (6.14). Cauchy exhibited this development in lectures at the Collège de France in May and June of 1830. He was preparing the publication of these lectures when the July Revolution of 1830 broke out (48).

In the case of an isotropic system of molecules (Eq. (6.15) with $F_i = 0$), Cauchy showed that the wave surface separates into two spherical waves, with the velocities $\sqrt{R+G}$ and $\sqrt{3R+G}$. The first spherical wave is the wave of light: The displacements of molecules are transverse, as Fresnel has assumed. The second one is longitudinal. It vanishes if the initial displacements of the molecules are transverse. In the case of a uniaxial system of molecules whose initial state is stress-free (i.e., if G, H and I are null), the three sheets of the wave surface are one sphere and two ellipsoids. The sphere and the first ellipsoid yield Huygen's construction for the ordinary and extraordinary rays. The second ellipsoid is assumed to be invisible. Cauchy also considered the case of a system of molecules with three rectangular axes (Eq. (6.14) with $F_i = 0$), in order to deduce the optical surface in a biaxial chrystal. As a result, by assuming some specific relations between the coefficients L, M, N, P, Q, R, he obtained a wave surface which is only an approximation of Fresnel's. Moreover, Cauchy's theory required the polarization plane to be the plane which contains the direction of wave propagation and that of the displacements of molecules. This last definition differed from Fresnel's. But the main difficulty of Cauchy's theory of birefringence is that one of the two optical waves is not exactly transverse.

In 1830, Cauchy also developed a theory of reflection and refraction of light on the boundary between two isotropic bodies. He obtained Brewster's law on the complete reflection and Fresnel's laws for reflection and refraction of polarized waves, by assuming that the density of ether is constant in all bodies.

Finally, Cauchy undertook, on Coriolis' advice, to examine the problem of light dispersion, which Fresnel had not explained. His theory consisted of using a Fourier series representation for the displacement of the molecules. Then, the displacement u_i of a given molecule at the point whose ray vector is $\mathbf{r}$ becomes

$$u_i = \sum_{\mathbf{k}} (a_{\mathbf{k}}(\mathbf{r}, \mathbf{t}) \cos \mathbf{k} \cdot \mathbf{r} + b_{\mathbf{k}}(\mathbf{r}, t) \sin \mathbf{k} \cdot \mathbf{r}).$$

Cauchy chose an arbitrary simple displacement $a_{\mathbf{k}}(\mathbf{r}, t) \cos \mathbf{k} \cdot \mathbf{r} + b_{\mathbf{k}}(\mathbf{r}, t) \sin \mathbf{k} \cdot \mathbf{r}$. He calculated the finite difference Δu_i in this simple displacement and substituted its expression into Eq. (6.12). By assuming that the molecules are symmetrically distributed around each other, he was able to obtain a symmetrical ordinary differential system

$$\partial_0^2 u_i = b_{ij}(\mathbf{k}) u_j, \tag{6.17}$$

which could be solved by classical methods of diagonalization.

Cauchy deduced from Eq. (6.17) the same results as he did from Eq. (6.16). But in the theory of light dispersion, the wave velocity V depends on the magnitude k of the wave vector $\mathbf{k}$, and thus on the wavelength $\lambda = \frac{k}{2\pi}$. Cauchy assumed that color depends on the wavelength and, in consequence of this assumption, he effectively showed the index of refraction depends on the color of the light ray and on the nature of the system of molecules. Cauchy intended to develop the theory of light dispersion in a paper 'Sur la dispersion', but its publication was suddenly interrupted by his departure from France in September 1830 (49).

By the end of the 1820s, Cauchy had used his molecular theory of elasticity in applications. However, he had still not completely abandoned the Eulerian point of view that he held in 1822. In May 1830, in an article of the *Exercices de Mathématiques*, he presented a general continuum theory of elasticity for the first time (50). This occurred after Poisson presented a paper to the Académie, in which he suggested that in the general case the continuum theory leads to 36 coefficients of elasticity. In his article, Cauchy compared his own 2 theories, the continuum one, with 36 coefficients, and the molecular one, with only 15 coefficients. In the special case of isotropy, two coefficients are obtained by the first theory [Eq. (6.9)] and only one by the second theory [Eq. (6.16)]. Cauchy then showed that the molecular theory could be derived, by reduction of the coefficients, from the continuum theory. Such, then, are the origins of the famous discussion on the 'Cauchy relations', which connect these two theories.

By 1830, Cauchy seems to have chosen the molecular theory, not for doctrinal reasons but on the basis of experiments Savart had undertaken on isotropic bodies. Three years later, in his lectures given at Turin, Cauchy unhesitatingly opted for the molecular theory (51). He then justified his choice, not on the basis of experiments, but on the basis of dogmatic arguments of a philosophical and theological nature.

From 1815 to 1830, Cauchy was relentless in his pursuit of a single project common to all the great physicomathematicians of that era, Laplace, Poisson, Ampère, Fresnel, and many others. This project was the inquiry into the mathematical laws governing the propagation of physical phenomena within a medium, whether for liquid waves, as in 1815, or for light waves, as in 1830. On that, in the tradition of Eulerian mechanics, Cauchy elaborated the basic concepts of continuum mechanics in 1822; as a mathematician, Cauchy thought that he had found in his continuum theory of elasticity a universal model from which it would be possible to deduce laws governing such diverse phenomena as sound, heat, and light. When, in the second half of the 1820's, Cauchy substituted a molecular theory for his continuum theory, this project became, in a sense, united with the Laplacian goal to create a universal molecular physics. However, Cauchy treated the molecular theory merely as a mathematical tool, albeit that it also became a metaphysical dogma for him, and he neglected its physical implications. These features were especially evident in his research on the theory of light.

Chapter 7

From the Theory of Singular Integrals to the Calculus of Residues

Cauchy's crowning achievement in analysis was unquestionably his theory of complex functions, a branch of mathematics that, except for some interruptions, commanded his attention from 1814 until his death. Few mathematicians concerned themselves with Cauchy's theory before the late 1840s. Almost all of the progress made in this area until that time was due to Cauchy. The principal lines followed by his work are well known. Cauchy first embarked on what would become his complex function theory by way of studying the integration along closed paths in the complex plane; he did not undertake to investigate analytic functions of a complex variable until 1831 and only much later, in 1846, did he begin to work out the fundamental notions that would govern his complex function theory.

Actually, Cauchy constructed this theory in order to develop new methods for calculating integrals, roots of algebraic and transcendental equations, and solutions of linear ordinary and partial differential equations with constant coefficients. These mathematical problems were closely connected with the study of mechanical topics. Nevertheless, the historians of mathematics who have examined Cauchy's papers relative to this branch of analysis have generally failed to consider this context. Accordingly, in this chapter we explain how the development of the complex function theory fit into the general framework of Cauchy's overall scientific concerns.

Cauchy presented his first paper on analysis, a major study entitled 'Sur les intégrales définies', to the Académie on August 22, 1814 (1). Accordingly to Cauchy, this paper was the first work of reference concerning his complex function theory. The original version of the paper, without the footnotes Cauchy added in 1825, however, was in the tradition of Cauchy's great mentors, Laplace, Legendre, and Poisson (2). In his 'Recherches sur les approximations des formules qui sont fonctions de très grands nombres' of 1785, quoted by Cauchy in the introductory section of his paper, Laplace, just as Euler before him, had made use of complex numbers in integrating certain real functions of a real variable, for instance, $\cos bxe^{-x^2}$, between 0 and $+\infty$ (3).

Laplace's method, however, was based on a simple induction that followed from the then-accepted principle of the generality of analysis. On such a basis, Laplace applied the usual methods of integration to complex functions of a complex variable. 'These methods', he admitted, 'although used with great care and due caution, nevertheless need proof for the results they generate' (4). Laplace examined this method again when, in 1812 and 1814, he published the first two editions of *Théorie Analytique des Probabilités.* At that time, he interested his protégé Cauchy in the investigation of this question.

'I have conceived', wrote the youthful scholar in the introduction to his paper, 'the hope of basing the passage from the real to the imaginary domain on a direct and rigorous analysis'.

Cauchy first considered a real function $f(u)$, where u was supposed to be itself a real function of two real variables x and y. He easily established the equation

$$\frac{\partial}{\partial y}\left[f(u)\frac{\partial u}{\partial x}\right] = \frac{\partial}{\partial x}\left[f(u)\frac{\partial u}{\partial y}\right]. \tag{7.1}$$

Thus, Cauchy assumed that f and u were complex functions, which he wrote

$$u = M(x, y) + N(x, y)\sqrt{-1} \quad \text{and} \quad f(u) = P'(x, y) + P''(x, y)\sqrt{-1}.$$

Substituting these expressions in Eq. (7.1) and setting

$$S = P'\frac{\partial M}{\partial x} - P''\frac{\partial N}{\partial x} \qquad U = P'\frac{\partial M}{\partial y} - P''\frac{\partial N}{\partial y}$$

$$T = P'\frac{\partial N}{\partial x} + P''\frac{\partial M}{\partial x} \qquad V = P'\frac{\partial N}{\partial y} + P''\frac{\partial M}{\partial y},$$

he stated the imaginary differential equation

$$\frac{\partial S}{\partial y} + \frac{\partial T}{\partial y}\sqrt{-1} = \frac{\partial U}{\partial x} + \frac{\partial V}{\partial x}\sqrt{-1}$$

equivalent to the two real differential equations

$$\frac{\partial S}{\partial y} = \frac{\partial U}{\partial x} \quad \text{and} \quad \frac{\partial T}{\partial y} = \frac{\partial V}{\partial x}, \tag{7.2}$$

from which he deduced the classical equations

$$\frac{\partial P'}{\partial x} = \frac{\partial P''}{\partial y} \quad \text{and} \quad \frac{\partial P'}{\partial y} = -\frac{\partial P''}{\partial x}$$

in the special case $[M(x, y) = x$ and $N(x, y) = y]$. The differential equations, (7.2), as he put it, 'summarize the whole of the theory relating to the passage from the real to the imaginary [domain]'. After having formed the double integrals

$$\iint \frac{\partial S}{\partial y} dxdy = \iint \frac{\partial U}{\partial x} dxdy$$

and

$$\iint \frac{\partial T}{\partial y} dxdy = \iint \frac{\partial V}{\partial x} dxdy \tag{7.3}$$

over a rectangle $R = [x_0, X] \times [y_0, Y]$, where S, T, U, and V were supposed 'to keep a determinate value', he obtained two fundamental equations between two simple integrals by permuting the order of integration:

$$\int_{x_0}^{X} [S(x, Y) - S(x, y_0)] dx = \int_{y_0}^{Y} [U(X, y) - U(x_0, y)] \, dy$$

and

$$\int_{x_0}^{X} [T(x, Y) - T(x, y_0)] dx = \int_{y_0}^{Y} [V(X, y) - V(x_0, y)] dy. \tag{7.4}$$

These formulas are equivalent to the famous Cauchy's integral theorem applied to various closed paths, which depends on the mapping u of x and y. By means of this theorem, Cauchy calculated a number of definite integrals.

The second part of the study was longer and still more innovative. Cauchy assumed up to that point that the functions to be integrated were regular (i.e., had a 'determinate value') between the limits of integration. He focused his attention on certain cases where this regularity failed to hold at certain isolated points. It was then no longer possible to permute the order of integration in Eq. (7.3), so, he posed, in section II.2, the general problem of determining a double integral:

$$\int_{a'}^{a''} \int_{b'}^{b''} \frac{\partial K}{\partial z} dx \, dz$$

when the primitive $K = \phi(x, z)$ assumes an indeterminate value at a point (X, Z) in the rectangle $[a', a''] \times [b', b'']$. Then, the value of the integral depended on the order of integration. Cauchy obtained the difference A between the two values depending on the order of integration in the form of one or several singular integrals:

$$\lim_{\substack{\varepsilon \to 0 \\ \zeta \to 0}} \int_0^{\varepsilon} \phi(X \pm \xi, Z \pm \zeta) d\xi.$$

Singular integrals were nonnull definite integrals taken over infinitely small intervals. This class of integrals, which Cauchy introduced into analysis, was entirely new.

Likewise, in section II.3, Cauchy examined the simple integral

$$\int_{b'}^{b''} \phi'(Z)\,dz$$

in the case where the primitive $\phi(Z)$ of $\phi'(Z)$ is discontinuous at a point Z in the interval $[b',b'']$. If Z was a pole of $\phi'(z)$, he obtained a determination of the integral that he called his principal value in 1822. For instance,

$$\int_{-2}^{+4} \frac{dz}{z} = \operatorname{Log} 2$$

and

$$\int_{-2}^{+2} \frac{dz}{z} = 0.$$

In section II.4, Cauchy investigated the double integrals [Eq. (7.3)] in the case of a function $f = \frac{g}{h}$. The conditions Cauchy stated for g and h were rather imprecise, but actually he dealt only with functions f real valued on the real axis and with simple poles $u_i = u(x_i, y_i)$ on the border or inside $u(R)$. Equations (7.4) are thus generally false. In order to calculate the differences between both sides, respectively, A and A', Cauchy substituted the first-order expansions

$$M(x_i, y_i) + \xi \frac{\partial M(x_i, y_i)}{\partial x} + \eta \frac{\partial M(x_i, y_i)}{\partial y}$$

of $M(x, y)$ and

$$N(x_i, y_i) + \xi \frac{\partial N(x_i, y_i)}{\partial x} + \eta \frac{\partial N(x_i, y_i)}{\partial y}$$

of $N(x, y)$ into the two singular integrals

$$\lim_{\substack{\varepsilon \to 0 \\ \eta \to 0}} \int_0^{\varepsilon} S(x_i \pm \xi, y_i \pm \eta)\,d\xi$$

and

$$\lim_{\substack{\varepsilon \to 0 \\ \eta \to 0}} \int_0^{\varepsilon} T(x_i \pm \xi, y_i \pm \eta)\,d\xi.$$

In this way, he obtained the values

$$\begin{aligned} A &= 2\mu\pi, \\ A' &= 2\lambda\pi, \end{aligned} \tag{7.5}$$

where

$$\mu = -\operatorname{Im}\left[\sum \frac{g(u_i)}{h'(u_i)}\right]$$

and

$$\lambda = \operatorname{Re}\left[\sum \frac{g(u_i)}{h'(y_i)}\right]$$

if all the u_i are inside $u(R)$. This result was equivalent to the residue theorem applied to closed paths depending on the mapping of u (rectangle if $u = id_{\mathbf{R}}$, semicircle, triangle, etc.). If some u_i were on the border of $u(R)$, the values of A and A' relative to these poles were $\mu\pi$ and $\lambda\pi$. This first version of the method of residues was called by Cauchy the theory of singular integrals.

The final section of the second part of the paper, as well as the supplementary parts written on the request of the evaluating commisioners, were devoted to applications. Cauchy used the theory of singular integrals to calculate integrals between $-\infty$ and $+\infty$ of real-valued functions with simple poles in the half-plane $\operatorname{Im} z \geqslant 0$, assuming only that the functions vanish at infinity in the same half-plane. In the case of functions with poles on the real axis, Cauchy obtained the principal value of the integrals. Unfortunately, he also considered functions with an infinity of poles on the real axis [for instance, if $h(x) = \cos xp(x)$]. Such an applications could, of course, lead to incorrect results.

The paper 'Sur les intégrales définies' was only the most remarkable part of a whole set of research studies on questions relative to the definite integrals: calculation techniques, transformation properties, and applications to other areas of analysis. It opened the way for a less famous study, 'Sur diverses formules relatives à la théorie des intégrales définies' which was presented to the Académie on January 2, 1815 (5). The key feature of this paper was a discussion of methods of integration used in Laplace's *Théorie Analytique des Probabilités* and, to a lesser extent, in Legendre's *Exercices de Calcul Intégral*. In the first part of his paper, Cauchy recalculated certain integrals that Laplace had deduced by passing from the real to the complex domain. In performing these reevaluations, Cauchy used a new method that was based on the consideration of double integrals. This method differed entirely from the one he had used in his paper of August 22, 1814.

The third part of the paper of January 2, 1815, was devoted to the transformation of finite differences into definite integrals, one of Laplace's favorite topics. Here, using new methods, Cauchy derived several important formulas of the *Théorie Analytique des Probabilités*. Moreover, he introduced a new class of integral (just as he had done the preceding year with singular integrals) that he labeled 'extraordinary integrals': if $f(x)$ is a function such that $f(0) \neq 0$, the integral $\int_0^x \frac{f(x)}{x^{a+1}}\,dx$ takes an infinite value, but the integral $\int_0^x \frac{f(x) - \phi(x)}{x^{a+1}}\,dx$, where $\phi(x)$ is the nth-order expansion of $f(x)$ in the neighborhood of 0, generally takes a finite value if n is the integer part of a. This last integral, which Cauchy designated by $\int' \frac{f(x)}{x^{a+1}}\,dx$, is an extraordinary integral.

During the following years, Cauchy developed the applications of the theory of singular integrals. He used it in his paper 'Sur la théorie des ondes' of October 1815 to establish, for $a \geqslant 0$, the inversion formulas

$$\phi_1(m) = \sqrt{\frac{2}{\pi}} \int_0^{\infty} \cos m\mu F_1(\mu)\, d\mu$$

and

$$\phi_2(m) = \sqrt{\frac{2}{\pi}} \int_0^{\infty} \sin m\mu F_2(\mu)\, d\mu \tag{7.6}$$

of the Fourier transforms

$$F_1(a) = \sqrt{\frac{2}{\pi}} \int_0^{\infty} \phi_1(m) \cos am\, dm$$

and

$$F_2(a) = \sqrt{\frac{2}{\pi}} \int_0^{\infty} \phi_2(m) \sin am\, dm \tag{7.7}$$

by showing that

$$\int_0^{\infty} \int_0^{\infty} \cos am \cos m\mu F_1(\mu)\, dm\, d\mu = \frac{\pi}{2} F_1(a)$$

and

$$\int_0^{\infty} \int_0^{\infty} \sin am \sin m\mu F_1(\mu)\, dm\, d\mu = \frac{\pi}{2} F_2(a). \tag{7.8}$$

His proof, based on the use of the convergence-producing factor $e^{-\alpha m}$ and on permutations of the signs $\int$ and lim was not rigorous (6). Cauchy argued that

$$\int_0^{\infty} \cos am \cos m\mu e^{-\alpha m}\, dm = \frac{1}{2}\left(\frac{\alpha}{\alpha^2 + (\mu - a)^2} + \frac{\alpha}{\alpha^2 + (\mu + a)^2}\right).$$

Since $\lim_{\alpha \to 0} e^{-\alpha m} = 1$, the left sides of Eq. (7.8) become, respectively,

$$\lim_{\alpha \to 0} \frac{1}{2} \int_0^{+\infty} F_1(\mu) \frac{\alpha\, d\mu}{\alpha^2 + (\mu - a)^2},$$

and

$$\lim_{\alpha \to 0} \frac{1}{2} \int_0^{+\infty} F_2(\mu) \frac{\alpha\, d\mu}{\alpha^2 + (\mu + a)^2},$$

which are singular integrals. By the change of variable $\mu = a + \alpha\xi$, Cauchy obtained for these singular integrals the expressions

$$\frac{1}{2} F_1(\mu) \int_0^{+\infty} \frac{d\xi}{1 + \xi^2},$$

and

$$\frac{1}{2} F_2(\mu) \int_0^{+\infty} \frac{d\xi}{1+\xi^2},$$

equal, respectively, to $\frac{\pi}{2} F_1(\mu)$ and $\frac{\pi}{2} F_2(\mu)$.

Thereafter, between 1817 and 1821, Cauchy's productivity in mathematics declined sharply. Absorbed in teaching at the École Polytechnique, he presented only six papers to the Académie during this four-year period, and none at all for the two-year interval from November 1819 to October 1821. However, he continued to work on applications of singular integrals.

Cauchy first presented his findings at the Collège de France in 1817 (7). In his lectures, he extended the use of his method of 1814 to the calculation of improper integrals:

$$\int_{-\infty}^{+\infty} \phi(x) \frac{u(x)}{v(x)} dx,$$

where $\phi(x)$ is the real part of a complex-valued function $\phi(x) + \chi(x)\sqrt{-1}$ of a real variable, and applied the formula to the special case

$$\phi(x) + \chi(x)\sqrt{-1} = \cos rx + \sin rx \sqrt{-1}.$$

He also investigated the properties of the Fourier transforms, Eq. (7.7), and of the inversion formulas, Eq. (7.6), called by him 'reciprocal functions'; he proposed various applications of them, which he had already discussed at the Société Philomatique on March 20, 1816 (8). From these lectures, Cauchy extracted an article on reciprocal functions for the *Bulletin* of the Société Philomatique (9). On the other hand, Cauchy developed some considerations about the theory of singular integrals and the reciprocal functions in his analysis teaching at the École Polytechnique during the year 1817–1818 (10).

Fourier now made an objection, claiming prior discovery (11). Actually, he had already set forth the properties of his transforms in 1807 and, using them, had developed a method for solving the linear partial differential equations related to the theory of heat. This method was identical to Cauchy's. Fourier's paper had been deposited in the archives of the Académie since 1811. However, Cauchy was completely ignorant of its existence, for it was not yet published, and proclaimed his good faith in a second article, which appeared in the December 1818 issue of the *Bulletin* of the Société Philomatique (12). Without questioning whether or not Cauchy had indeed acted in good faith in this matter, we are nevertheless inclined to believe that Fourier's work indirectly influenced Cauchy's research.

In fact, Poisson and probably Laplace, unlike Cauchy, knew of Fourier's unpublished study on the theory of heat. Poisson made use of formulas that were identical to those appearing in Fourier's paper in his own 1815 paper on the theory of waves. Cauchy was on very friendly terms with Laplace and Poisson in 1815, and as we have seen, he had discussed the theory of waves with both of them at this time. Accordingly, through Laplace and Poisson, Cauchy might well have been aware of Fourier's results. Such an hypothesis might explain why in his research he failed to use a basic method that had already enabled him to integrate the equations of the theory of waves and instead employ the method used by Fourier and Poisson (13).

After an interruption of two years, Cauchy came back to the theory of singular integrals in 1819. On November 22, 1819, he presented a paper to the Académie on the determination of the roots of an algebraic or transcendental equation in integral form (14). The method was based on the theory of singular integrals. Cauchy considered a function $f(u)/uF(u)$ with real simple poles 0, a, a', a'', etc., between -1 and $+1$ and complex simple poles $\alpha+\beta\sqrt{-1}$, $\alpha'+\beta'\sqrt{-1}$, etc., inside the upper unit semicircle. Integrating $f(u)/uF(u)$ along this circuit and using for the first time integrals of a complex function, he applied the residue theorem and wrote

$$\begin{aligned}\int_0^\pi \frac{f(\cos p+\sin p\sqrt{-1})}{F(\cos p+\sin p\sqrt{-1})}\,dp &= \sqrt{-1}\int_{-1}^{+1}\frac{f(r)}{rF(r)}\,dr \\ &+\pi\left[\frac{f(0)}{F'(0)}+\frac{f(a)}{aF'(a)}+\frac{f(a')}{a'F'(a')}+\cdots\right] \\ &+2\pi\left[\frac{f(\alpha+\beta\sqrt{-1})}{(\alpha+\beta\sqrt{-1})F(\alpha+\beta\sqrt{-1})}+\frac{f(\alpha'+\beta'\sqrt{-1})}{(\alpha'+\beta'\sqrt{-1})F(\alpha'+\beta'\sqrt{-1})}+\cdots\right].\end{aligned}$$

Unfortunately, he never published the paper.

This work gave an impulse to new investigations relative to the theory of singular integrals. During the following months, Cauchy wrote a valuable paper, entitled 'Recherches sur les intégrales définies qui renferment des exponentielles imaginaires' (14). In this paper, he systematically used the residue theorem of 1814 applied to circular paths. Moreover, he definitively gave up the idea of separating the real and imaginary parts of complex equations, which characterized the paper of 1814. Therefore, he expressed his findings by means of imaginary integrals:

$$\int_{x_0}^{\chi} u(x)+v(x)\sqrt{-1}\,dx$$

containing imaginary exponentials

$$e^{rx\sqrt{-1}}=\cos rx+\sin rx\sqrt{-1}.$$

Thus, we know that he exhibited important results in this paper, such as the integral expression of the nth derivative:

$$\frac{1}{2}\int_{-\pi}^{+\pi} e^{-np\sqrt{-1}} f(b + e^{p\sqrt{-1}})\, dp = \frac{\pi}{n!}\frac{d^n f(b)}{db^n}, \tag{7.9}$$

Cauchy's integral formula in the special case of the unit circle:

$$\frac{1}{2}\int_0^{\pi}\left[\frac{f(e^{p\sqrt{-1}})}{1 - ae^{p\sqrt{-1}}} + \frac{f(e^{-p\sqrt{-1}})}{1 - ae^{-1\sqrt{-1}}}\right] dp = \pi f(a) \quad (a < 1), \tag{7.10}$$

and the average property:

$$\frac{1}{2}\int_0^{\pi} [f(e^{p\sqrt{-1}}) + f(e^{-p\sqrt{-1}})]\, dp = \pi f(0).$$

Probably, it is not certain, he also obtained the integral formula

$$\phi(u_0) + \phi(u_1) + \cdots + \phi(u_{m-1}) = \frac{1}{2\pi}\int_{-\pi}^{+\pi} re^{p\sqrt{-1}} f(re^{p\sqrt{1}})\, dp, \tag{7.11}$$

where ϕ is a continuous function, u_i the (simple) roots of the equation $F(u) = 0$ inside the disc with center 0 and radius r, and $f(u)$ the function $\phi(u)\dfrac{F'(u)}{F(u)}$. Strangely enough, Cauchy did not present this paper to the Académie until September 16, 1822, nearly three years after it was begun, and he never published it.

One month after, on October 28, Cauchy presented another study on the theory of singular integrals to the Académie (15). In this paper, entitled 'Sur les intégrales définies, où l'on fixe le nombre et la nature des constantes arbitraires et des fonctions arbitraires que peuvent comporter les valeurs de ces mêmes intégrales quand elles deviennent indéterminées', he summarized all the improvements he had introduced into the theory of singular integrals since the presentation of his memoir 'Sur les intégrales définies' of 1814, which was still unpublished. Moreover, he made important theoretical innovations: he defined the new concept of principal value and generalized and radically simplified the intricate formulas of 1814 by means of imaginary integrals. The paper was unpublished but a 'very short analysis' of it immediately appeared under the same title in the *Bulletin* of the Société Philomatique (16). In 1823, he twice summarized the theory with its improvements and new applications, first in an appendix to an article on the linear partial differential equations (17) and then in his *Calcul Infinitésimal* of 1823 (18).

Cauchy began by considering a real function of a real variable $f(x)$ with some poles $x_0, x_1, \ldots, x_{m-1}$ within the interval $[x', x'']$. Then, he defined the principal value of the integral $\int_{x'}^{x''} f(x)dx$ as the sum

$$\lim_{\varepsilon\to 0}\int_{x'}^{x_0-\varepsilon} f(x)dx + \int_{x_0+\varepsilon}^{x_1-\varepsilon} f(x)dx + \cdots + \int_{x_{m-1}+\varepsilon}^{x''} f(x)dx.$$

For instance, the principal value of $\int_{-1}^{+1} \frac{dx}{x}$ is 0, the value that Cauchy had given in 1814. If the integral is taken from $-\infty$ to $+\infty$, its principal value is

$$\lim_{\varepsilon\to 0} \int_{-1/\varepsilon}^{x_0-\varepsilon} f(x)\,dx + \int_{x_0+\varepsilon}^{x_1-\varepsilon} f(x)\,dx + \cdots + \int_{x_{m-1}+\varepsilon}^{1/\varepsilon} f(x)dx.$$

In order to obtain the general value of the integral, which is indeterminate in most cases, it is necessary to add singular integrals of the form

$$\lim \int_{-1/\varepsilon}^{-1/\varepsilon\mu} f(x)\,dx,$$

$$\lim \left[\left(\int_{x_0+\varepsilon}^{x_i-\varepsilon\mu_i} f(x)\,dx + \int_{x_i+\varepsilon\nu_i}^{x_i+\varepsilon} f(x)\,dx\right)\right],$$

and

$$\lim \int_{1/\varepsilon\nu}^{1/\varepsilon} f(x)\,dx,$$

where μ, μ_i, ν_i, and ν are arbitrary positive constants. If $\mathbf{f}^- = \lim_{x\to-\infty} xf(x)$, $\mathbf{f}_i = \lim_{x\to x_i} (x-x_i)f(x)$, and $\mathbf{f}^+ = \lim_{x\to+\infty} xf(x)$ exist and are finite, Cauchy obtained the formulas

$$\lim_{\varepsilon\to 0} \left[\left(\int_{x_i-\varepsilon}^{x_i-\varepsilon\mu_i} f(x)\,dx + \int_{x_i+\varepsilon\nu_i}^{x_i+\varepsilon} f(x)\,dx\right)\right] = \mathbf{f}_i \log\frac{\mu_i}{\nu_i},$$

$$\lim_{\varepsilon\to 0} \int_{-1/\varepsilon}^{-1/\varepsilon\mu} f(x)\,dx = \mathbf{f}^- \log\mu,$$

and

$$\lim_{\varepsilon\to 0} \int_{1/\varepsilon\nu}^{1/\varepsilon} f(x)\,dx = \mathbf{f}^+ \log\frac{1}{\nu}$$

by using the integral mean-value theorem Eq. (5.2), with $\chi(x) = \frac{1}{x-x_i}$ and $\chi(x) = \frac{1}{x}$. For instance,

$$\int_{-1}^{+1} \frac{dx}{x} = \lim_{\varepsilon\to 0}\left(\int_{-\varepsilon}^{-\varepsilon\mu} \frac{dx}{x} + \int_{\varepsilon\nu}^{\varepsilon} \frac{dx}{x} = \lim_{\varepsilon\to 0}\left[\log(\varepsilon\mu) + \log\frac{1}{\varepsilon\nu}\right] = \log\left(\frac{\mu}{\nu}\right)\right.$$

and the general value of $\int_{-1}^{+1} \frac{dx}{x}$ is indeterminate, since it depends on μ and ν.

Cauchy then examined the case of two real functions of two real variables $\phi(x, y)$ and $\chi(x, y)$ satisfying the relation

$$\frac{\partial \phi(x, y)}{\partial x} = \frac{\partial \chi(x, y)}{\partial y}. \tag{7.12}$$

He proceeded in the same manner as in 1814. From the equations

$$\int_{x_0}^{x} \int_{y_0}^{y} \frac{\partial \phi(x, y)}{\partial x} \, dx \, dy = \int_{x_0}^{x} \int_{y_0}^{y} \frac{\partial \chi(x, y)}{\partial y} \, dx \, dy,$$

he deduced the relation

$$\int_{x_0}^{x} \phi(x, Y) - \phi(x, y_0) \, dx = \int_{y_0}^{y} \chi(X, y) - \chi(x_0, y) \, dy, \tag{7.13}$$

which is to be compared with Eq. (7.4), on the assumption that $\phi(x, y)$ and $\chi(x, y)$ are finite and continuous in $R = [x_0, X] \times [y_0, Y]$. If $\phi(x, y)$ and $\chi(x, y)$ have one pole (a, b) in R, Eq. (7.13) deals only with the principal values of the integrals, and we have, therefore, the equation

$$\int_{x_0}^{x} \phi(x, Y), - \phi(x, y_0) \, dx = \int_{y_0}^{y} \chi(X, y) - \chi(x_0, y) \, dy - \Delta, \tag{7.14}$$

where

$$\Delta = \lim_{k \to 0} \int_{y_0}^{Y} \chi(a + k, y) - \chi(a - k, y) \, dy. \tag{7.15}$$

Cauchy applied these formulas to complex functions. This extension was justified in his own opinion by the definition of imaginary expressions as symbols equivalent to two real expressions, which he had provided in his *Analyse Algébrique* of 1821. Cauchy argued as follows: let f be a complex function of a complex variable and u a complex function of two real variables. The relation Eq. (7.12), is obviously satisfied by $\phi(x, y) = f(u)\frac{\partial u}{\partial x}$ and $\chi(x, y) = f(u)\frac{\partial u}{\partial y}$, and therefore, we have Eq. (7.13), if the functions are finite and continuous, or Eq. (7.14), if they have one pole $u_0 = a + b\sqrt{-1}$ inside $u(R)$.

For instance, if $f = \frac{h}{g}$, with g continuous and finite and h having simple poles inside $u(R)$, Eq. (7.15) yields

$$\Delta = 2\pi\sqrt{-1}(\lambda + \mu\sqrt{-1}),$$

which is equivalent to Eqs. (7.5). In the special case $u(x, y) = x + y\sqrt{-1}$, it follows from Eqs. (7.14) and (7.15), by substituting $f(x + y\sqrt{-1})$ for $\phi(x, y)$, and $\sqrt{-1}f(x+y\sqrt{-1})$ for $\chi(x,y)$, that

$$\int_{x_0}^{X} f(x+Y\sqrt{-1})-f(x+y_0\sqrt{-1})\,dx$$
$$=\sqrt{-1}\int_{y_0}^{Y} f(X+y\sqrt{-1})-f(x_0+y\sqrt{-1})\,dy-\Delta, \tag{7.16}$$

where

$$\Delta=\sqrt{-1}\lim_{k\to 0}\int_{y_0}^{Y} f(a+k+y\sqrt{-1})-f(a-k+y\sqrt{-1})dy. \tag{7.17}$$

In order to determine the singular integral Δ, Cauchy used the substitution $y=b+kz$ and the mean-value theorem Eq. (5.2). Thus, he obtained the formula

$$\Delta=2\pi\mathbf{f}\sqrt{-1}, \tag{7.18}$$

where

$$\mathbf{f}=\lim_{k\to 0} kf(u_0+k). \tag{7.19}$$

The limiting value $\mathbf{f}$ is, of course, the residue of f at the simple pole u_0, and accordingly, the residue theorem can be expressed in the form

$$\int_{x_0}^{X} f(x+Y\sqrt{-1})-f(x+y_0\sqrt{-1})\,dx$$
$$=\sqrt{-1}\int_{y_0}^{Y} f(X+y\sqrt{-1})-f(x_0+y\sqrt{-1})dy$$
$$-2\pi\sqrt{-1}(\mathbf{f}_0+\mathbf{f}+\cdots\mathbf{f}_{m-1}), \tag{7.20}$$

where $\mathbf{f}_0, \mathbf{f}_1, \ldots, \mathbf{f}_{m-1}$ are the residues of f at the simple poles $u_0, u_1, \ldots, u_{m-1}$ inside R.

For $a=x_0$ or $a=X$ (with $Y>b>y_0$), however, one must take the principal value of

$$\int_{x_0}^{X} f(x+Y\sqrt{-1})-f(x+y_0\sqrt{-1})\,dx$$

and attribute the value $\pi\mathbf{f}\sqrt{-1}$ to Δ in Eq. (7.15) and, likewise, if $b=y_0$ or $b=Y$ (with $X>a>x_0$).

Cauchy also discussed the case $u(r,p)=re^{p\sqrt{-1}}$. If $r=1$, he obtained the equation

$$\int_{-\pi}^{+\pi} e^{p\sqrt{-1}}f(e^{p\sqrt{-1}})\,dp=2\pi\sqrt{-1}(\mathbf{f}_0+\mathbf{f}_1+\cdots+\mathbf{f}_{m-1}), \tag{7.21}$$

where the residues $\mathbf{f}_0, \mathbf{f}_1, \ldots, \mathbf{f}_{m-1}$ are at the poles inside the unit circle. The formulas of the paper of November 22, 1819, are immediate consequences of this equation.

In 1822, just as in 1814, Cauchy investigated only functions with simple poles inside R. In a footnote to his article of July 1823 (19), he also gave the value of the residue $\mathbf{f}$ of f at a multiple pole u_i of order p in the form

$$\lim_{k\to 0} \frac{d^{p-1}}{dk^{p-1}} \frac{k^p f(u_i+k)}{(p-1)!}, \tag{7.22}$$

which generalizes formula (7.19). Cauchy gave no proof of Eq. (7.22), but further articles allow us to restore his arguments: he made the change of variable $y=b+kz$ in the singular integral Eq. (7.17) and then he expanded $\mathbf{f}[u_0+k(1+z\sqrt{-1})]=k^p(1+z\sqrt{-1})^p f[u_0+k(1+z\sqrt{-1})]$ to the $(p-1)$th order with respect to k, obtaining:

$$\mathbf{f}(u_0)+k(1+z\sqrt{-1})\mathbf{f}'(u_0)+\cdots+\frac{k^{p-1}(1+z\sqrt{-1})^{p-1}}{(p-1)!}\mathbf{f}^{(p-1)}(u_0)$$
$$+\,\omega[(u_0+k(1+z\sqrt{-1})].$$

The substitution of this expansion in the singular integrals yields a sum of p integrals, all vanishing except $\dfrac{\mathbf{f}^{(p-1)}(u_0)}{(p-1)!}\displaystyle\int_{-\infty}^{+\infty}\frac{1}{1+z^{\sqrt{-1}}}+\frac{1}{-1+z^{\sqrt{-1}}}dz.$ Thus, he once more obtained the residue formula, Eq. (7.18), but with $\mathbf{f}=\dfrac{\mathbf{f}^{(p-1)}(u_0)}{(p-1)!}$, that is

$$\mathbf{f}=\lim_{k\to 0}\frac{d^{p-1}}{dk^{p-1}}\frac{k^p f(u_0+k)}{(p-1)!}$$

Therefore, Eqs. (7.20) and (7.21) also deal with functions having multiple poles.

Cauchy published all these results on the theory of singular integrals while working on the resolution of the equations by means of integrals. In fact, it was by way of his research into the problem of solving linear ordinary and partial differential equations with constant coefficients that Cauchy was led to perfect his theory of singular integrals. For this reason, we will examine the work that Cauchy did between 1815 and 1823 relative to this topic before pursuing the historical development of his complex function theory.

In his paper 'Sur la théorie des ondes' of 1815, Cauchy exhibited the general solution of the equations he had examined, Laplace's equation and the equation of propagation,

$$\frac{\partial^4 y}{\partial z^4}+g^2\frac{\partial^2 y}{\partial z^2}=0$$

in the form of Fourier integrals containing an arbitrary function that was determined *a posteriori* from the initial and boundary conditions. He did not set forth in precise terms the methods by which he had arrived at the formulas. It was probably by induction. However, he tried to establish their generality by giving their series expansions and by comparing them to the results he obtained by the method of undetermined coefficients, 'a method', he wrote, 'that possesses all generality possible' (20).

Over the following years, Cauchy did not get an opportunity to develop his solution method to problems of mathematical physics. His appointment to the chair in mechanics at the Faculté des Sciences of Paris in 1821 led him back to this research area. Poisson, Cauchy's arch rival in analysis, had just published an important study in which he solved equations relating to vibrating elastic surfaces, to the distribution of heat in solid bodies, as well as the motion of fluids (21). Cauchy also presented several papers to the Académie in these same subject areas. In these papers, he developed a general method involving Fourier transforms for solving linear partial differential equations with constant coefficients and applied it to the main equations of mathematical physics.

In the first paper, 'Sur l'intégration générale des équations linéaires à coefficients constants', which he presented to the Académie on October 8 1821 (22), Cauchy substituted a single Fourier transform with an imaginary exponential for the two transforms with circular functions for the first time. Thus, he expressed a function $f(x, y, z, \ldots)$ of n variables contained in $[\mu', \mu''] \times [\nu', \nu''] \times [\omega', \omega''] \times \cdots$ by a multiple integral:

$$\left(\frac{1}{2\pi}\right)^n \int_{-\infty}^{+\infty} \int_{-\infty}^{+\infty} \int_{-\infty}^{+\infty} \cdots \int_{\mu'}^{\mu''} \int_{\nu'}^{\nu''} \int_{\omega'}^{\omega''} \cdots e^{\alpha(\mu - x)\sqrt{-1}} e^{\beta(\nu - y)\sqrt{-1}}$$
$$e^{\gamma(\omega - z)\sqrt{-1}} \cdots f(\mu, \nu, \omega \cdots)\, d\alpha\, d\beta\, d\gamma \cdots d\mu\, d\gamma\, d\omega \cdots.$$

Then, he showed that the integral formula

$$\left(\frac{1}{2\pi}\right)^n \int_{-\infty}^{+\infty} \int_{-\infty}^{+\infty} \int_{-\infty}^{+\infty} \cdots \int_{\mu'}^{\mu''} \int_{\nu'}^{\nu''} \int_{\omega'}^{\omega''} \cdots e^{\theta_i t} e^{\alpha(\mu - x)\sqrt{-1}} e^{\beta(\nu - y)\sqrt{-1}}$$
$$e^{\gamma(\omega - z)\sqrt{-1}} \cdots f(\mu, \nu, \omega \cdots)\, d\alpha\, d\beta\, d\gamma \cdots d\mu\, d\nu\, d\omega \cdots$$

represents a solution $\phi(x, y, z, \ldots, t)$ satisfying the initial condition $f(x, y, z, \ldots)$ for $t = 0$ of the homogeneous linear partial differential equation (of mth order relative to t) with constant coefficients:

$$F\left(\frac{\partial}{\partial x}, \frac{\partial}{\partial y}, \frac{\partial}{\partial z}, \ldots, \frac{\partial}{\partial t}\right)\phi(x, y, z, \ldots, t) = 0, \tag{7.23}$$

if $\theta_i(\alpha, \beta, \gamma, \ldots)$ is one of the m roots of the algebraic equation

$$F(\alpha\sqrt{-1}, \beta\sqrt{-1}, \gamma\sqrt{-1}, \ldots, \theta) = 0.$$

Finally, he sought to determine the solution of the equation, knowing the values of ϕ and of its derivatives relative to t up to order $m - 1$ for $t = 0$ and from this to deduce the solutions of the main linear equations of mathematical physics satisfying given initial conditions in integral form. In an unpublished portion of this work, he generalized his method to inhomogeneous equations.

While this study was being written, Cauchy remained unquestioning in his belief in the generality of the solutions he had obtained by his method, since he could always set them in infinite series expansions equal to the expansions deduced by the method of undetermined coefficients. However, in the study 'Sur le développement en série et sur l'intégration des équations différentielles' he presented some three and a half months later to the Académie (23), he showed that the method of undetermined coefficients is deficient. Indeed, from the existence of the function

$$f(x) = e^{-\frac{1}{x^2}},$$

whose series expansion in the neighborhood of 0 is identically the null series, he deduced the fact that a given MacLaurin expansion can actually correspond to several functions and, consequently, the methods of undetermined coefficients do not guarantee the generality of a solution of an ordinary or partial differential equation. For this reason, in an addendum to this paper, Cauchy expressed his misgivings about 'the discussion in its present state' relative to the generality of the formulas representing the solutions of partial differential equations of mathematical physics (24).

A year later, on September 16, 1822, Cauchy presented a new paper, 'Sur l'intégrations des équations linéaires aux différences partielles à coefficients constants et avec un dernier terme variable' to the Académie (25). In this paper, which was an expansion of his October 8, 1821, study, he set forth the main results given in his lectures at the Faculté des Sciences. Cauchy began by expressing $F\left(\frac{\partial}{\partial x}, \frac{\partial}{\partial y}, \frac{\partial}{\partial z}, \ldots \frac{\partial}{\partial t}\right)$ in Eq. (7.23) in the form $\sum_{p=0}^{p=m} f_p\left(\frac{\partial}{\partial x}, \frac{\partial}{\partial y}, \frac{\partial}{\partial z}, \ldots\right)\frac{\partial^p}{\partial t^p}$. Then, developing an idea that was present in the paper of October 1821, he reduced the problem of solving a linear partial differential equation, with given initial conditions for ϕ and for its derivatives relative to t up to order $m-1$, to the simpler problem of solving the linear ordinary differential equation of mth order,

$$\sum_{p=0}^{p=m} f_p(\alpha\sqrt{-1}, \beta\sqrt{-1}, \gamma\sqrt{-1}, \ldots)\frac{\partial^p}{\partial t^p} S(t) \tag{7.24}$$

satisfying the initial conditions $S(0) = 1$, $\frac{d}{dt}S(0) = u \ldots,$ $\frac{d^{m-1}}{dt^{m-1}}S(0) = u^{m-1}$, where u is some variable. Cauchy solved this equation by the classical method he expounded in his first-division course at the École Polytechnique.

Moreover, in a paper that he presented to the Académie on May 26, 1823 (26), Cauchy gave a new method of solving linear ordinary differential equations with initial conditions, such as Eq. (7.24). This method was based on his theory of singular integrals. It consisted of applying the residue theorem, Eq. (7.20), to the function $\frac{F(\theta)}{(u-\theta)F'(\theta)}$, where $F(\theta)$ is the characteristic equation. Thus, the solution was expressed in integral form. It was in this context that Cauchy extended the residue theorem to functions with multiple poles in order to treat the case of characteristic equations with multiple roots.

At the end of 1823, the construction of a unified theory of singular integrals, a task that had been begun 10 years before, seemed to have been achieved. When his study 'Sur la théorie des ondes' was registered for publication on May 17, 1824, Cauchy appended a note to it on the theory of singular integrals in which he made no additions to his results of 1822 and 1823 (27). But, suddenly, early in 1825, just as he was teaching mathematical physics at the Collège de France, he announced discoveries of fundamental importance relative to the basis and applications of his theory of singular integrals. In fact, between January 31, and February 18, 1825, he presented three papers on analysis to the Académie, in which he investigated the calculus of residues and the theory of curvilinear integrals of complex functions. The circumstances that led Cauchy to write these papers remain a mystery to this day, since the study of February 14 has not been published and the other two works are known only by way of what appears to be later versions. Although we are unable to completely explain this affair, we will nevertheless attempt to shed some light on it. In order to come to grips with the mystery surrounding these discoveries, we have to consider the influence of three quite different mathematicians.

The first of these mathematicians was Poisson. In 1820, Poisson published a paper on definite integrals, which was perhaps, as is frequently stated, the origin of Cauchy's innovation (28). Poisson proposed to integrate a real function having a real pole between the limits of integration by passing through a neighborhood of the pole along a path in the complex plane. The value thus obtained, he asserted, was equal to the difference of the values of the primitive of the function at the two limits of integration. Twice, in 1822 and 1823, Cauchy criticized this method of integration (29). He was especially annoyed with Poisson for attempting to preserve the definition of definite integral deduced from the primitive. He also rejected the idea of 'passing through a sequence of imaginary variables' in order to integrate a function between real limits, arguing that this conception led one to giving imaginary values to integrals taken between real limits. In spite of these criticisms, Cauchy was probably prompted by the paper of Poisson to develop his own ideas on complex integration.

The second mathematician, Barnabé Brisson, is less famous. Some 12 years older than Cauchy, he was a brilliant chief engineer at the Ponts et Chaussées. He was also interested in pure and applied mathematics, to which he seems to have devoted his spare time, in particular to the theory of linear partial differential equations (30). In 1804, 1821, and 1823, Brisson presented papers on this subject to the Académie. Cauchy was appointed reporter for the evaluation commission on the paper of 1823, a work that was never published and seems to have been lost. In this study, Brisson appears to have developed a symbolic calculus derived from the Leibnizian analogy of powers and differences. He had already stated the elements of this calculus in his 1821 study. In the 1823 paper, he used these first results to obtain solutions for linear ordinary and partial differential equations in symbolic form. From these solutions in symbolic form, he proceeded to obtain series expansions and integral representations. Because of a disagreement between the members of the evaluation commission, Cauchy delayed submitting his evaluative report to the Académie (31). But, without delay, on December 27, 1824, he introduced Brisson's symbolic calculus, as revised and corrected by himself, in a long memoir devoted to his own method of solving the equations using Fourier's formula (32). The aim of this paper was to simplify this last method by using a concise notation. Thus, toward the end of 1824, Brisson's work had prompted Cauchy to focus attention on the possibilites of a new calculus for linear differential equations.

Elsewhere in his study, Brisson used an integration formula 'that rests', Cauchy critically wrote in his evaluative report, 'on the consideration of definite integrals that would assume imaginary values while the function under the $\int$ sign retains a real value' (33). In a new study that he was working on (34), Brisson, perhaps mindful of Cauchy's criticisms, used integrals taken between imaginary limits of integration. Cauchy, who quickly grasped the fruitfulness of Brisson's idea, became interested in this new class of integrals.

We turn now to the role of yet another mathematician, Michael Ostrogradski. Ostrogradski had come from Russia to Paris in 1822 to study mathematics. Though still very young, he was nonetheless, Cauchy wrote, 'blessed with great abilities and well versed in infinitesimal analysis' (35). Ostrogradski presented two highly significant papers to the Académie during the summer of 1824. In his paper of July 24, 'Sur la difficulté que se rencontre dans le calcul des intégrales définies lorsque la fonction à intégrer est discontinue entre les limites d'intégration', he criticized Poisson's 1820 study on definite integrals (36). Ostrogradski's criticisms coincided with Cauchy's to the extent that he reproached Poisson for his notion of definite integrals based on the consideration of primitives. But, it was Ostrogradski's paper of August 7, 'Remarques sur les intégrales définies', that was particularly instrumental in drawing Cauchy's attention to the young Russian (37). In this study, Ostrogradski proved—albeit in a clumsy but fundamentally correct way—that the residue of a function at a multiple pole assumed the value Cauchy had declared without proof in 1823. In a note to the Académie on February 14, 1825, Cauchy alluded to this study (38). But, rather curiously, while we see this paper as a study closely related to the calculus of residues, Cauchy merely mentioned this work incidentally in discussing his own research on integration along arbitrary paths in the complex plane. Moreover, he even pointed out (quite incorrectly) that in this study Ostrogradski had used integrals taken

Mikhail Ostrogradski (1801–1862). This Russian mathematician took part in the invention of the calculus of residues. Photograph by J. L. Charmet, permission of the Académie des Sciences.

between imaginary limits. Ostrogradski, in fact, used only integrals between real values and merely indicated that, although the function was real between the limits of integration, it might, under certain conditions, become imaginary.

On January 31, 1825, Cauchy presented the paper 'Sur un nouveau genre de calcul analogue au calcul infinitésimal' to the Académie. One year later, on February 27, 1826, the manuscript was deposited with the Secretariat, and several days later the work was published in *Exercices de Mathématiques* (39). We can reasonably assume that the original version of the work—that is, the version of January 1825—scarcely differed from the February 1826 version. In this study, directly prompted by Ostrogradski's paper of August 7, 1824, Cauchy announced a new calculus, which he called the calculus of residues. He

Nouveau Mémoire sur le Calcul des Résidus et sur les intégrales Définies, 1825. Manuscript by Cauchy. Published by permission of the Académie des Sciences of Paris.

drew a parallel between the differential coefficient of a function f at the regular point u_0, which can be defined as the coefficient of $u - u_0$ in the expansion of the function in the neighborhood of u_0, and the residue of the same function relative to a pole u_0 of order m, which he defined as the coefficient of $(u - u_0)^{m-1}$ in the expansion of the function $\mathbf{f}(u) = (u - u_0)^m f(u)$ in the neighborhood of u_0, that is, $\frac{\mathbf{f}^{(m-1)}(u_0)}{(m-1)!}$ or, by setting $k = u - u_0$, $\lim_{k \to 0} \frac{d^{m-1}}{dk^{m-1}} \frac{k^m f(u_0 + k) f^{(m-1)}(u_0)}{(m-1)! \quad (m-1)!}$ [to be compared with Eq. (7.22)]. Thus, he transformed Ostrogradski's demonstration into a definition.

The calculus of residues was not really a novelty since Cauchy had already developed the mathematical tools necessary to calculate the residue of a function at a simple pole, which he denoted by **f**; moreover, in 1823, he had even determined the value of the residue of a function at a multiple pole. However, up to then these residues had been obtained by means of singular integrals, so that Cauchy had not thought it necessary to develop a special terminology and notation. It is not clear, however, whether Cauchy already possessed the notation for the extraction of residues at the beginning of 1825. This uncertainty follows from the fact that, prior to 1826, Cauchy did not make use of this notation which resulted from his collaboration with Ostrogradski.

On February 7, 1825, one week after having presented his first paper on the calculus of residues, Cauchy wrote a note announcing a new paper 'Nouveau mémoire sur le calcul des résidus et sur les intégrales définies', to the Académie (40). In this note, he indicated a generalization of his calculus of residues to the case in which the order of multiplicity of a pole was replaced by a fractional power. In the end, however, Cauchy gave up on this particular generalized calculus of residues, and he never published his paper. His mistake in this matter clearly followed from the failure to define the class of functions to which his new calculus should apply. Between February 7 and February 14, Cauchy inserted several lines in his note announcing new research on definite integrals between imaginary limits. This research, he said, was contained in the last paragraph of his paper. Citing Brisson and Ostrogradski, Cauchy articulated the crux of what would be his outstanding paper, 'Sur les intégrales définies prises entre des limites imaginaires', which he presented to the Académie on Feburary 28, 1825.

This latter study was the most important step from the crude theory of singular integrals of 1814 toward the sophisticated articles on meromorphic functions of the 1850s. Aware of the significance of this paper, which was the outcome of 10 years of research, Cauchy had the entire paper edited in August 1825 (41).

The essential innovation was the definition and the use of integration in the complex plane. Just as he had done in *Calcul Infinitésimal* in 1823, Cauchy first defined the integral

$$\int_{x_0 + y_0\sqrt{-1}}^{x + y\sqrt{-1}} f(u)\, du$$

as the limit, as the intervals $[x_0, X]$ and $[y_0, Y]$ are indefinitely subdivided, of the finite sums:

$$[(x_1 - x_0) + (y_1 - y_0)\sqrt{-1}] f(x_0 + y_0\sqrt{-1})$$

$$+[(x_2-x_1)+(y_2-y_1)\sqrt{-1}]f(x_1+y_1\sqrt{-1})+$$

$$\cdots+[(X-x_{n-1})+(Y-y_{n-1})\sqrt{-1}]f(x_{n-1}+y_{n-1}\sqrt{-1}),$$

where $x_0 \leqslant x_1 \leqslant \cdots \leqslant X$ and $y_0 \leqslant y_1 \leqslant \cdots \leqslant Y$. Then, for any path from $x_0+y_0\sqrt{-1}$ to $X+Y\sqrt{-1}$, he introduced the parametric representation $\phi(t)+\chi(t)\sqrt{-1}$ (with $\phi(t_0)+\chi(t_0)\sqrt{-1}=x_0+y_0\sqrt{-1}$ and $\phi(T)+\chi(T)\sqrt{-1}=X+Y\sqrt{-1}$).

In the case of a function that is finite and continuous, Cauchy proved that the value of its integral is independent of the path of integration between $x_0+y_0\sqrt{-1}$ and $X+Y\sqrt{-1}$ by using the calculus of variation. Following this, he examined the case in which the function had (simple or multiple) poles in the domain between two given paths. In an early stage of his investigation, Cauchy made use of the theory of singular integrals, which he had been developing since 1822, in order to obtain the residue theorem for the case of two nearby curves (he also used a method based on the consideration of the whole part of the function); and in a later stage, he generalized this result to the case of two arbitrary curves. He represented a path as a curve in the complex plane and then considered a 'variable and moving curve whose form is such that over any two distinct time intervals it can be made to coincide successively with two fixed curves' representing the two paths under consideration. However, he restricted himself to paths in the rectangle $\{x+y\sqrt{-1}:(x,y)\in[x_0,X]\times[y_0,Y]\}$ and indicated without proof that the theorem 'holds even when such contours (i.e., paths) are not confined to the rectangle' (42). From the general residue theorem, Cauchy easily derived all the formulas that had been obtained over the preceding years. In concluding, he gave the values of a number of improper integrals.

Up until now, Cauchy had formulated all his research on definite integrals in solidly analytical terms and scarcely paid any attention to formulations based on geometrical considerations; for, in his view, such formulations lacked rigor. However, in the present situation, the fact that paths in the complex plane could be represented geometrically enabled Cauchy to explain simply their relevant topological properties, even if he also analytically defined the homotopy of two paths by means of a parameter. Later on, when this paper was out of the way, Cauchy did not in any sense renounce his view of imaginaries as symbolic expressions, because such a view seemed to him to be the only one that could be rigorously justified.

The simultaneous creation of the calculus of residues and the theory of integration in the complex plane corresponds, in Cauchy's works, to two radically different directions. The calculus of residues is fundamentally a formal calculus that arose from the theory of singular integrals, but was expounded in its own right in an abstract manner. It seems that Cauchy had been presenting this new calculus in his physics courses at the Collége de France during 1824–1825, when he was striving to perfect Brisson's symbolic calculus. He perceived that the calculus of residues could be of use to him in simplifying the calculations arising in connection with the solution of differential equations. On the other hand, the theory of integration in the complex plane represented an unusual attempt to reinterpret and generalize all the results Cauchy had obtained between 1814 and 1823 by using his theory of singular integrals. Cauchy, of course, was perfectly aware of these opposing tendencies: he relegated his studies on the calculus of residues to the broad area of 'calculi'; because, as concerns residues, the focus was on algorithms that would render his theory of singular integrals more amenable. On the contrary, he consigned his paper 'Sur les intégrales définies prises entre des limites imaginaires' to the theoretical realm, thus completing the goal he had set for himself in 1814, namely, to endow the method of integrating by passing from the real to the imaginary with a rigorous theoretical foundation. Therefore, he made no allusions at all to the calculus of residues in the notes that he appended to his 1814 paper 'Sur les intégrales définies', which in its updated version was handed over to the Secretariat of the Académie on September 14, 1825, for publication (43). For that matter, they were not mentioned in an important article published on October 1825 in the *Annales de Mathématiques* of Gergonne (44), in which he again proved the residue formula he had given in 1814 for the calculation of

$$\int_{-\infty}^{+\infty} f(u)\,du,$$

specifying more rigorously the conditions that must be imposed on the function $f(u)$ and generalizing the formula to the case of a function with multiple poles in the half-plane $\operatorname{Im} u > 0$.

It was not until 1826 that he began to publish his first works on the calculus of residues in *Exercices de Mathématiques*, which had begun to appear at this time. He proposed the specific notation

$$\sideset{_{x_0}^{X}}{_{y_0}^{Y}}{\mathcal{E}}((f(u)))$$

for denoting all the residues of f inside the rectangle $R = \{x + y\sqrt{-1}, (x, y) \in [x_0, X] \times [y_0, y]\}$. The expressions

$$\sideset{_{x_0}^{X}}{_{y_0}^{Y}}{\mathcal{E}}\phi(u)((\chi(u))), \sideset{_{x_0}^{X}}{_{y_0}^{Y}}{\mathcal{E}}\frac{((\phi(u)))}{\chi(u)}$$

and

$$\sideset{_{x_0}^{X}}{_{y_0}^{Y}}{\mathcal{E}}\frac{\phi(u)}{((\chi(u)))}$$

were used to denote the residues of $\phi(u) \cdot \chi(u)$, $\dfrac{\phi(u)}{\chi(u)}$ and $\dfrac{\phi(u)}{\chi(u)}$, respectively, at the poles of $\chi(u)$, $\phi(u)$, and $\dfrac{1}{\chi(u)}$ contained in R. Moreover, Cauchy called the integral residue, the sum of all the residues of the function $f(u)$. One of the first

SUR UN NOUVEAU GENRE DE CALCUL

ANALOGUE AU CALCUL INFINITÉSINAL.

On sait que le calcul différentiel qui a tant contribué aux progrès de l'analyse, est fondé sur la considération des coefficients différentiels ou fonctions dérivées. Lorsqu'on attribue à une variable indépendante x un accroissement infiniment petit ε, une fonction $f(x)$ de cette variable reçoit elle-même en général un accroissement infiniment petit dont le premier terme est proportionnel à ε, et le coefficient fini de ε dans l'accroissement de la fonction est ce qu'on nomme le coefficient différentiel. Ce coefficient subsiste, quelque soit x, et ne peut s'évanouir constamment que dans le cas où la fonction proposée se réduit à une quantité constante. Il n'en est pas de même d'un autre coefficient dont nous allons parler, et qui est généralement nul, excepté pour des valeurs particulières de la variable x. Si, après avoir cherché les valeurs de x qui rendent la fonction $f(x)$ infinie, on ajoute à l'une de ces valeurs, désignée par x_1, la quantité infiniment petite ε, puis, que l'on développe $f(x_1+\varepsilon)$ suivant les puissances ascendantes de la même quantité, les premiers termes du développement renfermeront des puissances négatives de ε, et l'un d'eux sera le produit de $\frac{1}{\varepsilon}$ par un coefficient fini, que nous appellerons le *résidu* de la fonction $f(x)$ relatif à la valeur particulière x_1 de la variable x. Les résidus de cette espèce se présentent naturellement dans plusieurs branches de l'analyse algébrique et de l'analyse infinitésimale. Leur considération fournit des méthodes simples et d'un usage facile, qui s'appliquent à un grand nombre de questions diverses, et des formules nouvelles qui paraissent mériter l'attention des géomètres. Ainsi, par exemple, on déduit immédiatement du calcul des résidus la formule d'interpolation de Lagrange, la décomposition des fractions rationnelles dans le cas des racines égales ou inégales, des formules générales propres à déterminer les valeurs des intégrales définies, la sommation d'une multitude de séries et particulièrement de séries périodiques, l'intégration des équations linéaires aux différences finies ou infiniment petites et à coefficients constants, avec ou sans dernier terme variable, la série de Lagrange et d'autres séries du même genre, la résolution des équations algébriques ou transcendantes, etc. ...

Definition of the residue of a function at a pole. 'Sur un nouveau genre de calcul analogue au calcul infinitésimal', *Exercices de Mathématiques*, **1**, 1826, p. 11. Published by permission of the École Polytechnique.

applications of the calculus of residues consisted of representing the singular part of a [meromorphic] function f by the integral residue

$$\mathcal{E}\frac{((f(u)))}{x-u}. \tag{45}$$

Developing a notion that originated in the concept of the extraordinary integral, he then investigated the whole part of the function f, that is,

$$\omega(x) = f(x) - \mathcal{E}\frac{((f(z)))}{x-z}$$

and thus obtained a method for decomposing rational fractions into simple fractions and later, more generally, for expanding functions into series (46). Cauchy also obtained a simple proof of the residue theorem, first for rectangular contours by the same procedure and shortly after this, for circular contours (47). All calculations discussed were applied to functions for which the number of poles increased infinitely with the length of the contour. In 1827, he showed that it was necessary to consider in the calculations the integral residue corresponding to a circle with center 0 and a radius that tends to infinity. This he called the principal residue of the function (48).

Important areas of application of the calculus of residues were in the research on the solutions of algebraic and transcendental equations as well as on the solution of linear ordinary and partial differential equations with constant coefficients and—in some cases—with variable coefficients. As we saw earlier, it was within the context of the solutions of such equations that Cauchy had created his new calculus. Furthermore, he generalized the formula he had obtained in 1819 for the sum of similar functions $\phi(u_i)$ of the roots u_i of $F(u) = 0$ [see Eq. (7.11)] to the case of multiple roots, and presented this generalized result in the form

$$\sum \phi(u_i) = \mathcal{E}\frac{\phi(u)F'(u)}{((F(u)))}.$$

in *Exercices de Mathématiques* (49). Likewise, he gave the general solution of the linear homogeneous differential equation

$$\frac{d^n y}{dx^n} + a_1\frac{d^{n-1}y}{dx^{n-1}} + a_2\frac{d^{n-2}y}{dx^{n-2}} + \cdots + a_{n-1}\frac{dy}{dx} + a_n y = 0$$

by the formula

$$y = \mathcal{E}\frac{\phi(r)e^{rx}}{((F(r)))},$$

where $F(r)$ is the characteristic equation and $\phi(r)$ is an arbitrary function that can be determined from the initial conditions (50). Finally, in several other papers, Cauchy tried to apply the calculus of residues to linear partial differential equations by combining it with the Fourier transform method.

Thus, he was able to obtain solutions for some types of equations, especially when a variable t could be separated from the other variables (51). In 1829 and 1830, Cauchy made use of this method in his theory of light (52).

In his research work on the application of the calculus of residues to mathematical physics, Cauchy obtained representations of functions by means of residues in the form of series expansions, especially Fourier's series expansion. He thought that he had proved the convergence of such series (53). However, this was not the case, because in 1829 Dirichlet showed the insufficiency of Cauchy's argument (54).

Aside from the calculus of residues and the first definition of a complex function of a complex variable, which he gave in 1829 in his *Leçons sur le Calcul Différentiel*, Cauchy did not develop any new results in his theory of functions until 1831. Indeed, during this period he seems to have been in a sort of retreat from the position he had held in 1825; he made no use of integration in the complex plane for arbitrary regions, but limited himself to rectangular and circular areas. With the exception of Ostrogradski, no mathematicians seem to have been interested in Cauchy's theory of functions during these years. This lack of interest, however, was not due to any lack of effort on Cauchy's part, because he certainly tried to stimulate an interest and understanding of this theory by his publications and in his teaching at the École Polytechnique and at the Collège de France. But, for a considerable time, his methods were considered as being too complicated, and it was only after 1840 that the theory began to find acceptance in the French, German, and Italian mathematical communities.

Chapter 8

A Mathematician in the Congrégation

Augustin-Louis Cauchy was now 28 years old and still living at his parents' home, near the Palais du Luxembourg, where he had returned in 1812. Louis-François decided that it was now high time for his eldest son to get married, since he now had a secure place in life. The elder Cauchy's choice of bride for his son was the only daughter of the bookseller Marie-Jacques de Bure, Aloïse de Bure, 23 years of age. The de Bures were an old, solidly bourgeois family. They had been in the book trade since the 17th century and were connected with quite a few publishers, particularly with Didot and Saugrain. The two de Bure brothers, Marie-Jacques and Jean-Jacques, had followed their father Guillaume in the book trade in 1813 and were later associated with the King's Library, where they compiled many catalogs. Serving in this capacity until 1838, they became well known as collectors of books and prints (1).

The marriage took place amid great pomp and ceremony on April 4, 1818, at the Church of Saint-Sulpice in Paris. Aside from her personal effects, which had an estimated value of 5000 francs, Aloïse brought a considerable dowry to the marriage: a perpetuity of 1300 francs; another perpetuity and income of 1000 francs, to be paid annually in cash by her parents; 2500 francs in silver; and 10 shares of common stock in the Bank of France. In addition to the foregoing, she brought a gift of 3000 francs from her uncle, Jean-Jacques de Bure. Augustin-Louis brought a more modest financial contribution to the marriage, including 15,000 francs that his parents had advanced as part of his inheritance. He also brought 900 francs in state credits; 4000 francs in hard cash, and his personal effects, including scientific instruments valued at some 3000 francs. In effect, almost this entire amount came from Augustin-Louis' own earnings and thrift, which confirms the fact (as if it needed to be confirmed) that during those days one would be unlikely to enter the field of mathematics to make a fortune.

The marriage had been announced under the most favorable conditions, with Louis XVIII and the entire royal family attesting the goodwill and esteem they bore toward young Cauchy by signing the marriage contract (2). A

Monsieur le Directeur Général,

J'ai l'honneur de vous prévenir que, d'accord avec mes parents, j'ai formé le projet de m'unir en mariage à Mlle Aloyse Debure, fille unique de M. Debure de Milly, l'un de MM. Debure frères, libraires de la Bibliothèque du Roi. Les qualités personnelles de Mlle Debure, l'honorable réputation dont jouit sa famille, distinguée depuis deux siècles par son instruction et sa probité dans le commerce de la librairie, les principes religieux et politiques dont cette famille a toujours fait profession, me donnent lieu d'espérer, Monsieur le Directeur Général, que vous daignerez accorder votre approbation au projet de mariage dont j'ai l'honneur de vous faire part.

Daignez aussi, je vous prie, agréer l'hommage du respect avec lequel je suis

Monsieur le Directeur Général

Votre très humble et très obéissant serviteur

A. L. Cauchy

Ingénieur des Ponts et chaussées, membre de l'académie Rle des Sciences, Profr à l'éc. Rle Polytechnique

Paris le 25 Mars 1818

M. le Conseiller d'État, Directeur Général des Ponts et Chaussées et des Mines.

Petition by Augustin-Louis Cauchy to the Director General of the Ponts et Chaussées to be licensed to marry Aloïse de Bure. Archives Nationales, F^{14} 2187^2, Cauchy file. Published by permission of the École Polytechnique.

number of Peers of France attended the ceremonies, among them several Congregationalists, such as the Count of Polignac, the Marquis of Semonville, and the Viscount of Dambray. Also present were the Director of the École Polytechnique; mathematicians, such as Laplace, Reynaud, Dinet, and Binet; and several ecclesiastics, among whom were Teysseyrre, Genoude, Desjardins, and Legris-Duval. The book trade was represented by Firmin-Didot and others. A little pamphlet was published in celebration of the marriage (3). This publication opened with an address by Augustin-Louis Cauchy to the Duchess of Angoulême requesting that she and her husband attend the wedding ceremonies. It contained some verses written in honor of the young couple by Louis-François and Alexandre Cauchy, Pierre Didot, and Charles Magnin and a poem Augustin-Louis addressed to his bride. This last poem is the sole remaining testimony on how he felt toward Aloïse. It reads:

> If ever there was a blessed moment in my life,
> A happy day
> It was most assuredly, my sweet friend,
> The day you gave your love to me.
> God Himself, has just blessed
> The marriage that crowns all my desires.
> His blessed law wills that I should love you
> Ah! It is but a command that I am happy to obey.
> . . . .
> And I will love you, my beloved friend,
> Right to the end of my days;
> But even then there is another life
> Since your Louis will love you forever.

According to Valson, 'Cauchy's first years of married life were smooth, untroubled by any rancors and problems' (4). The couple lived in the Hôtel de Bure in the rue Serpente, No. 7. The house belonged to Marie-Jacques and Jean-Jacques de Bure. Starting in 1822, Augustin-Louis habitually spent summers in Sceaux, near Paris, at the Maison Trudon, a peaceful, shaded property that had been purchased by Marguerite de Bure, Aloïse's grandmother.

In 1819, the couple's first daughter, Marie Françoise Alicia, was born; and, in 1823, the second and last daughter, Marie Mathilde, was born. In any event, it appears that Augustin-Louis Cauchy accorded his wife and two children no important place in his life: his scientific work and other preoccupations took all of his time. In fact, we will see that in 1830 Cauchy did not hesitate to leave his family for several years and go into voluntary exile. It is, of course, a very dangerous thing to speculate on the motives that compel the behavior of so complex a person as Augustin-Louis Cauchy, particularly when so little relevant evidence is available. Yet, it would seem highly likely that Cauchy's marriage was a marriage of convenience, a union entered into in order to satisfy his parents. Since early youth, his sensitive and passionate nature had been strained to the utmost by his somewhat arrogant und uncompromising search for truth. In short, then, domestic life was simply a bother to him; it was stifling and placed an intolerable strain on his fragile nerves.

In August 1826, Cauchy contemplated taking a month's vacation in the Pyrenees Mountains for his health. Lamennais, who also suffered from a nervous condition would also go with him on the trip (5). In any event, the vacation never took place. As the years passed, Cauchy's condition did not improve (6). However, he refused to rest, to slacken his pace, but rather continued his regular participation in meetings and working sessions at the Académie. In addition to this, he produced a prodigious amount of research—treatises, studies, articles, and mathematical studies.

Although he lived with his in-laws in the rue Serpente, he was still a regular visitor at the Palais du Luxembourg, where he spent time with his parents and his sisters and brothers. Until his retirement in 1825, Louis-François continued to support his eldest son, advancing his career whenever possible. Through his influence, Augustin-Louis was awarded the Légion d'Honneur in August 1819 and an appointment as engineer-ordinary first class in March 1820 and then as chief engineer second class in May 1825 (7).

In spite of the number and variety of mathematical and scientific works he produced between 1816 and 1830—a list that, aside from the many unpublished courses he taught at the École Polytechnique, the Collège de France, and the Faculté des Sciences, included 92 papers read, presented, or registered at the Académie, 41 articles published in scientific reviews, and 10 longer published works—he devoted a significant part of his time to other activities. In particular, he was a regular participant in the various charitable works founded by the Congrégation, which he had entered in 1808, later being joined in this association by his two brothers, Alexandre and Eugène.

The Congrégation had officially begun to hold meetings at the start of the Restoration in 1814. It was managed now by a Jesuit, Father Legris-Duval. Taking full advantage of the new political climate, it prospered under the protection of the authorities. Little by little, the Congrégation increased its activities and spheres of operation through its affiliates. These were specialized organizations that the Congrégation created to take care of particular areas of concern. The first affiliate was the Société des Bonnes Oeuvres, which was established in 1816. This organization's area of concern basically consisted of three charitable practices: visiting hospitals, visiting prisons, and instructing young chimney sweeps in the catechism. Cauchy was a member of a commission that, in April 1816, was charged by Father Legris-Duval with the responsibility of forming the Société. In this capacity, he served alongside the extremely conservative leading figures, such as Mathieu de Montmorency and Alexis de Noailles (8). He worked tirelessly. In fact, Lamennais, to whom Cauchy was closely bound, stated in a letter written in July 1818 that he (Cauchy) was absorbed from 'dawn to dusk' with charitable works (9). Cauchy's close association with Lamennais had no doubt come by way of his (Cauchy's) prior friendship with Teysseyre.

Later on, he also worked with the Société Catholique des Bons Livres, an organization that had been created in Paris in August 1824. Its goal was to finance the publication of good books, to propagate their message, and to distribute them as cheaply as possible. The Duke Mathieu de Montmorency was head of this organization until he died, when the Duke de Rivière replaced him. Cauchy was one of its five directors. He served in this capacity along with the Abbé Perreau, Grand Almoner of the Chevaliers de la foi; the Abbé Dufriche-Desgenettes, Curator of the Church for Foreign Missions; the Abbé de Salinis, a close relative of Lamennais; and Pierre Sebastien Laurentie, an ultraroyalist journalist on *La Quotidienne*. In July 1827, a serious dispute broke out between the Société and the Grand Master of the Université, Monseigneur Frayssinous.

At the behest of Laurentie, the Société Catholique des Bons Livres planned to create and publish the Catholic Encyclopedia of the Sciences, a major work that was to have extreme conservative leanings. Abbé Clausel, a member of the Royal Committee on Public Instruction and an ardent Gallican, prompted by Frayssinous, violently attacked the encyclopedia project and the Société itself in three very sharply worded pamphlets. Clausel laid particular blame on Laurentie, whom he regarded as the great enemy of the Université and of censorship. Cauchy, Salinis, Perreau, and Laurentie then handed in their resignations as directors of the Société to the Duke of Rivière, as an expression of protest against the publication of the pamphlets. Following various negotiations and compromises, Laurentie was forced to give up his post as a director; the other directors, including Cauchy, resumed their duties. Following this dispute, the Société lost its ministerial protection, and as a result, it slowly declined; finally, in August 1830, it was disbanded altogether (10).

In the meantime, Cauchy became further involved in the extreme right-wing battle that was now raging between the Jesuits and the Université, with Cauchy on the side of the Jesuits and against the Université monopoly. In order to understand the nature of this dispute, it is necessary to briefly recall a few points relative to the evolution of the political climate in France during the years from 1821 to 1828 and the role Cauchy played at the Académie des Sciences on the side of the clerics.

Ever since the assassination of the Duc de Berry, eldest son of the future Charles X, in February 1821, the government had been breaking away from the more liberal political policies that it had followed during the preceding period and pursued an increasingly reactionary road. At the end of the reign of Louis XVIII, the Prime Minister Villèle instituted the repression against the Charbonnerie (a liberal secret association) and used all means at his disposal to reduce the influence of the liberal parliamentarians in the two houses. Charles X came to the throne in 1824, a few months after the election (skillfully prepared by Villèle) of a Chambre retrouvée with the same leanings as the famous Chambre introuvable of 1815.

The ascension of Charles X further strengthened the reaction. Measures were instituted in 1825 in favor of the landed aristocracy (the Milliard for

Cauchy as young academician. Half-length portrait by Boilly, 1821. Published by permission of the École Polytechnique.

the indemnification of the émigrés). The reestablishment of birthrights was projected, but it was rejected by the Chamber of Peers in 1826. As for the bourgeoisie, they saw their interests threatened, in 1824, by a scheme for the conversion or funding of stocks, and again in 1827, by the dissolution of the National Guard in Paris. The guard was guilty of having protested, during a review, against the notorious 'Law of Justice and Love,' which curtailed the freedom of the press.

More than anything else, however, it was the so-called 'alliance between the throne and the altar' that aroused deep-seated passions in France during this time. Influenced by the ideas of Maistre, Bonald, and Lamennais, a large part of the French right regarded the Great Revolution of 1789 as nothing less than 'Satan's handiwork', in the fullest sense of the term. Therefore, in order to reestablish political order and the traditional civil arrangements—both of which were undermined by the catastrophe of 1789—it was necessary to support religion. Bit by bit, a faction of the counterrevolutionary groups longed for a kind of theocratic purification (11). The king, a very pious man, wanted to see the reestablishment of the prerogatives that the Catholic religion had enjoyed prior to the Great Revolution. The law on sacrilegious persons, the law on religious communities for women (which, though instituted merely by royal ordinance, had the weight of a legal statute), and the solemn coronation at Rheims all seemed, in 1825, to foretell a 'government by priests'. In the provinces, the various missions that had been organized by the many congregations and were supported by the civil authorities redoubled their efforts to lead the masses back to the cult and faith of Catholicism. Augustin-Louis Cauchy fully approved of these policies and echoed this conservative religious mood in the Académie, where, on several occasions, he did not hesitate to raise objections and denounce doctrines that were contrary to religion. This behavior, of course, exasperated and angered the Académie's liberal wing.

On July 19, 1824, he presented a note pertaining to Geoffroy Saint-Hilaire's verbal report on Serres' work entitled 'Anatomie comparée du cerveau dans les quatre classes d'animaux vertébrés'. Going outside the framework of a purely scientific debate and inquiry, he severely condemned Dr. Gall's 'very eminent philosophical principle, a principle that rejects both the true philosophy and the vital doctrines on which rest the peace and well being of society' (12). This action by Cauchy, however, met with sharp and determined resistance from the Académie. In his note, he pointed out that the Académie had itself never placed Gall on its list of candidates for the medicine and anatomy sections because of his theory of cerebral protuberances. The President of the Académie intervened in the dispute, forcing Cauchy to withdraw this intrusive observation (13).

A short time later, at the Académie's session of October 4, 1824, Cauchy gave the evaluative report on a study of Souton about the theory of light. In this report, he reproached Souton for having erroneously asserted (following Voltaire) that Newton doubted the existence of the soul. Cauchy took advantage of this opportunity to attack Voltaire and to reaffirm the superiority of the Christian religion (14). This was sufficient to set off whispers and murmurs among the academicians and to ignite a sharp polemic in the press. In the October 6, 1824, issue of *Le Corsaire*, a liberal periodical, there appeared an article in which was denounced 'this little discourse that rather more resembles a homily than a scientific report'. In a bantering tone, the journalist continued:

> Now, it was certainly a curious thing to see an academician who seemed to fulfill the respectable functions of a missionary preaching to the heathens. It is not a novelty, except when it happens at the Académie, to hear a geometer busily trying to prove—not by way of equations and solidly logical arguments but by declamations that are both hackneyed and completely out of place in an enlightened age—that materialism has taken over the domains of science, a dubious proposition that ought not to be difficult to disprove. Equally strange to hear was that the names

> of Voltaire and other philosophers are found mentioned—by what standards we do not know—in this peculiar little sermon, which as a sermon, ended up very amusing indeed.

Le Mémorial, a Catholic publication, replied to *Le Corsaire's* 'indecent article' by citing Cauchy's outstanding qualities:

> When he was no older than the age at which most youths have scarcely completed their elementary school training, he was called to take his seat in the most illustrious academy in the kingdom, a singular distinction that was due to nothing more than the brilliance of a first-class mind, a brilliance that outstripped his years. That much so, one should not pay much mind to the insipid carpings of a literary corsaire (15).

In the meantime, Cauchy had earned a reputation in liberal circles as an inquisitor in the breast of the Académie.

In the June 1, 1825, issue of the *New Monthly Magazine*, Stendhal captured the perceived personality of the controversial savant in a definitive phrase, writing:

> Nowadays, M. Cauchy of the Institut, a veritable Jesuit in short frock, is charged with the honorable task of ferreting out and bottling up physiology, a science that of late has been advanced so far smartly by the experiments of MM. Flourens, Magendie, and Edwards (16).

A short time later, in the November 1826 issue of the same publication, Stendhal reported to his English-language readers an incident that befell the Académie:

> Some time ago, a naturalist whose name I will not mention out of fear of doing him injury, read a paper on the various phenomena that can be observed in the life of certain insects. Of fundamental importance in its own right, the subject was here treated in a very witty manner. At the end of the reading, whispers and murmurs of approbation could be heard, whereupon M. Cauchy took the floor and made the remark that the Academy should not honor this curious little discussion of animal life by applauding it. Said M. Cauchy: 'Even if it were admitted that the things just told us were true—in my opinion they are false—it would still not be useful to communicate such truths to the masses, given the devilish state into which our misbegotten Revolution has hurled public opinion. Any such talk can only harm our holy religion. They clearly show the influence of physical causes and tend to confirm Cabanis' mischievous doctrines'. These words from Mr. Cauchy, who is himself a dignitary of the Academy of Sciences, were greeted with great bursts of laughter.

Although this little anecdote, which combines the ridiculous with the downright odious, was particularly stinging, an examination of the minutes of the Académie's meetings shows this incident was merely a clever, malicious invention that was doubtlessly inspired by Cauchy's actions against Gall's system. Using the same ironic tone, Stendhal kept up his attacks on the scientist, writing:

> This courageous man, who is so enterprising and who apparently wants to be a martyr to contempt, requested at the meeting on Monday last that the Academy remove from its library all books that were tainted with philosophical leanings. Up to now, the Academy has not dared reply to M. Cauchy's request. Further to the point, several men of science, whose livelihoods depend on salaries from some insignificant employment, have been obliged to refuse to publish their most recent discoveries and findings in physiology out of fear that they might be accused of throwing some new light on the interplay between man's physical and moral aspects' (17).

This was all baseless gossip, pure and simple, and Cauchy paid no attention whatsoever to these malicious criticisms. In Dupin's *Observations sur les Recherches Statistiques sur l'Instruction et sur la Moralité Comparée des Différents Départements de la France*, which was presented on January 15, 1827, he found arguments that allowed him to set forth once more a justification for the Church's role in education, particularly that of the Frères des Écoles Chrétiennes in Brittany (18).

Such clericalism—and Cauchy was the champion of clericalism at the Académie—naturally angered the liberal movement; upset the bourgeoisie, which was still largely attached to the principles of 1789; and caused the lower classes—particularly in the town and cities—to become disaffected. The issue was clear, and the stakes were high: would the clergy go against the educational establishment, the press, and the legislature in order to reassert, with the support of the State, its ideological control over society? Oddly enough, the liberals found allies in the ranks of the extreme royalists, who were attached to Gallican traditions and were hostile to the ever-increasing role played by the congregations that acted outside of any control by the Church of France. As was seen in the crisis that shook the Société des Bonnes Oeuvres in 1827, the advocates of Gallican liberties had powerful supporters in the government.

The publication of a pamphlet against the Jesuits by the Comte de Montlosier, an old reactionary royalist with strong Gallican leanings, in February 1826, was the signal for a vast anticlerical campaign. This time around, the liberals were allied with the Gallicans in denouncing the secret role of the 'priestly party', whose driving force was the Jesuits, whose tool was the Congrégation, whose doctrine was ultramontanism, and whose ultimate aim was nothing less than the establishment of a theocracy. The liberal press, the deputies of the opposition parties, and many sensationalist pamphlets all joined forces to popularize the idea of a clerical plot and clamored for the dissolution of the Congrégation, as well as the dispersion of the Jesuits.

In spite of the official illegality of the noncertified Jesuit residences, the Minister of Ecclesiastical Affairs, Monseigneur Frayssinous, who had nevertheless become suspect in the eyes of the ultramontains, defended the Society of Jesus. Thus, the Jesuits were not only able to continue as before, but even to expand their activities. For example, in 1826, there were 476 Jesuits in France, as well as 2 noviciates and 8 small seminaries that the bishops had entrusted to the order. In reality, these seminaries were colleges, the most famous being that of Saint-Acheul. The Jesuits' power was extended by the Congrégation on which they exerted an unquestionable influence. At this time, some 1200 persons, many of whom occupied leading positions in society, belonged to the Congrégation (19).

Meanwhile, Villèle's government, increasingly isolated by its reactionary clerical policies, was seeking to curb the opposition groups, especially in the press, by applying repressive administrative measures and police tactics, all of which were unpopular and ineffective. Moreover, Villele's enemies—that is, the liberals and the Gallicans—expected to win the next elections that were announced November 1827, because the enfranchised classes were busily disassociating themselves from the extreme conservative party and its unrealistic reactionary policy. In order to calm the liberals, Martignac, the new Chief of the Ministry, obtained two ordinances from the King in June 1828. According to these ordinances, all church schools were placed directly under the control of the Université and members of the unauthorized congregations were prohibited from teaching. In effect, these regulations amounted to a ban on the Jesuits, and they, quietly submitting to the royal ordinances, dispersed.

A few days before this decision was made, the directors of the Société des Bons Livres, among whom was Cauchy, came together and founded the Association pour la Protection de la Religion Catholique. The aim of this organization was 'to unite the efforts of all persons of good will in order to defend the Catholic religion'. The association consisted of 45 members and a board of directors, of which Cauchy was one of the members. The real aim of the leaders of this organization was twofold: to establish a true ultramontain party that would protect the interests of the Church and to organize Catholic and antiliberal opinion by means of propaganda. The Gallicans regarded this undertaking with horror (20). By merely participating in the creation of such an organization, Cauchy became even more involved in politics. At the Académie, in the company of the scholars with whom he associated, he attempted to recruit others for the cause: Jacques Binet joined the association, and André-Marie Ampère agreed to sponsor, along with Cauchy, preparatory classes at the Collège de Juilly, an institution that had been established for the benefit of students who had been attending the Jesuit secondary schools that were closed in 1828. The Almones of the Collège de Juilly, the Abbé de Salinis, was a director of the Association (21).

In March 1828, the Association began publication of a weekly, *Le Correspondant, Journal Religieux, Politique. Philosophique*, with Bailly as editor. Lamennais was hostile to this publication, which was 'plowing the same field' as the *Mémorial Catholique*, the paper that he had founded. However, as the Abbé de Salinis explained to him in a letter dated February 25, 1829, '*Le Correspondant* was an idée fixe with the majority of the Association' (22). In truth, however, Lamennais was disappointed by the Association's timidity. Renouncing the theocratic ideal he had adhered to in the preceding period, he published a work in early 1829 entitled *Des Progrès de la Révolution et de la Guerre contre l'Église*, in which he associated democracy and liberty with the progress of Catholicism. However, notions such as these could not satisfy reactionary ultramontains such as Cauchy. In his letter to Lamennais, the Abbé Salinis observed:

> Cauchy and the Vicar of Missions [Dufriche-Desgenettes], who is basically a fine man, approve of the things you are doing to the extent that they understand them.

But Salinis was an optimist, because their degree of comprehension was, in truth, very scant. In fact, the Association so was paralyzed by its own internal divisions that, on balance, its actual accomplishments were quite small: a few brochures in 1828 and the journal that vacillated between Lamennais' and the congregationalists' positions right down to the Revolution of July 1830.

While condemning liberal agitation, *Le Correspondant* watched the policies followed by Polignac during the final months of the Restoration with increasing alarm and assailed the Revolution of July 1830 as an all too foreseeable misfortune. Adopting 'civil and religious liberty' as its motto, it continued publication after the Revolution, disappearing only when Lamennais began publication of his new journal, *L'Avenir*.

Taking advantage of the discomfort of the reactionary congregationalists, in November 1830, Lamennais, in conjunction with Montalembert, founded a new association, the Agence Générale pour la Défense de la Religion Catholique. This organization, which replaced the Association pour la Protection de la Religion Catholique, can be regarded as the first liberal Catholic organization in France.

Augustin-Louis Cauchy, of course, never changed his views as regards liberal Catholicism and throughout his life remained deeply attached to the reactionary clerical ideals of the Congrégation. Deeply involved in the antiliberal struggle, he regarded the events of 1830 not only as a national catastrophe, but also as a personal drama.

Chapter 9
Exile in Turin

The Revolution of 1830 marked a turning point in Cauchy's life and a break in his mathematical productivity. On August 2, 1830, Charles X abdicated in favor of his grandson, the Duke of Bordeaux. One month later, Cauchy left Paris and France and went into voluntary exile, which would last eight years. His exile was a consequence of the events that took place during the famous July days.

The Revolution of 1830 had its origins in the crisis between the royal power and the liberal movement, a crisis that divided the propertied classes and shook the country for several years. In August 1829, a right-wing cabinet was formed with La Bourdonnaye as the Minister of the Interior and Polignac as the Minister of Foreign Affairs. This cabinet represented a veritable provocation to the liberal majority in the Chamber. One year later, on July 26, 1830, following the liberals' victory in the elections, a set of official proclamations appeared in *Le Moniteur* suspending the freedom of the press; announcing the dissolution of the Chamber, which had just been elected; and modifying the election laws. This was nothing less than a coup d'état, and its effect was to provoke Paris into insurrection. On July 27, the first confrontations took place between the insurgents and the royal troops under the command of Marmont. For three days, the 'Glorious Three', Paris bristled with barricades as the insurrection gained strength. On July 29, the Tuileries fell to the insurgents, and Marmont was forced to evacuate Paris, leaving the capital in the hands of the revolution. The wealthy liberal bourgeoisie and its leaders Laffitte, Thiers, and La Fayette now brought the Duke of Orléans forward. The deputies had appointed him Lieutenant Général du Royaume on July 31 and King of the French on August 7. During these hectic days, Charles X fled from Paris and went to Cherbourg where he and his family embarked for England. Cauchy watched these events unfold from close range. During the weeks immediately preceding the Revolution of 1830 his antiliberal hatred seems to have reached unparalleled heights. On July 13, the day on which the election of deputies for the Department of the Seine was held, he provoked an incident at the Sorbonne. The incident was reported that same day by the *Journal du Commerce* in the following terms:

> M. Cauchy, a member of the Académie des Sciences, is an elector and votes in the 2nd Section of the 7th College, which meets at the Sorbonne. This morning, when M. Cauchy was called to deposit his paper ballot, he found two lists on the desk, one consisting of the names of the members of the provisional office and the other the list of candidates for final office. M. Cauchy was quite loud in expressing his indignation at seeing a list of liberal candidates posted up right in front of the royalist electors. He seized the list and crumpled it up in his hand.
>
> Premier President Séguier, who was standing near M. Cauchy, spoke sharply to him and reminded him that just as he was free to vote as he saw fit, he must leave to the other electors the possibility of correctly writing the names of the candidates of their choice. But M. Cauchy, paying no mind to M. Séguier's observation, continued to speak and gesticulate vehemently as he crumpled the list even more. M. Cropelet, the President of the 2nd Section, had to call M. Cauchy to order and request that he put the list back on the desk and withdraw as soon as he had cast his vote.

A few days later, Paris was gripped by insurrection. Barricades sprang up in the Latin Quarter. In the Place Saint-Michel, a few steps away from the rue Serpente where Cauchy lived, there was a violent confrontation on July 28, 1830. Students from the École Polytechnique took an active part in the uprising. Indeed, on July 28, 1830, the polytechnicians left the École, where they were billeted and, in uniform, took command of small bands of insurgents. Their action was decisive everywhere, especially in the storming of the Babylon Barracks, during which the polytechnician Vanneau was killed.

Cauchy was deeply shaken by the collapse of the government that he had so ardently supported. He did not attend a single session at the Académie between July 26 and August 30. Never before—not since his appointment in 1816—had he been absent for so long a time (1). On August 30, 1830, he attended the Académie's weekly meetings for the last time, and during the first days of September, he left Paris to go abroad, taking neither wife nor children with him (2).

What reasons could have compelled Cauchy to abandon everything, his family and his position, during the summer of 1830? Although definitive evidence is lacking, we can nevertheless advance a likely hypothesis. First, however, it should be noted that the most common explanation for his behavior does not fit in very well with the chronological order of the facts:

Polytechnicians climb over the wall on July 28, 1830 and rejoin the insurgents. Published by permission of the École Polytechnique.

Cauchy would have refused to swear allegiance to the new political arrangement in France out of loyalty to the now deposed king, he would have been faced with immediate loss, by forfeiture, of all the positions he had come to occupy, and in consequence he had to decide to leave France (3).

As a matter of fact, the law requiring the oath of loyalty to the King of the French from all persons occupying a position in the public service was not passed until August 30, 1830, after Cauchy had made his decision to leave the country. Penalties against him by the administrations of the various institutions with which he was affiliated—the Ponts et Chaussées, the École Polytechnique, and the Faculté des Sciences of Paris—were imposed several days after his actual departure (4). Thus, Cauchy did not go abroad after having lost his positions for refusing to take the oath, as Valson claims. A careful examination of the available facts and documents suggests other explanations. In truth, when Cauchy left Paris at the beginning of September 1830, he probably had no idea that he was now embarking on a long exile. But, several compelling reasons required that he leave the capital for a time.

For quite a while, Cauchy had been considering going on a trip in order to rest. As we have already seen, he had planned on traveling to the Pyrenees in 1826, but had been unable to do so at that time. His exceptional creative activity during the years prior to the Revolution of 1830 had left him physically and emotionally drained. Indeed, during the two years immediately preceding the turmoils of July 1830, he seems to have even accelerated the 'furious pace of his creative work' (5). In 1828 he published the last installment of Volume II of *Exercices de Mathématiques* as well as the 12 installments of Volume III. During the same year, he also published the second volume of *Leçons sur les Applications du Calcul Infinitésimal à la Géométrie* and a paper in the *Bulletin des Sciences* of Férussac. Moreover, he presented 10 papers to the Académie during this same period. The following year, 1829, he published the first 8 installments of Volume IV of *Exercices de Mathématiques*, the *Leçons sur le Calcul Différentiel*, as well as 8 articles in the *Bulletin des Sciences* of Férussac. Aside from this, he presented 15 papers to the Académie. Finally, during the first 8 months of 1830, he published the last 4 installments of Volume IV of *Exercices de Mathématiques* along with 3 installments of Volume V, 4 articles in the *Bulletin des Sciences* of Férrussac, and 2 studies on the theory of light, in addition to presenting 12 papers to the Académie. While it is certainly true that none of the works were produced during the 2 years immediately preceding the Revolution of 1830, they can be counted among his great masterpieces. In terms of sheer quantity, his productivity was impressive, to say the least. Referring to this period, Cauchy declared in 1831, in a letter to his family, that:

> I dare not work any longer in the evenings, as I was once silly enough to do over a period of time, in order to preserve the good health I am now enjoying. In this regard, I have gained much, because it is certain that when I left Paris, I felt so weak and exhausted that I had started to believe that I could not very well continue keeping the same pace (6).

Intellectually exhausted and physically weak, he had to revive his strength far from Paris. Cauchy's assertions in this matter are confirmed by letters from his brother and from his wife, written in September, October, and November 1830. On September 11, 1830, Alexandre Cauchy wrote to the Director General of the Ponts et Chaussées, explaining Augustin-Louis' absence:

> For the past two or so weeks, my brother, not being scheduled for any active duties at the Ponts et Chaussées, has used the leisure time afforded by the recess periods at the École Polytechnique and at the Faculté des Sciences to take the short trip to Switzerland and Italy that his health has so long demanded (7).

Alexandre Cauchy thus hoped to soothe and put off the Ponts et Chaussées administration, which was now pressing for Cauchy to take the oath of allegiance.

In other letters, letters written right up to the end of November 1830, Augustin-Louis' family continued to pretend against all evidence that his poor state of health necessitated the extension of his trip and that this necessity was,

Augustin-Louis Cauchy. Painting by J. Roller, undated. Published by permission of the École Polytechnique.

in fact, the only reason for his absence (8). But, if this was a plausible explanation in September, it became a poor excuse as time passed, and the Cauchy family was unable to keep Cauchy from losing all his public positions: on November 26, 1830, he lost his adjoint professorship at the Faculté des Sciences, in February 1831 he was stripped of his professorship at the École Polytechnique, and in early March of that year, he lost his rank as an engineer in the Ponts et Chaussées.

The extension of his leave of absence beyond the month of September indicates that Cauchy's decision to leave France was not based solely on the need for rest to improve his health, but that his decision was also the direct consequence of the July Revolution. Cauchy, his physical and emotional system already weakened by overwork, simply could not bear the shock of events brought on by the political turmoil that came to a head in July. Passionate to the point of blindness, with an emotional sensitivity that was hidden behind his outward rigidity and coldness, but which often betrayed itself, as in his tendency to stammer and gesticulate whenever he lost control, and traumatized since childhood by the very idea of revolution, he resolved at the beginning of August to leave Paris, to depart with no more of a definitive plan in mind than simply to avoid turmoil and agitation.

Two facts, no doubt, have a definite bearing on his decision: the massive participation of his students from the École Polytechnique in the revolution and the violent anticlericalism, whose main victims were the Jesuits, his friends. On July 28, 1830, the Jesuit residence in Montrouge was sacked and plundered, and it is very likely that Cauchy sheltered in his home Jesuits who were attempting to flee from the popular wrath. It seems even more likely that he accompanied some of them to Switzerland as they made their way into exile (9).

The exact route that he followed on his departure from France is not known, but he seems to have gone directly to Fribourg, Switzerland (10). A small colony of émigrés were settled there, having been drawn to this particular location because of its proximity to the French frontier and the friendly protection of the canton's patrician government. The Jesuits were very influential there and operated a college that, since 1827, had welcomed numerous French students, who were lodged in a private boarding school there. Cauchy did not remain in Fribourg long, but rather continued on to northern Italy. By early October, he had reached Turin, and by the end of the month he was in Modena (11). During the last day of November, he reached Genoa, where he met the King of Sardinia. By now, however, his trip abroad had already become an exile.

In effect, at the end of September 1830, while he was still in Fribourg, Cauchy had to decide whether he would take the oath of allegiance. Supporters of the now-dethroned Charles X were divided on whether or not the oath should be taken, with a minority of the royalists regarding themselves as bound to their sovereign by a feudal code of honor; to take the oath would, in their eyes, be to commit a felony. This was the way Cauchy saw the matter, and disregarding the advice of his family, who continued to hope that he would return, he finally rejected the idea of returning to France and decided to remain abroad, where he expected to be able to continue teaching.

Cauchy's father, Louis-François, and his two younger brothers, Alexandre and Eugène, who had succeeded their father in the Chamber of Peers in 1825, took the oath of allegiance to the new regime and, having done so, served the new government with all the zeal that they had shown in serving its predecessor. Moreover, this latest political contortion, so adeptly performed by the Cauchys, father and sons, earned them an honorable mention in the *Dictionnaire des Girouettes*, a satirical work that appeared in 1832 and gave the following definition:

> *Cauchy*—Honorary Keeper of Archives in the Chamber of Peers. An official with the Intendancy of Rouen before the great Revolution. Keeper of the Archives and of the Seal of the Senate during the Consulate. Secretary to the Curator of Archives of the Senate under the Empire. Keeper of the archives and Editor of the Verbal Proceedings in the Chamber of Peers during the Restoration. Honorary Keeper of the Archives in the Chamber of Peers before, during, and after the July

> Revolution. His son Alexandre, Counselor to the Royal Court and Titular Keeper, and his son Eugene, Adjoint Keeper. These three are very cosy at the Palais du Luxembourg... Napoleon made him a knight of the Legion of Honor... Louis XVIII made him an officer in the same order. Charles X, whose praises he sang, advanced his [Cauchy's] children. He has not yet sung the praises of Louis-Philippe. Let's wait a while! (12)

The Cauchy family, of course, fully shared Augustin-Louis' political views, but while the scientist of the family held to his convictions in a passionate, uncompromising way, the other members of the family preferred to hold their views with a prudent reserve.

In early October 1830, while he was still at Fribourg, Cauchy—with the support of the Jesuits—threw himself into work on the Académie Helvétique project. This project, in the words of the prospectus, which was written by Cauchy, aimed to establish courses so that 'young people, who shall have completed their studies, whether at the college of the city [Fribourg] or at some other institution, may come and pursue instruction in the philosophical and literary sciences, in oriental languages, and in the mathematical sciences...' In the prospectus, Cauchy specified that courses at the Académie Helvétique would be taught by 'scholars from France, who, after having carefully considered the conditions under which the new laws in their own country would allow them to retain their professorships, have resolved to remain faithful to the oath they swore to Charles X' (13). This passage was omitted when the prospectus was printed.

During the next few weeks, Cauchy devoted all of his energies to bringing this project to a successful conclusion (14). A major problem facing him at this point was that of finding the necessary financial means. In October 1830, he wrote (and had printed) a prospectus that was to launch his fund-raising efforts. In order to obtain the title of founder of the Académie Helvétique, it was necessary to take one or several issues of stock worth a thousand francs per annum. But, this would not bring in sufficient funds, and Cauchy tried to get the financial support of Europe's most reactionary rulers: he wrote to both the Emperor of Austria and the Csar of Russia, and he personally met the Duke of Modena and the King of Sardinia, Carlo-Felice.

Cauchy also developed a working list of prospective teachers for the Académie, a list of legitimist scholars who, like himself, had gone into exile. Aside from himself, the list contained the names of Charles-Louis Haller, former professor at the École des Chartes; Joseph Récamier, a former professor at the École de Médecine of Paris and at the Collège de France; Henri-Francois Gaultier de Claubry and Jacques Clarion, both formerly of the Collège de Pharmacie; Charles-Francois Leroy and Camille Menjaud, from the École Polytechnique staff; Meyraux, from the Collège de Bourbon; the engineer Hippolyte d'Haranguier de Quincerot; and d'Horrer, a diplomat (15). The project seemed to be on the verge of successful completion, the financial problem having been straightened out and the necessary authorization papers obtained when, on December 2, 1830, the 'Day of the Sticks' saw the reactionary patrician regime in Fribourg overthrown and the liberals swept into power. This spelled the end of the Académie, and the entire project had to be abandoned.

The months that followed this setback, up until the Autumn of 1831, are one of the least known periods in Cauchy's life. After his trip to Italy and the abandoning of the Académie Helvétique project, Cauchy seems to have settled in Switzerland (16). Events now unfolding in Paris kept him from returning home: on February 14, 1831, a mob sacked the church of Saint-Germain l'Auxerrois, and the following day, the Archbishop's palace was also sacked. A brief trip back to Paris in early March convinced him that, despite the pleas of his family, it would be better to remain abroad so as to avoid the turmoil in Paris (17). Thus, he returned to Switzerland, again without his wife and two children, whom he left with his parents-in-law. This separation would continue until 1834, and we are forced to wonder if indeed Cauchy was not fleeing from the responsibilities of family and marriage quite as much as from the revolutionary turmoil....

Be that as it may, Cauchy used these months of solitude to regain his strength. In July, he informed the President of the Académie in Paris that the 'precarious state of his health, weakened as it was by intensive study, obliged him to extend his leave of absence a while longer' (18). Cauchy's scientific productivity had now been considerably reduced. During this time, from December 1830 to June 1831, he published three articles in Italian entitled *Sui metodi analytici* in the *Biblioteca Italiana* of Milan. In these three articles, he presented an introduction to the methods of his courses at the École Polytechnique (19).

The reason for the publication of these articles was a report by Giuseppe Cossa that examined Cauchy's *Exercices de Mathématiques* in the *Biblioteca Italiana.* In his report, Cossa advised readers of negligent errors and oversights in Cauchy's presentations of the foundations of analysis and further observed that the author of a treatise who fails to attain his end should at least have the honesty to acknowledge the fact and not hide his own inabilities. In a note appended to this report, Cossa specifically declared that he was not referring to Cauchy. Nevertheless, Cauchy felt that he and his work were the subjects of Cossa's report, and he wrote the three articles to show what he meant by "the need for rigor" in mathematics (20). Given their didactic character, it might be thought that they were the summary of a course from this period.

Cauchy soon went to Turin, where a small colony of legitimists were thriving under the protection of the King of Sardinia, Carlo Alberto. Unfortunately, it has not been possible to determine the exact date of his move, however, it appears to have been sometime during the month of August. In any event, the Academy of Sciences of Turin devoted its meeting of October 11, 1831, to an address given by Cauchy on his paper 'Sur la mécanique céleste et sur un nouveau calcul qui s'applique à un grand nombre de questions diverses'. A short while later, on November 27, 1831, he presented a second paper to the Academy, entitled 'Sur les rapports qui existent entre le calcul des résidus et le calcul des limites, et sur les avantages qu'offrent ces deux calculs dans la résolution des équations algébriques ou transcendantes'. These two studies represent Cauchy's real return to the world of mathematical research after an almost total absence, which had lasted for a full year.

The paper presented on October 11 was a "composite" study (21). In the introduction, Cauchy indicated the occasion for which he had undertaken this study. The methods of calculation used in astronomy since Laplace first presented them in his monumental *Mécanique Céleste* were based on series expansions with convergences that had not been rigorously shown. Cauchy had been aware of this problem for quite some time and, prior to 1830, had begun to think about ways that would enable him to strengthen the analytical foundations of mathematical astronomy. Thus, Cauchy's paper of October 1831, in its most significant aspects, was the result of research that had been begun several years before the paper was written (22).

It is sometimes erroneously concluded from a remark made by Cauchy that this paper was written following a conversation that he had with the astronomer Jean Plana in Turin. Plana, of course, had spoken with Cauchy several weeks before about the time needed for calculating the coefficients in the series expansions related to the various perturbations, and Cauchy later gave him certain formulas by which the numerical calculations could be performed more simply and more quickly, he himself having already made use of these formulas. Cauchy observed, however, that in order to obtain these formulas, it sufficed merely to 'apply to the expansion of the function that, in the *Mécanique Céleste*, is denoted by R certain well-known results, such as Taylor's theorem and Lagrange's theorem on the expansion of the functions of the roots of algebraic or transcendental equations'. In fact, by citing Plana, Cauchy dealt tactfully with the most outstanding astronomer in Turin, whose scientific monopoly could be threatened by the presence of a brilliant and illustrious mathematician.

The most important part of the study dealt with the general principles of the calculus of limits. First, Cauchy proved the integral formula

$$f(x) = \frac{1}{2\pi}\int_{-\pi}^{+\pi} \bar{x}\frac{f(\bar{x})}{\bar{x}-x} \quad dp \quad (\bar{x} = Xe^{p\sqrt{-1}}), \tag{9.1}$$

where f is a finite and continuous function and x is a complex value in the open disc D with center 0 and radius X. He immediately obtained this formula, known today as Cauchy's formula, by replacing the finite and continuous function $f(\bar{x})$ by

$$\bar{x}\frac{f(\bar{x})-f(x)}{\bar{x}-x}$$

in the mean-value formula

$$\int_{-\pi}^{+\pi} f(\bar{x})dp = 2\pi f(0) \qquad (\bar{x} = Xe^{p\sqrt{-1}}).$$

Cauchy noted that the mean-value theorem, as well as his integral formula, can also be deduced from the residue theorem applied to the function

$$\frac{\phi(x)}{x}$$

taken along a circle with a center 0 and radius X:

$$\int_{-\pi}^{+\pi} \phi(\bar{x})dp = 2\pi \, {}_0^x\mathcal{E}_{-\pi}^{+\pi} \frac{\phi(x)}{x} \qquad (\bar{x} = Xe^{p\sqrt{-1}}).$$

By this method, he had established his integral formula for the first time about 1820 (see Eq. 7.11).

Then, Cauchy expanded $\dfrac{\bar{x}}{\bar{x}-x}$ into the MacLaurin series $\sum_0^{\infty} \left(\dfrac{x}{\bar{x}}\right)^n$, which is convergent in D, and by permuting the signs $\sum$ and $\int$, he showed that any function of a complex variable x has a convergent MacLaurin series expansion, provided the modulus ξ of the variable maintains a value less than the value for which the function ceases to be finite and continuous. This statement is sometimes called the Turin theorem.

Cauchy also majorized the modulus of the nth term,

$$\frac{1}{2\pi}\int_{-\pi}^{+\pi} \left(\frac{x}{\bar{x}}\right)^n f(\bar{x})dp,$$

of the series expansion of $f(x)$ in Eq. (9.1) by $\left(\dfrac{\xi}{X}\right)^n \wedge f(\bar{x})$ and the modulus of the remainder of order n

$$\frac{1}{2\pi}\int_{-\pi}^{+\pi} \frac{x^n f(\bar{x})}{\bar{x}^{n-1}(\bar{x}-x)} dp$$

by $\dfrac{\xi^n \wedge f(\bar{x})}{X^{n-1}(X-\xi)}$ $\left(\text{and also by } \wedge\left[\dfrac{x^n f(\bar{x})}{\bar{x}^{n-1}(\bar{x}-x)}\right]\right)$. The symbol $\wedge$ designates the limit of the function, that is, the maximum of the modulus of the function in its domain of definition; in particular, we have

$$\wedge f(\bar{x}) = \sup_p |f(Xe^{p\sqrt{-1}}).$$

These majorations are now called the Cauchy inequalities. Cauchy used them in order to determine upper bounds of errors when evaluating functions whether explicit or implicit by means of series expansion. This method, which is known today as the method of majorants, was called by Cauchy the calculus of limits.

In his paper, Cauchy presented also a new theory on the variation of arbitrary constants and, in a very extended second part, applications of this theory and of the calculus of limits to celestial mechanics especially for the expansion of the function R.

The second paper, which was presented on November 27, 1831, to the Academy of Turin, remained almost unknown, in spite of the publication of several abstracts, until its republication in 1974 (23). It is unquestionably one of Cauchy's most beautiful studies, and less because of the importance of the mathematical results it contains than because of the polish and elegance of its proofs, which are based on the systematic use of integration in the complex plane, which he had created some six years earlier. Moreover, this new study marks a theoretical advance over his paper of 1825, since he now introduced closed paths. For instance, he stated the residue theorem in the synthetic form

$$\mathcal{E}((f(z))) = \frac{1}{2\pi\sqrt{-1}} \int_C f(z) \frac{dz}{ds} ds \tag{9.2}$$

where C is a closed path of length c, s is the curvilinear abscissa of C, and $\mathcal{E}((f(z)))$ is the integral residue of f inside C. Moreover, he gave the majoration $c/2\pi \wedge f(z)$ for the modulus of $\mathcal{E}((f(z)))$.

By using Eq. (9.2), Cauchy resumed using the residue method he had developed to investigate the roots of the equation $f(z) = 0$. Thus, he obtained the formula

$$\sum_i F(z_i) = \frac{1}{2\pi\sqrt{-1}} \int_C \frac{f'(z)}{f(z)} F(z) \frac{dz}{ds} ds \tag{9.3}$$

for the summation of the similar functions of the roots z_i of the equation lying inside the contour C, to be compared with Eqs. (7.11) and (7.25); the modulus of $\sum F(z_i)$ is majorized by $\frac{c}{2\pi} \wedge \left[\frac{f'(z)}{f(z)} F(z)\right]$. For $F(z) = 1$ and $F(z) = z$, Eq. (9.3), yields, respectively,

$$m = \frac{1}{2\pi\sqrt{-1}} \int_C \frac{f'(z)}{f(z)} \frac{dz}{ds} ds,$$

where m is the number of roots inside C, and

$$\sum z_i = \frac{1}{2\pi\sqrt{-1}} \int_C z \frac{f'(z)}{f(z)} \frac{dz}{ds} ds.$$

Then, Cauchy developed a new calculus, called the calculus of indices of functions. The index of the real function $f(s)$ of a real variable relative to the root σ is the number

$$\lim_{\varepsilon \to 0} \frac{1}{2} \left\{ \frac{f(\sigma + \varepsilon)}{\sqrt{[f(\sigma + \varepsilon)]^2}} - \frac{f(\sigma - \varepsilon)}{\sqrt{[f(\sigma - \varepsilon)]^2}} \right\},$$

which can take the values $-1, 0$, or $+1$. The integral index of f, denoted by Cauchy's $\mathscr{I}((f(s)))$, is the sum of the indices of f relative to all of its roots; $\mathscr{I}_{s_1}^{s_2}((f(s)))$ is the sum of the indices of f relative to all of its roots between s_1 and s_2. The calculus of indices was based on the formula

$$\frac{1}{2\pi\sqrt{-1}}\int_C \frac{f'(z)\,dz}{f(z)\,ds}\,ds = \mathscr{I}_0^c\left(\left(\frac{\chi(s)}{\phi(s)}\right)\right) \tag{9.4}$$

where $\chi(s) + \sqrt{-1}\,\phi(s)$ is the complex function $f(z)$ taken on the contour C of length c. From this formula, Cauchy elegantly deduced many theorems on the roots of algebraic equations, especially the Sturm theorem.

Moreover, using the calculus of limits, he set forth methods for determining the series expansion of a single root of an algebraic equation or of a sum of such roots. A short time later, on September 10, 1832, he presented a study to the Academy of Turin in which he exhibited a new method for expanding the real and complex roots of an equation depending on one parameter and, in certain cases, for determining a bound for the error that is committed by neglecting a given number of terms in the series (24).

Cauchy was now far removed, physically, from the French mathematical community, and it took him almost 10 years to make the results of his first Turin paper known. As for the second Turin paper, Cauchy refused to publish it at all, for he soon found new methods for attacking the problem it treated. Well aware of the importance of the results he had obtained, results that not only generalized his work in 1813 on the roots of equations, but also Fourier's results and Sturm's famous theorem of 1829, which dealt with the number of real roots (of an algebraic equation) between two given numbers, Cauchy attempted to create a formal calculus of indices of functions that was not based on the calculus of residues (25).

Cauchy's arrival was an important event for the progress of the sciences in Turin. The exiled French scientist had the support of the Jesuits, who were very influential at court, and was thus well received in Turin. As has been seen, the doors to the Academy were wide open to him; but even more important, he was offered a professorial chair at the university so that he could resume his work as a teacher. On December 19, 1831, Count Gloria, the President of Studies, petitioned King Carlo Alberto to reestablish the chair in mathematical physics at the University of Turin, which the illustrious Avogadro had formerly held. This chair had been eliminated by King Carlo Felice on account of the events of 1821. Thus, on January 5, 1832, the king established the chair in physics for Cauchy (26), who was thus able to inform his family of his new position on January 11, declaring:

> At the request of the University of Turin, a chair in sublime physics [that is to say, mathematical physics, which I taught at the Collège de France] has been established, and I am appointed to teach this subject for a beginning salary of 1000 écus (27).

On 15 January, after he had been appointed to the chair, Cauchy visited the court in order to express his gratitude to his royal benefactor. While there, he read a paper, concerning which King Carlo Alberto wrote that 'the views he expressed seem, in my opinion, to be very wise, and I expect to reflect deeply on the matter' (28).

We do not know exactly when Cauchy began teaching his courses, and it is not even clear whether he taught during the winter of 1832. However, it is known that during the month of March he went on a trip that took him to Paris, among other places. While in his home city, Cauchy attended the meetings of the Académie on March 5, March 12, and March 19. Taking advantage of the fact that he was in Paris, he presented the two papers he had written in 1831 to the Académie. He seems not to have headed directly back to Turin, because, during the same month, he spent two weeks in Rome, where he resided at the Hotel Cesari. While in Rome, he met the mathematician Tortolini (29) and was given the honor of an audience with Pope Gregory XVI.

After his return to Turin, Cauchy delayed committing himself to returning to France, even though his family was urging him to do so. A month earlier, his youngest brother Amédée had died; and, no doubt, his wife and children were worrying about him. Moreover, his second brother, Eugène, soon arrived in Turin as deputy for the family and asked him to hasten his return, which was so heartily desired. Indeed,

> Those reasons most likely to touch him were set forth. Aside from considerations for his family, he was made aware of the fact that France was now neither so tumultous nor so distressed as he assumed. It was explained [to him] that the revolutionary movement had become progressively more calm—and, moreover, by not letting himself get involved in any kind of official position, he would surely be able to find the peace and quiet necessary for his studies (30).

Unfortunately, just as he seemed to have reached the point of relenting and returning home, new troubles broke out in Paris. The funeral of General Lamarque, who had died of cholera, was the occasion for a republican uprising on June 5 and 6, 1832. The fighting, though brief, was very violent, and it left many dead. The ensuing repression, which was a measure of the depth of the great fear that had shaken the bourgeoisie, was extremely harsh, and Cauchy quickly seized on these events and used them as a pretext for not returning home (31).

Cauchy did return to Paris for a two-week stay in October 1832, before he started teaching his courses at the University of Turin. On October 8, 1832, Cauchy took part in the commission that was charged with selecting a problem for the Grand Prix de Mathématiques of 1832, and the fact that he did so suggests that he was then contemplating a longer stay in Paris. However, he did not return to the Académie after October 29. Aside from presenting a paper that had been published using lithography, a study entitled 'Sur la rectification des courbes et la quadrature des surfaces courbes', dated October 19, 1832, he had enough time to present a note entitled 'Sur le versement des voitures publiques', in which, drawing on his own experiences as a traveler, he advanced proposals for reducing the number of serious accidents (32). He seems to have returned to Turin at the end of October 1832.

Once Cauchy had returned to Turin, his main concern was in teaching his courses. This remained the case until his departure in July 1833. According to Joseph Bertrand, Cauchy first planned to give his lectures in Latin, but gave up on the idea and, instead, lectured in Italian, a language he had first studied years before at the École des Ponts et Chaussées. This substitution of Italian for Latin was made on the basis of the belief that those attending his courses could more clearly understand the former that the latter. Though he lectured in Italian, the lecture notes and other material for his courses were no doubt in French, as were his *Sept Leçons de Physique Générale* of 1833 (33) and the *Résumés Analytiques*, which were published (thanks to a subsidy of 200 lira that King Carlo Alberto granted on request from Cauchy) by the royal printing service between 1833 and 1835 (34). The first of these works presented (albeit in the very general language of natural philosophy) the principles of molecular physics. The second work, which, according to the foreword, was a composite 'of a series of articles aimed at presenting a resume of the most important theorems in analysis, older results as well as newer ones, and especially the theories that include algebraic analysis and those methods that make for an easier exposition of it [algebraic analysis]', seemed rather like a compromise between a miscellany along the lines of *Exercices* and a work like his *Analyse Algébrique* of 1821.

The only account we have today bearing on Cauchy's performance as a teacher in Turin is a rather unflattering picture by Louis-Frédéric Menabrea. The courses Cauchy taught were not required. However, attracted by his tremendous reputation, many students enrolled in his classes. At that time, education was carried out under fairly precarious conditions. The government was afraid of student agitation and turmoil, and accordingly, students were not even authorized to reside in Turin itself. In light of this restriction, professors only gave private lessons (35). As to Cauchy's teaching, Menabrea wrote:

> [His courses] were very confused, skipping suddenly from one idea to another, from one formula to the next, with no attempt to give a connection between them. His presentations were obscure clouds, illuminated from time to time by flashes of pure genius. The students found them to be exhausting, and only a few were able to endure them to the end. In fact, of the thirty who were enrolled in his course with me, I was the only one to 'see it through' (36).

Thus, if Menabrea is to be believed, it is clear that Cauchy failed as a teacher in Turin. This failure may, of course, be explained in terms of Cauchy's known lack of teaching ability as well as in terms of covert meddling and obstructions by certain students who were close to Plana, the titular of the chair in analysis at the University of Turin, for it is known that Plana had been hostile to Cauchy since his arrival in Turin.

As we have seen, Cauchy made reference to Plana in his study of October 11, 1831. As he read the introductory remarks of his paper to the Academy of Turin, Plana interrupted him at several different points without, however, contesting the assertion that he had conversations with Cauchy relative to the topic under investigation. Plana alleged that he had dealt with one of the questions investigated in Cauchy's study in papers of his own, papers that he had registered with the Academy in sealed envelopes on September 2 and 6, 1832. In any event, under questioning by Cauchy, Plana had to acknowledge that he had not solved the problem of determining the general expansion of the perturbation function (37). Moreover, according to Menabrea, a short time later, Plana got the notion that Cauchy was attempting to obtain one of his [Plana's] positions (38). Whether or not Cauchy did indeed seek to undermine Plana's position is open to question; however, this affair illustrates the mistrust that governed relations between the two men. Thus it was that Plana hardly bothered to conceal his hostility toward the 'new hypertranscendent analysis' his great contemporary had developed (39).

Cauchy had a much better relationship with Bidone, who years later recalled 'with genuine pleasure' the talks he had on scientific matters with 'the great scientist' (40). Finally, it should be noted that Cauchy was very close to Jesuit mathematicians in Turin, to Father La Chèze, a professor of mathematics and physics at the College of the Holy Martyrs, and to Abbé Moigno, whom he had known for a long time. After his departure from Turin, Cauchy proposed that one of these two men should be appointed to fill his chair in sublime (mathematical) physics at the university (41).

The Revolution of 1830 and his consequent departure and exile from France unquestionably marked a turning point in Cauchy's life. Deeply involved in the extreme conservative movement and weakened by intensive intellectual activity, he could not bear to see a regime that he wholeheartedly supported collapse in a few days. The legitimate king went into exile; the Jesuits went into hiding to avoid being slaughtered, and the liberals triumphed at the École Polytechnique. Cauchy felt each of these events to be a personal affront. He once again experienced the nervous illness that had afflicted him in 1812 when he was in Cherbourg, the symptoms being depression with moments of excitement and physical weakness. When he left Paris early in September 1830, his aim was merely to take a much needed rest, but this had changed by early October and what had begun as a trip to recuperate became voluntary exile, which Cauchy justified on grounds of his refusal to take the oath, but which his family regarded as a flight pure and simple.

The events of 1830 also signaled a break in Cauchy's mathematical productivity. Fourier's death, Legendre's death a few years later, and Galois' tragic fate: all foretold the end of an era. Cauchy's absence from Paris was now painfully felt. He was, to be sure, temporarily replaced (and without any apparent regrets) at the École Polytechnique by Coriolis, his friend and former répétiteur. But, it was a long time indeed before anyone could even pretend to take his place at the Académie. Writing to Libri in 1831, Sophie Germain spoke in darkly prophetic terms when she observed that:

> [There is] decidedly a kind of fate or spell hovering over everything that has to do with mathematics. Your own difficulties, Cauchy's problems, M. Fourier's death, as well as that of the student Galois, who, for all his impertinence, suggested certain exciting developments and tendencies... (42).

A few months after this letter was written, she herself passed from the scene.

Chapter 10

The Education of the Duke of Bordeaux

At the beginning of the summer of 1833, Cauchy received a letter dated June 22 from the Baron de Damas in Toeplitz. De Damas was the tutor of the Duke of Bordeaux, the grandson of Charles X, the former King of France, who was now living in Bohemia. The letter stated:

> Sir, you can be of great service in the education of the Duke of Bordeaux, and the king himself ordered me to write to you. His Majesty would like you to come, if it is at all possible, and work with my student and be in charge of his education in the sciences, which up to now has been the responsibility of M. Barande.
>
> M. Barande has just been removed for reasons that grieved us greatly, but which reflect neither on his honor nor on his abilities. I have thought it fitting to mention this fact to you as I want to avoid giving grounds for any unfounded suspicions that might conceivably arise.
>
> With my highest regards and continued affection,
>
> Yours truly,
>
> Baron de Damas

In addition to this official letter, Cauchy received another letter, which was filled with expressions of goodwill and high esteem for him and which closed with the following words:

> While I cannot speak of your personal position, I do know what your sentiments are: the heir to our king needs your services, and I have been given the responsibility of asking for them. I now assure you of the deep attachment that I swore to you in times better than these and that will not change.
>
> Baron de Damas

In a postcript to this letter, the baron declared:

> The king would like you to come to Prague and to proceed thence to the countryside, where His Majesty resides, as soon as possible. Write me of your travel plans and the presumed date of your arrival.
>
> The king will leave Toeplitz during early July and will return to Buschtierad, near Prague. His Majesty is quite pleased with his stay here, since the waters helped him very much (1).

Cauchy gave his last examination at the University of Turin on July 22 and left Turin for Prague, traveling by way of Geneva, Switzerland, and Bavaria (2). Thus, a new life began for Augustin-Louis Cauchy, which would last for six years.

In order to understand Charles X's decision, it is necessary to examine the complex situation in the little court of exiles presided over by the old king. Following the July Revolution, Charles X and his family had first taken refuge at Lullworth, in England and then Holyrood, in Scotland. On August 2, 1830, at Rambouillet, Charles, as well as his son, the Duke of Angoulême, abdicated in favor of his grandson, the Duke of Bordeaux. However, on November 27, he was declared regent until the Duke of Bordeaux reached his majority; that is to say, until September 29, 1833, when the Duke would celebrate his fourteenth birthday. This decision alienated the Duke of Bordeaux from his mother, Duchess of Berry, Marie-Caroline, who was not trusted by the royal family at all. In any event, the duchess had firmly decided to act on her young son's behalf. Settled in Italy and surrounded by an entourage of frivoulous-minded aristocrats, she contrived a grandiose but ill-conceived plan. The plan essentially consisted of organizing a vast uprising in France to reestablish the legitimate heir to the throne. When Charles X was informed of what was afoot, he completely disapproved of the undertaking. However, encouraged by a number of leading Parisian legitimists, Marie-Caroline paid no attention to Charles' disapproval of her scheme, just as she paid no attention to the defeat of the first conspiracy in February 1832, and continued to pursue her designs. Thus, on April 29, 1832, she secretly landed in Provence. In June, the uprising miscarried, and after a grotesque escapade, the duchess went into hiding in Nantes in order to escape Louis-Philippe's police. But, on November 7, 1832, she was finally arrested and imprisoned in the fortress of Blaye.

Marie-Caroline became now a heroine and martyr for the Parisian legitimists of the Faubourg Saint-Germain. But an even greater shock came in February 1833 when it was learned that the duchess, having secretly remarried while in Italy, was now pregnant. Charles, who in the autumn of 1832, had left his refuge in Scotland and taken his family to Prague, where Emperor Francis II placed the Hradschin Palace at his disposal, completely disowned his daughter-in-law as soon as he learned of her remarriage and pregnancy. The absolutist faction at the court of the old king was lead by Cardinal de Latil,

Baron de Damas, the tutor of the young Duke of Bordeaux, and especially by the Duke of Blacas, the old king's confidant. It now gained the upper hand over the liberal faction, which was led by Madame de Gontaut, the governess of the Duke of Bordeaux; the Duke of Gramont; and Joachim Barande, the director of education of the young duke, and which still supported Marie-Caroline. Even in France, the legitimists were split; many were simply outraged by the duchess's conduct, but the most active of the legitimists, those who now went by the name 'Young France', continued to support her. Chateaubriand came to Prague to plead her cause. However, his efforts were of no avail.

The dispute between the two factions actually went beyond the duchess's personality, for the real issue had to do with the kind of education that would be given to the young man, who might, one day, occupy the throne of France. Marie-Caroline's supporters thought that the Duke of Bordeaux should be given a modern education, which would be relevant to the times. The aim of the education they contemplated would at last be a reconciliation of the monarchist ideal with the principles of liberty. Adhering to this view, Chateaubriand had notions of becoming the tutor of the future Henri V. However, the absolutists had a wholly different view of things. As they saw things, the Duke of Bordeaux should be educated in the traditional way. Accordingly, they reproached Barande who, in spite of his known merits, was regarded as being too liberal minded. In January 1833, the decision to dismiss him was made, and a replacement had to be found. In early June, the Marquis of Foresta, the subtutor of the Duke of Bordeaux, returned from Rome and brought with him two Jesuits, Fathers Druilhet and Deplace, both of whom had formerly been professors at the Collège de Saint-Acheul. Damas's plan was simple: he wanted to retain Barande, a brilliant polytechnician and engineer of Ponts et Chaussées, to teach mathematics and the natural sciences to the young duke and to entrust instruction in other subjects to the two newly arrived Jesuits. But, Barande refused to go along with this plan, and Charles X sent him packing (3). This explains why, at the time Barande was dismissed, Billot, a professor of law, and Cauchy were called (4). In August, when Cauchy arrived in Prague, matters had just taken a new turn. The announcement of the two Jesuits as tutors to the Duke of Bordeaux aroused a storm of protests back in France. The Jesuits' unpopularity in France was a known fact, and this appointment was seen as an outright provocation. Under heavy pressure from legitimists in France and after the Emperor of Austria had intervened in the matter, Charles X decided to dismiss the Jesuits. In August, Father Druilhet left Prague, and in November, Father Deplace followed. Now, disowned by Charles X, de Damas tendered his resignation and the liberal faction prevailed. On September 9, 1833, the Duke of Bordeaux reached his majority and several aristocrats of the Young France persuasion came to Prague. By recognizing him as king, they hoped to weaken the convervative influence of Charles X and his family; once more, Chateaubriand arrived in Prague to make a final plea to the old king on behalf of the duchess.

Under such conditions, Cauchy was not all sure that he would be able to remain in Prague. In fact, on September 24, writing from Buschtierad, he informed the Count of L'Escarène of the situation he faced in Prague:

> Unfortunately, we still have not gone beyond provisional arrangements here so that it is really impossible for me to say definitely whether or not I shall remain here or whether I will return to Turin to teach the courses entrusted to me by the King (5).

Nevertheless, taking part in the intrigues of the absolutist party, Cauchy had a brochure published in Prague, entitled *Quelques Mots Adressés aux Hommes de Bon Sens et de Bonne Foi* (6). In this publication, he declared his intentions of participating in the education of the prince, writing:

> The love which authors have for their own works and the attraction they feel for the theories they invent are so well known that it is easy to imagine how much it has cost me to interrupt the scientific works that I had undertaken. But, what my own weakness kept me from doing while my king was on the throne and in the Tuileries, I cannot refuse to do for my king now that he is in exile, and for the child of the miracle who now wears the double crown of glory and misfortune. A stranger to the language of courts and to the art of flattering the mighty, I can only bring a true heart and unblemished life to the service of my king. Many of my friends have preceded me here to this land, which is so new an experience to me. I no longer see them. When I look at what I have to offer, I see new grounds for fearing to involve myself in so difficult and challenging a career. But, I am well aware that the great ideal that controls the education of the prince is both a religious and moral ideal, which can inspire even the most exalted persons and which is the only ideal that can form great princes and great kings... I dare, therefore, to reply to the voice that calls on me. I will not recoil in the face of any obstacles.

After the dismissal of the Jesuits, Charles X appointed the absolutist Bishop of Hermopolis, Monseigneur Frayssinous, tutor to the duke, during final days of September, the Bishop appointed two instructors, Cauchy and Billot, the two scholars who had been first summoned to Prague to assist Fathers Druilhet and Deplace. At the same time, Charles summoned the Marquis of La Tour-Maubourg, a man of fairly liberal tendencies, as replacement for the Baron de Damas, the dismissed tutor. Citing the state of his health as a reason, La Tour-Maubourg declined the appointment and, as a substitute for himself, sent General d'Hautpoul, who arrived in Prague in early October. From the very first, General d'Hautpoul encountered the underhanded opposition of Billot and especially of Cauchy.

In his memoirs, d'Hautpoul discusses the reception he received, recalling that:

> I learned that M. Cauchy had been named to replace M. Barande, and I knew of his reputation as an outstanding scholar. He was totally committed to the faction that was supporting the Jesuits against the prevailing popular mood in France. I observed that Cauchy showed me a very reserved attitude.... He often visited the Bishop of Hermopolis, but he rarely came to see me, and when he did, he seemed embarrassed by it (7).

General d'Hautpoul officially undertook his duties as subtutor on November 1, 1830, and an official ceremony was organized for the occasion with Monseigneur Frayssinous, the Baron de Damas, d'Hautpoul, Cauchy, Billot, and other members of the young Duke of Bordeaux's household attending. Cauchy took advantage of this gathering to read a declaration in which he did not bother to hide his hostility toward the decision taken against the Jesuits. D'Hautpoul recalled the scene in the following terms:

> M. Cauchy took a sheet of paper out of his pocket and read a farewell speech that he had prepared in advance. Such a solemn way of going on leave seemed rather bizarre to me, and moreover, the little speech was given in such a voice that it was quite impossible for me to completely grasp the sense of what was said. However, I did manage to catch some words expressing regret at the departure of M. de Damas and the Jesuits, which was completely acceptable. Aside from this, however, I believe I heard him say that he deplored the changes that these dismissals had brought about and that he was repelled at the thought of having to participate in them.

Thus, Cauchy, who did not even once mention General d'Hautpoul by name in his statement, put himself squarely on the side of the Jesuits and against the 'obscure prejudices' by which they had been victimized (8).

As the weeks passed, relations between Cauchy and General d'Hautpoul, instead of improving, became increasingly worse. Their notions on teaching and pedagogy were diametrically opposite, for the latter wanted to endow the young prince with a modern education, a training and outlook suitable for the times they were living in, while the former filled his teaching with religious considerations and indulged the misconduct of his high-born pupil to an extent that d'Hautpoul found downright culpable (9). Moreover, Cauchy never ceased to decry loudly the dismissal of the Jesuits. The final break between the two men came in February 1834 when d'Hautpoul learned that Cauchy had made a farewell statement to de Damas in the *Gazette du Lyonnais*. Outraged, he had a meeting with Cauchy and demanded that he give an explanation for his action. Cauchy, of course, replied to d'Hautpoul's demands in a very vague, evasive way (10). Dissatisfied, d'Hautpoul demanded of Monseigneur Frayssinous that Cauchy be dismissed. This dismissal did not occur, and General d'Hautpoul handed in his resignation and left Prague on February 20, 1834.

Thereafter, tutors came and went, one after the other. But, the young prince's real education was taken charge of by Monseigneur Frayssinous and his assistant, the Abbé Trébuquet, aided by the more or less permanent tutors, Billot, Clouet, Montbel, etc., and, of course, Cauchy in the exact sciences. With the departure of d'Hautpoul, Cauchy was assured that he would continue to play a role in the education of the Duke of Bordeaux; and, that being so, he had his wife and children come to Prague, to the Hradschin, in 1834 (11).

Cauchy's duties as royal tutor consumed most of his time. He took the mission that had been entrusted to him very seriously; and, with great ardor and great clumsiness, he set about trying to teach his high-born pupil the rudiments of science. A short while after she arrived in Prague, Aloïse Cauchy wrote of her husband:

> It would be hard to imagine a position that is more agreeable than the one he has. However, at the same time, I must say, in order to justify some of the reproaches we had voiced about him for his having failed to write to us, that he does not have a moment to call his own. I hardly see him except for an hour at dinner and later, for a few moments in the evening. He is busy all day: in the morning there are lessons that he must give; these are followed by other lessons that he assists and by walks and strolls, which scarcely leave him enough time to add a few words and algebraic symbols to a study that he is now writing and which he hopes to send to the Académie just as soon as it is finished (12).

According to d'Hautpoul, Cauchy taught mathematics each day (at least in 1833) for one hour, from 8:15 to 9:15 a.m., and other sciences, especially physics and chemistry, for one hour, from 5 to 6 p.m. At the end of each week, on Saturday, there were exams, and the prince's close companions were allowed to participate in them. Cauchy prepared his courses very carefully (13). In fact, according to Moigno, he wrote elementary treatises on arithmetic and geometry. Cauchy's mathematical imagination and creativeness compelled him to look for original methods to his teaching (14). In line with this creativity, he developed a new decimal notation which would enable mental calculations to be performed more easily (15).

Were Cauchy's efforts successful with the prince? The results suggest a negative answer. The young prince showed neither a taste nor talent for mathematics and the exact sciences; and, moreover, Cauchy once again gave proof of his own mediocre teaching abilities. We have only two pieces of relevant evidence, and they agree completely on this point. The first of these comes from d'Hautpoul, a man who had no liking for Cauchy personally, but nevertheless seems worthy of belief (16). D'Hautpoul presents us with a picture of Cauchy in 1833, shortly after the mathematician arrived in Prague. Here, we see Cauchy busily working with the young prince in an attempt to alienate him from the general's influence and from the memory of Barande.

Indifferent to his pupil's ridicule and stoically suffering the child's intolerable snubs and misconduct, Cauchy once more proved himself to be a man of heroic patience. During the strolls that he took with the young prince, the boy would frequently play tricks on him that passed the bounds of simple jokes. 'For example', says d'Hautpoul, 'one day, when the young prince came back from a walk, he brought with him a snowball and hit his tutor with it' (17). Moreover, the child very often addressed Cauchy in 'stinging, disrespectful' terms, and again Cauchy would tolerate such misconduct, showing an extraordinary degree of understanding and compliance.

Two anecdotes illustrate the trials Cauchy endured in the service of his pupil:

> M. Cauchy, having seen that the story of my campaigns was of interest to the prince [d'Hautpoul wrote] asked me permission to sit with him sometimes in the evening in order that he might also tell him about his own. He had been some time in the Ponts et Chaussées, and he called "campaigns" the works that were executed each year during good weather. Earlier in his career, M. Cauchy had been responsible for repairing some of the sewers of Paris. The prince, with great malice, seized upon this information and went about saying that M. Cauchy's first campaign had been in a sewer (18).

On the Day of the Feast of the Kings, the Duke of Bordeaux was chosen king by the person who found the bean. The prince then formed his council and bestowed his favors. 'One person', d'Hautpoul tells us, 'was overlooked—perhaps intentionally: M. Cauchy. "And I?" the latter asked, seeing that everybody else had been given an appointment. "And", the prince replied, after a moment of reflection, "You will be the minister of stars". This little flash of wit caused a great deal of laughter, because everyone had observed that M. Cauchy too often brought astronomical discussions into his conversations' (19). But, if Cauchy failed to show firmness with the young prince outside the classroom, pure chaos reigned in the courses. The prince had no liking or talent for mathematics and would often put on violent temper tantrums. 'But what good are all these numbers and figures that M. Cauchy always makes me do?' the boy would complain to d'Hautpoul. Ignored and abused, Cauchy was simply incapable of making the boy respect his authority. One day, the young man took his tutor by the collar and literally pushed him out the door, 'and,' recalled d'Hautpoul, 'M. Cauchy went along with this, muttering excuses and asking what he could have done to displease Monseigneur'. On another occasion, the young man was beside himself with anger and grabbed the table cloth, jerking it so hard that the books lying on it went flying off, as did a basket of coins that was lying on the table. Without budging, the child stood glaring at poor Cauchy, who, on hands and knees, scurried about across the floor gathering up the coins (20).

As for the courses, Cauchy never succeeded in developing a course at a level suitable for the prince. According to d'Hautpoul:

> [One of the things] that inspired the young man with such a dislike for mathematics was that he was never able to understand what was being taught him. One day, I attended one of the sessions and during this class, the professor was proving one of the most simple propositions in geometry. I watched as the prince became more and more agitated and confused, and I swore to myself that although I remembered this proposition very well, I myself could hardly follow Professor Cauchy's argument (21).

In order to soothe his pupil's hostilities, Cauchy began substituting other activities for mathematics: one day he might tell stories to the boy in order to amuse him; and on the next day, he and his pupil would sing songs and homilies from Saint-Sulpice en duo (22).

As time passed, the prince began to pick up on the rudiments of mathematics. However, according to an account by the Marquis of Villeneuve, who came to see the prince at Goritz, Cauchy had a conspicuous lack of success in getting mathematics across to his pupil. Villeneuve recalled:

> At Goritz, as elsewhere, I paid close attention to the exams that were given each Saturday. Unfortunately, these always began with mathematics. When questioned by Cauchy on a problem in descriptive geometry, the prince was confused and hesitant. The king [the Duke of Angoulême] then stood up, explained the problem, and was praised by Cauchy for having done so. There was also material on physics and chemistry. But, as in the case with mathematics, the prince showed very little interest in these subjects. Cauchy became annoyed and screamed and yelled. The queen [the Duchess of Angoulême] sometimes said to him, soothingly, smilingly, 'too loud, not so loud!' (23)

Thus, we see that Cauchy had to have an extraordinary faith and sense of mission to tolerate the isolation in which he found himself for five years—first at the Hradschin Palace in Prague and then, after May 1836, at the Kirchberg Castle in Toeplitz, and finally at Goritz—in a petty, illiberal, cynical court in exile. Moreover, it all seems to have been for no good reason, because, when the Duke of Bordeaux finished his formal education in October 1838, he had acquired an abiding dislike for mathematics.

Worn out by his efforts to educate the young prince, Cauchy did not do much research between 1834 and 1839. Until the end of 1834, it seems that he worked on the publication of the five installments of his *Résumés Analytiques*, which was being printed in Turin and which he had begun the preceding year (24). On the other hand, resuming the work he had first started in Turin in 1833, he published in lithographic form—this time in Prague—an important study entitled *Sur l'Intégration des Équations Différentielles.* In it, he presented a new application of his calculus of limits. Thus, he carried through to fruition a goal he had set for himself in 1831: to use the calculus of limits to determine

18

et les formules (96) jointes aux formules (107) (108) donnent

(112)

$$\xi = \rho \cos\left(\vartheta + \frac{K'+K''}{2} z - st\right) \cdot \cos\left(\frac{K'-K''}{2} z\right)$$

$$-2\rho \frac{(\Theta''^2 - \Theta'^2) \sin\left(\vartheta + \frac{K'+K''}{2} z - st\right) + 2\Theta'\Theta'' \sin(\theta' - \theta'') \cos\left(\vartheta + \frac{K'+K''}{2} z - st\right)}{\Theta'^2 + \Theta''^2 - 2\Theta'\Theta'' \cos(\theta' - \theta'')} \sin\left(\frac{K'-K''}{2} z\right),$$

$$\eta = 2\rho\Theta'\Theta'' \frac{\Theta' \sin\left(\vartheta + \theta' + \frac{K'+K''}{2} z - st\right) - \Theta'' \sin\left(\vartheta + \theta'' + \frac{K'+K''}{2} z - st\right)}{\Theta'^2 + \Theta''^2 - 2\Theta'\Theta'' \cos(\theta' - \theta'')} \sin\left(\frac{K'-K''}{2} z\right).$$

Pour une valeur nulle de z, les formules (111) donnent, comme on devait s'y attendre

(113) $\xi = \rho \cos(\vartheta - t), \quad \eta = 0$

c'est-à-dire que le rayon est polarisé parallèlement à l'axe des x. De plus les formules (109), comme on devait s'y attendre encore, se réduisent à

(114)

$$\xi = \rho \cos\left(\vartheta + \frac{K'+K''}{2} z - st\right) \cos\left(\frac{K'-K''}{2} z\right),$$

$$\eta = \mp \rho \cos\left(\vartheta + \frac{K'+K''}{2} z - st\right) \sin\left(\frac{K'-K''}{2} z\right);$$

et alors on obtient la polarisation [illegible], puisque pour une valeur donnée de z le déplacement d'une molécule tourne autour de l'axe des x de l'angle $\pm \frac{K'-K''}{2} z$.

Research work on the theory of light. Manuscript by Cauchy, January 1836, Sorbonne Library, ms 1762. Published by permission of the École Polytechnique.

the solution $(x(t), y(t), z(t), \ldots)$ of a system of ordinary differential equations:

$$\frac{dx}{F_1(x, y, z, \ldots, t)} = \frac{dy}{F_2(x, y, z, \ldots, t)} = \frac{dz}{F_3(x, y, z, \ldots, t)}$$

$$= \cdots = \frac{dt}{F_0(x, y, z, \ldots, t)}, \tag{10.1}$$

satisfying the initial conditions $x(\tau)=\xi$, $y(\tau)=\eta, z(\tau)=\zeta$, by means of series. In this system, the functions $F_0, F_1, F_2, \ldots$ where supposed to be continuous and finite.

Hamilton's study, 'On a general method in dynamics', the first part of which was published in 1834 (25), had led Cauchy back to the idea of reducing the resolution of Eq. (10.1) to the equivalent problem of determining the prime integrals $X(x,y,z,\ldots,t,\tau)=\xi$, $Y(x,y,z,\ldots,t,\tau)=\eta$, $Z(x,y,z,\ldots,t,\tau)=\zeta,\ldots$ of Eq. (10.1), such that $X(x,y,z,\ldots,t,t)=x$, $Y(x,y,z,\ldots,t,t)=y$, $Z(x,y,z,\ldots,t,t)=z,\ldots$ (26). More generally, a prime integral of Eq. (10.1) can be expressed by the equation $U=\upsilon$, where $U=u(X(x,y,z,\ldots,t,\tau)$, $Y(x,y,z,\ldots,t,\tau)$, $Z(x,y,z,\ldots,t,\tau),\ldots)$, and $\upsilon=u(\xi,\eta,\lambda,\ldots)$. Then, Cauchy showed that $U-\upsilon$ is a solution of the first-order linear partial differential equation

$$\frac{\partial s}{\partial t}+\frac{F_1}{F_0}\frac{\partial s}{\partial x}+\frac{F_2}{F_0}\frac{\partial s}{\partial y}+\frac{F_3}{F_0}\frac{\partial s}{\partial z}+\cdots,=0. \tag{10.2}$$

He constructed this solution by an iterative method, much like the one he had used in 1825. If ∇s is the integral

$$-\int_\tau^1\left(\frac{F_1}{F_0}\frac{\partial s}{\partial x}+\frac{F_2}{F_0}\frac{\partial s}{\partial y}+\frac{F_3}{F_0}\frac{\partial s}{\partial z}+\cdots\right)dt$$

and $\nabla^n s$ are the multiple integrals $\nabla(\nabla(\cdots(\nabla s)\cdots))$, the iteration of the integral equation $U-u-\nabla U=0$ yields the development

$$U=u+\nabla u+\nabla^2 u+\nabla^3 u+\cdots+\nabla^{n-1}u+\nabla^n U.$$

Therefore, $\Sigma(\nabla^n u)-\upsilon$ represents a solution of Eq. (10.2), if the series $\Sigma(\nabla^n u)$ is convergent.

Cauchy used the calculus of limits to establish the convergence of the series $\Sigma\nabla^n u$. Thus, he obtained not only approximate values of the solution $U-\upsilon$ of Eq. (10.2), but also an existence theorem for the solutions of the holomorphic differential system, Eq. (10.1), which differed from the existence theorem he had established in his lessons at the École Polytechnique. For instance, in the case of a single equation, $\dfrac{dx}{dt}=F(x,t)$ (with $x(\tau)=\xi$), he found that

$$\operatorname{mod}\nabla^n x\leqslant\operatorname{mod}\left\{2(\tau-t)\wedge\frac{F[x+\bar{x},t+\theta(\tau-t)]}{\bar{x}}\right\}^n\wedge(x+\bar{x})$$

where $\bar{x}=re^{p\sqrt{-1}}$ is on the circle $C(x,r)$ of center x and radius r and θ is a certain number in $[0,1[$ depending on r, by applying the inequalities

$$\operatorname{mod}f^{(n)}(x)\leqslant n!r^{-n}\wedge f(x+\bar{x}),$$

which he had given in his first memoir of Turin. $F(x)$ and $f(x)$ were supposed to be finite and continuous inside and on the circle $C(x,r)$. Therefore, the series $\Sigma(\nabla^n x)$ is convergent if

$$\operatorname{mod}\left\{2(\tau - t) \wedge \frac{F[x+\bar{x},t+\theta(\tau-t)]}{\bar{x}}\right\} < 1$$

and then the solution $x(t)$ such that $x(\tau) = \xi$ exists and can be expanded into a convergent series in a neighborhood of τ. Cauchy also gave the majorant

$$\frac{\operatorname{mod}\{2(\tau - t) \wedge F[x+\bar{x},t+\theta(\tau-t)]/\bar{x}\}^n}{1 - \operatorname{mod}\{2(\tau - t) \wedge F[x+\bar{x},t+\theta(\tau-t)]/\bar{x}\}} \wedge (x+\bar{x})$$

for the rest of order n of the series expansion of the solution.

It was to the theory of light that Cauchy devoted most of his efforts during the years 1835–1836. The publication of his study, *Sur la Dispersion de la Lumière*, which had been held up by the turmoils of 1830, came at the beginning of this period of new research. He presented this study to the Royal Society of Sciences of Prague, and that body, in effect, accepted the responsibility of financing the printing of this work as a separate study. However, when it began to appear in October 1835, it was not *en mémoire détaché*, as was originally foreseen; rather, it was included in a new series of *Exercices* entitled *Nouveaux Exercices de Mathématiques*, which followed the pattern of *Exercices de Mathématiques* of 1826–1830 and *Résumés Analytiques* of 1833–1834 (27).

Cauchy was not content with publishing his 1830 manuscript. He undertook new investigations that were only partly published at the time. In the case of an isotropic ether, he showed that the velocity V of a plane wave can be written

$$V = a_1 + a_2k^2 + a_3k^4 + \cdots,$$

where k is the magnitude of the wave vector $\mathbf{k}$. The coefficients a_n depend on the medium and decrease very quickly when n increases. Cauchy deduced from this series-expansion the value of k^2:

$$k^2 = b_1 + b_2s^2 + b_3s^4 + \cdots,$$

where

$$s = kV \quad \text{and} \quad b_1 = \frac{1}{a_1}, b_2 = \frac{a_2}{{a_3}^2}, b_3 = \frac{a_1a_3 - 2{a_2}^2}{{a_1}^4}, \cdots.$$

Using a new method for the calculation of errors which he had presented in the lithograph 'Sur l'interpolation' (28), he proved that his formula is compatible with the results of Frauenhofer's experiments on the indices of refraction in different media.

One highly significant paragraph of his study on dispersion was devoted to the physical properties of molecular ether. Cauchy considered the relation between s and k in the case of an isotropic ether that is not dispersive. The intermolecular forces were supposed to be attractive at a great distance in inverse ratio to the square of the distance and repulsive in the neighborhood of each molecule, in inverse ratio to the fourth power of the distance. With this hypothesis, Cauchy showed that the ether is necessarily an extremely dense system of molecules. Its density varies according to the material medium involved, maximal in empty space, where there is no dispersion, and minimal in the dispersive bodies.

In August 1836, just before his departure for Goritz and hard on the heels of the publication (in installments) of his study on dispersion, Cauchy had a lithograph of another of his works published.

Entitled *Sur la Théorie de la Lumière*, this study was an extract of the materials from the notebooks of 1836 (29).

The first part of this study was devoted to some mathematical preliminaries. In particular, Cauchy examined a new method for determining the boundary conditions for a body regarded as a system of molecules. The remaining sections treated various questions concerning the theory of light. Cauchy developed a new method for transforming the equation of motion of a system of molecules:

$$\partial_0^2 u_i = \sum_{\mathrm{p}} m_{\mathrm{p}} \left(f(r_{\mathrm{p}}) \frac{\Delta_{\mathrm{p}} u_i}{r_{\mathrm{p}}} + (r_{\mathrm{p}} f'(r_{\mathrm{p}}) - f(r_{\mathrm{p}})) \mu_{\mathrm{p}i} \mu_{\mathrm{p}j} \frac{\Delta_{\mathrm{p}} u_j}{r_{\mathrm{p}}} \right). \qquad (6.13)$$

Instead of representing the deplacements u_i by Fourier series expansions, as in 1830, he used Taylor's series. He simplified the notation by means of the symbolic operator $e^{r\mu_i \partial_i}$, which represents the symbolic series expansion relative to ∂_i:

$$1 + r\mu_i \partial_i + \frac{r^2}{2} (\mu_i \partial_i)^2 + \cdots.$$

Thus, according to the analogy between the powers and the differences, we have the symbolic equation

$$\Delta = e^{r\mu_i \partial_i} - 1.$$

Cauchy then defined two other symbolic operators:

$$\nabla = \sum_{\mathrm{p}} m_{\mathrm{p}} \frac{f(r_{\mathrm{p}})}{r_{\mathrm{p}}} (1 - e^{r_{\mathrm{p}} \mu_{\mathrm{p}i} \partial_i}),$$

and

$$\square = \sum_{\mathrm{p}} m_{\mathrm{p}} \frac{f(r_{\mathrm{p}})}{r_{\mathrm{p}}} \left(\frac{(\mu_{\mathrm{p}i} \partial_i)^2}{2} - \frac{e^{r_{\mathrm{p}} \mu_{\mathrm{p}i} \partial_i}}{r_{\mathrm{p}}^2} \right).$$

He obtained new operators $\square_{ij}$ by derivating $\square$ with respect to ∂_i and ∂_j, symbolically interpreted as variables; for instance, if $i = 1$, and $j = 2$,

$$\square_{12} = \sum_{\mathrm{p}} m_{\mathrm{p}} \frac{f(r_{\mathrm{p}})}{r_{\mathrm{p}}} \mu_{\mathrm{p}1} \mu_{\mathrm{p}2} (\partial_2 - e^{r_{\mathrm{p}} \mu_{\mathrm{p}i} \partial_i}).$$

Finally, he showed that Eq. (6.13) can be rewritten in the form

$$\partial_0^2 u_i = \square_{ij} u_j + \nabla u_i. \qquad (10.3)$$

If the system of molecules is isotropic, Cauchy obtained the equations

$$\partial_0^2 u_i = \diamond\, u_i + \bowtie \partial_i U_{jj} \tag{10.4}$$

and

$$\partial_0^2 U_{ii} = \diamond\, U_{ii} + \partial_{ii} \bowtie U_{jj}, \tag{10.5}$$

where U_{ij} is the cubical dilation $\partial_j u_j$ and the operators $\diamond$, $\bowtie$, and $\partial_{ii} \bowtie$ are defined by the symbolic equations

$$\diamond = \sum_{\mathrm{p}} m_{\mathrm{p}} \frac{f(r_{\mathrm{p}})}{r_{\mathrm{p}}} \left(\left(\frac{e^{r_{\mathrm{p}}\sqrt{\partial_{ii}}} - e^{-r_{\mathrm{p}}\sqrt{\partial_{ii}}}}{2r_{\mathrm{p}}\sqrt{\partial_{ii}}} - 1 \right) - \left(\frac{1}{3} - \frac{e^{r_{\mathrm{p}}\sqrt{\partial_{ii}}} - e^{-r_{\mathrm{p}}\sqrt{\partial_{ii}}}}{2r_{\mathrm{p}}^2 \partial_{ii}} + \frac{e^{r_{\mathrm{p}}\sqrt{\partial_{ii}}} - e^{-r_{\mathrm{p}}\sqrt{\partial_{ii}}}}{2r_{\mathrm{p}}^3 \partial_{ii}\sqrt{\partial_{ii}}} \right) \right),$$

$$\bowtie = \sum_{\mathrm{p}} m_{\mathrm{p}} \frac{f(r_{\mathrm{p}})}{r_{\mathrm{p}}^3} \left(\frac{r_{\mathrm{p}}^2 \partial_{ii}}{6} - \frac{(e^{r_{\mathrm{p}}\sqrt{\partial_{ii}}} - e^{-r_{\mathrm{p}}\sqrt{\partial_{ii}}})}{2r_{\mathrm{p}}\sqrt{\partial_{ii}}} \right),$$

and

$$\partial_{ii} \bowtie = \sum_{\mathrm{p}} m_{\mathrm{p}} \frac{f(r_{\mathrm{p}})}{r_{\mathrm{p}}} \left(\frac{e^{r_{\mathrm{p}}\sqrt{\partial_{ii}}} - e^{-r_{\mathrm{p}}\sqrt{\partial_{ii}}}}{2r_{\mathrm{p}}\sqrt{\partial_{ii}}} - 3\frac{e^{r_{\mathrm{p}}\sqrt{\partial_{ii}}} - e^{-r_{\mathrm{p}}\sqrt{\partial_{ii}}}}{2r_{\mathrm{p}}^2 \partial_{ii}} + 3\frac{e^{r_{\mathrm{p}}\sqrt{\partial_{ii}}} - e^{-r_{\mathrm{p}}\sqrt{\partial_{ii}}}}{2r_{\mathrm{p}}^3 \partial_{ii}\sqrt{\partial_{ii}}} \right).$$

Equation (10.4) generalizes Eq. (6.15). Cauchy also established the general equations of motion for a uniaxial system of molecules. Finally, he investigated the propagation of plane waves in isotropic and uniaxial systems of molecules.

Because of his departure, Cauchy could not pursue the publication of the conclusion of his two studies, *Sur la Dispersion* and *Sur la Théorie de la Lumière*, as well as the remainder of his 1836 notebooks. The problem concerned research on the propagation of spherical and cylindrical waves, on shades, on diffraction, and especially on a new theory of reflection and refraction on a transparent or opaque body (30). From letters that he sent to Libri and to Ampère in early 1836 (31), from an unpublished notebook manuscript from 1836–1837 (32) and from several notes from the *Comptes Rendus Hebdomadaires des Séances de l'Académie des Sciences*, which were published after 1839 it is possible to get an idea of this research (33). We will merely mention Cauchy's new theory of reflection and refraction here. Considering the variation of the density of ether according to the material medium involved, Cauchy examined the behavior of light at the boundary of a body. By applying the methods he had developed in the beginning of his lithographic study 'Sur la théorie de la lumière', he derived the laws that had been articulated earlier by Brewster and Fresnel. He now had to assume that molecular vibrations do not act parallel to the polarization plane (as he had supposed in his 1830 theory), but rather perpendicular to this plane in line with Fresnel's hypothesis.

While he was in Prague, Cauchy was scientifically isolated. His relations with the local scientific community seem to have been very tenuous. On April 20, 1836, he presented his lithographed paper *Sur l'Intégration des Équations Différentielles*, and as we have seen, a few months later, he obtained a subsidy for the publication of his paper *Sur la Dispersion de la Lumière* (34). It should be noted that around 1834 a meeting between Cauchy and Bolzano took place. There seems to have been no subsequent meetings between the two men, and the one that did take place appears to have been sought by Bolzano, who had sent Cauchy a tract on the problem of the rectification of curves, which he had written entirely in French for Cauchy's benefit (35).

Charles X and his court left the Hradschin Palace and the city of Prague in May 1836 for Toeplitz. This was necessary since the new emperor Ferdinand had come to Prague to receive his investiture as King of Bohemia. Charles planned to settle in Göritz, being attracted by the climate of the southern region. En route to Göritz, Charles and his entourage stopped at Kirchberg Castle, an estate that had been purchased by Blacas, because of a cerebral fever that affected the Duke of Bordeaux in August 1836. There, the old king remained until October 8, 1836, when he proceeded on to Göritz, where he died on November 6, 1836, a few days after his arrival. The Duke of Angoulême and the Duke of Bordeaux spent the following years traveling between Kirchberg and Göritz, with Cauchy accompanying the prince on these peregrinations (36).

Cauchy continued his research on the theory of light while he was at Göritz, and he also did work on a new study on analysis and on the calculus of limits. During 1837, he sent several pieces to Libri or directly to the Académie, works that dealt with the problem of the determination of the roots of an algebraic or transcendental equation and were to be published in the *Comptes Rendus* (37).

The Duke of Bordeaux reached his eighteenth birthday in September 1838, and this marked the conclusion of his formal education. It also ended Cauchy's duties with the exiled court. Would he now return to France or continue his life as an exile? It is clear that his friends, Moigno for example, urged him to return. But, it was Cauchy's mother, Marie-Madeleine Cauchy, who was now decisive. Her golden wedding anniversary had been celebrated the preceding year; and no doubt sensing that the end was now drawing near, she was loathe to have her eldest son so far away (38). Indeed, she was to die on May 5, 1839, less than seven months after Augustin-Louis had returned. At last, in October 1838, Cauchy and his family arrived back in Paris and took up residence once again in the de Bure townhouse on the rue Serpente. Cauchy's first act upon returning to Paris was to attend the October 22, 1838, meeting of the Académie des Sciences. He was now approaching 50, and a new period in his life was beginning.

Cauchy had little to show for the eight long years he had spent in exile: the title of Baron, which Charles X and bestowed on him for his years of faithful service—and to which Cauchy was very attached—and several notebooks that were filled with handwritten notes on various mathematical topics. It is from these notes that he would draw much of his scientific inspiration in the coming years.

Chapter 11
The Legitimist Mathematician

When he returned to France in the autumn of 1838 after an exile that had lasted eight years, Cauchy was firmly resolved to participate fully in contemporary scientific life. His refusal to take the oath of allegiance in 1830 had meant the loss of his professorial chairs. But, he still had his position at the Académie des Sciences, because it had not been required that he swear to the July Monarchy in order to keep his seat there. But, in spite of the many advantages that came with membership in the Académie—aside from attendance at the meetings and the rapid publications of notes and studies in the *Comptes Rendus Hebdomadaires des Séances*—Cauchy could not be content with a single position in the scholarly world.

Opportunity presented itself on July 29, 1839, with the death of Cauchy's old teacher, Prony, which created a vacancy in the geometry section of the Bureau des Longitudes. The Bureau des Longitudes occupied a unique position among the French scientific institutions of the time. Like the Académie des Sciences, it had the right to choose its new members, and the king merely approved its choice. In this respect, the Bureau tended to regard itself as something of an academy of the astronomical sciences in its own right (1). However, members of the Bureau—unlike members of the Académie—were under an obligation to take oaths of political allegiance. Within the Bureau, such oath taking was merely a formality to which no one paid any great attention. Accordingly, it was assumed that if Cauchy were elected, the matter of an oath would not assume the same importance that it assumed when the question was one of appointment to a professorial chair.

A few days after the announcement of Prony's death, Cauchy entered the competition. On August 5, 1839, he presented a paper on celestial mechanics to the Académie (2). This study was in reality only indirectly related to astronomy. In it, Cauchy gave a proof of his famous Turin theorem of 1831, dealing with the series expansions of functions. He pointed out the possible applications of the theorem to celestial mechanics, but did not detail or explore them in any way. During the same meeting, Cauchy paid homage to Prony, 'an illustrious colleague', he said, 'who, many years ago, seemed to have taken such pleasure in having me as one of his students and who willingly encouraged me with my first scientific works' (3).

The elections were to be held in the fall, and a commission was appointed on October 30, 1839. Aside from Poisson, who, since Prony's death, was the only member of the geometry section of the Bureau, the commission consisted of Biot and Arago. But, these three men were unable to agree on a common list, and on November 6, 1839, addressing a meeting of the Bureau's members, Arago declared:

> The majority, consisting of MM. Biot and Arago, recommends MM. Cauchy, Liouville, and Sturm as candidates; M. Poisson recommends M. Lacroix (4).

It seems that Biot and Arago were especially supportive of Cauchy's candidature. According to Valson, Arago was rather blunt and impatient with the other applicants to whose inquiries he invariably replied 'M. Cauchy is competing'. To those petitioners who persisted in their inquiries after having been given the standard answer, he would repeat, making a humorous little gesture: 'But, sir, I have just told you that M. Cauchy is competing' (5). Arago's influence in the learned society of that time is well known, and by supporting Cauchy's candidature, he was, in effect, showing a deep respect for a man that he did not like, either personally or politically, and with whom he had often clashed over scientific questions. We should not be surprised by Poisson's attitude, because, as we have seen, he and Cauchy had been on bad terms since 1825. On April 15, 1839, shortly after the latter's return to France, an altercation once again flared up between these two scientists. The disagreement centered around a study by Cauchy, 'Sur la quantité de lumière réfléchie sous les diverses incidences par les surfaces des corps opaques et spécialement des métaux'. Several notes were written, prolonging the dispute over the following few weeks (6).

On November 14, 1839, the Bureau held the election. Definitely hostile to Cauchy, Poisson made one last attempt to swing the election in favor of Lacroix, his candidate. The record indicates that:

> M. Poisson asked to speak regarding the nomination for which the presentation had already been made at the last meeting. It was pointed out that, at the Académie des Sciences, it has never been the practice to resume a discussion on a topic once the [formal] presentation had been made. M. Poisson replied that no such presentation was obligatory at the Bureau des Longitudes. Nevertheless, it was observed that even at the Bureau there is no instance on record where the discussion was resumed, but M. Poisson's motion was not opposed at this time. M. Poisson spoke of the great services that M. Lacroix's works had rendered to public instruction.

But this attempt failed, and:

> Following a discussion in which several members participated, a secret ballot was held. The outcome was that M. Cauchy received the majority and was accordingly nominated as a member of the Bureau. His nomination will be submitted to the king for approval (7).

Thus, a long and painful story began. A few days after the election, the Ministry of Public Instruction, which controlled the Bureau des Longitudes, sent Arago, the president of the Bureau, a letter about the recent nomination. Arago informed his colleagues accordingly:

> Being in doubt as to whether this letter had been addressed to him personally or to the Bureau itself, M. Arago read it to the staff and sought the Bureau's opinion as to the proper response. Basically, the Minister asserted that he would submit M. Cauchy's nomination to the king for approval when M. Cauchy had taken the oath. Is it up to the Bureau to make inquiries into M. Cauchy's frame of mind in this regard? This had never been done before, but if the Bureau takes the letter into account, then it would be obliged to take a position.
>
> On this matter, the Bureau is unanimous in its opinion that political considerations should have no part in this nomination and that the Bureau can in no way involve itself in the matters that the minister spoke of, nor can it concern itself with either M. Cauchy's intentions or opinions. Thus, in his own name, M. Arago replied to the Minister (8).

The situation at the Bureau became absurd: Lacking the royal approval of his election, Cauchy could neither receive payment from the Bureau nor participate in its meetings, even though he remained an elected member.

Good will intervened to clarify the situation (9). Villemain, Minister of Public Instruction, favored a compromise. This was also true of Victor Cousin, who succeeded Villemain in the ministry after March 1840. J. B. Biot, who was a member of the Bureau des Longitudes at that time and who perhaps played a role in the affair, declared:

> During the time that this problem was festering, the Ministry of Public Instruction was occupied successively by two persons of great literary renown, whose characters as well as inclinations placed them above petty and malicious aims. These two men labored, in ingenious ways, to reduce to the fewest possible conditions what their positions required them to obtain. They were willing to accept the minimum that would permit them to comply with the law. But, Cauchy shunned any such compromises... (10).

Still adhering to his attitude of 1830, Cauchy refused to take any oaths. Elected but not appointed, since his election by the Bureau was not approved by the king, he acted as though he were a full member of the Bureau, except that he could not participate in that body's business meetings. Accordingly, starting in

1839, he devoted an important part of his research to celestial mechanics. As for the men in authority at the Bureau, they did nothing to curtail Cauchy's activities in the sense that for four years they acted as though they were simply waiting on him to swear the oath, even though they knew that this was something he would never do.

On April 25, 1840, Poisson died, and this prompted Cauchy to devote even more of his research efforts to mathematical astronomy. Moreover, Poisson's death removed the last of the geometers from the staff of the Bureau des Longitudes, even though this was precisely the time at which the Bureau was beginning to recognize the need for staffing itself with as many theorists as the observational staff (whose number was also to be increased) (11). Since Laplace's death, mathematical astronomy had been a rather neglected area of research in France—and that in spite of the works of young French scientists, such as Liouville and Leverrier, compared to Germany, where a number of mathematicians, such as Gauss, Bessel, and Hansen, developed the methods of applied analysis as related to celestial mechanics.

Cauchy felt a need to prove again that he was worthy of holding the position to which he had been elected. The fact that he was not able to submit his papers to the Bureau des Longitudes did not hinder his research in mathematical astronomy, so that between June and November 1840, he presented a dozen papers to the Académie on celestial mechanics or on the applications of mathematical analysis to celestial mechanics (12).

On November 18, 1840, Liouville was elected to fill the vacancy that Poisson's death had created at the Bureau des Longitudes; Cauchy subsequently slackened his production. Nevertheless, in 1841, he still presented several important studies on celestial mechanics, which were inspired by Leverrier's research on the minor planet Pallas. During the following years, he continued to do research on mathematical astronomy, even after he was excluded from the Bureau des Longitudes in November 1843.

Cauchy's opposition to the regime was not expressed solely in his refusal to take the oath of loyalty to the king of the French. From the moment he returned to France in late 1838, he had favored the Jesuits in the struggle against the Université and for the 'freedom of teaching'.

Since 1837, the Jesuits had been training teachers and professors for their colleges at the École Normale Écclésiastique on the rue des Postes in the Latin Quarter. While it has not been possible to determine the truth of Hoeffer's assertion that Cauchy played an active role in mathematical teaching (13), it is nonetheless clear that he regularly went to the rue des Postes to consult, at the very least, with Abbé François Moigno, who was in charge of the young Jesuits' training and instruction in mathematics and physical sciences.

François Napoléon-Marie Moigno was born on April 20, 1804, at Guéméné in Brittany. His father, a member of the petty aristocracy of Brittany, had emigrated during the Great Revolution. In 1815, young François Napoléon-Marie enrolled in the College of Pontivy and, the next year, entered the Seminarie of Sainte-Anne, a small institution operated by the Jesuits. He left for Montrouge, having decided to become a member of the Society of Jesus in 1822, when his seminary days would be over. However, in September 1824, under orders from his superiors with a view to studying mathematics and physics, he entered the École Normale on the rue de Sèvres. There he remained until 1829, studying most of his mathematics under Charles Leroy, a professor of mathematics of the École Polytechnique. He also took classes under Cauchy at the Collège de France and at the Faculté des Sciences. Even at this time, Moigno seems to have been deeply attached to Cauchy and, in fact, in 1828, he even presented him with a new method for obtaining the equation of a tangent plane, a method that Cauchy set forth in his *Exercices de Mathématiques* of 1828 (14). In October 1829, he left for Saint-Acheul, where he studied dogmatic theology until the Revolution of July 1830. On September 6, 1830, he left France, going first to Brigg, Switzerland, and then to Turin. In September 1833, Cauchy recommended that Moigno be appointed as his sucessor to fill the chair in higher physics that he was now vacating at the University of Turin. Upon returning to France, Moigno taught theology at the Scholasticat of Vals, near Puy-en-Velay, and later, in 1835, at Saint-Acheul.

In October 1836, he returned to Paris where he had been appointed to fill a chair in mathematics at the École Normale Écclésiastique in the rue des Postes. Once at the École Normale, he established a physics department that soon attracted the attention of learned society. He began to acquire a name for himself in the scientific community, a reputation that connected him with a number of leading scientists of quite different persuasions, such as Arago, Liouville, Thénard, and, of course, Cauchy, with whom he kept in contact before 1838. From 1839 on, Cauchy frequently called on Moigno, regarding him as his disciple. The closeness of the relationship between these two men is indicated by the fact that Moigno's courses at the École Normale were inspired by courses that he had taken under Cauchy and by the fact that Cauchy entrusted manuscripts to him (15).

With Cauchy's full agreement and support, Moigno published the first volume of the *Leçons de Calcul Differentiel et Intégral* in 1840. In the introduction, Moigno invoked the authority of his distinguished teacher, declaring:

> M. Cauchy, whom I am honored to have had as a teacher and who I am even more honored to have as a friend, openly accepted me as intermediary and echo in his scholarly communications with the public. (16).

Several chapters of the *Leçons* were taken, almost word for word, from the manuscripts that Cauchy had entrusted to Father Moigno, this being particularly true of an unpublished treatise on the calculus of finite differences, and from a manuscript notebook that would have become the third volume of the *Applications du Calcul Infinitésimal à la Géométrie* (17). Moigno's work was quite successful.

Father Moigno was involved in many other activities. He directed the open reunion of the Society of Saint Francis-Xavier at Saint-Sulpice, the parish in which he planned to create a museum devoted to sculpture. He held retreats, preached, and took part in charitable works as well as writing articles for Catholic journals, particularly for the *Univers* and the *Union Catholique*. His superiors were very suspicious of all these activities, regarding them as agitation. Even more seriously, however, Moigno made the mistake of becoming involved in some unfortunate speculations and financial dealings. Along with several other persons, he had incurred debts amounting to tens of thousands of francs in order to finance the rather shady dealings of the Marquis de Jouffroy, the inventor of the palmipedes motor. He managed to get Cauchy to agree to make a favorable report to the Académie on an invention of Jouffroy's in November 1840 (18). A major financial scandal threatened to break out, and sensing the danger Father Boulanger, Moigno's superior, decided to send the imprudent professor to Laval so that he might teach a course in Hebrew there (19).

Moigno refused to obey, preferring to go into hiding in Paris; and, after a four-year battle, he withdrew from the Jesuit Order in October 1843. He continued his rather stormy career as an ecclesiastic over the following years. In 1844, he published the second volume of his *Leçons*, which, like the first volume, contained many chapters that were based on approaches that Cauchy had taken in the courses he had taught during the Restoration era (20).

Cauchy seems not to have been offended by Moigno's break with the Jesuits, and continued, at least until 1844, to visit his old friend regularly, coming to see him during retreats and occasionally writing him. Moreover, Cauchy agreed to participate in a commission charged with the responsibility of evaluating the study Jouffroy submitted on June 12, 1843, entitled 'Sur un nouveau système de chemin de fer'. Not only did Cauchy work on the commission, but he also wrote a favorable report about it in 1846 (21).

In 1839, while he was lending support to the École Normale Ecclésiastique with his prestige and knowledge, Cauchy took part in the founding of the Institut Catholique, an organization in which he played a key role (22). The purpose of this institution was to offer philosophical, literary, and scientific conferences and thereby soften the effects of the absence of Catholic university education. This organization really gained impetus in 1842, after the founding of the Cercle Catholique in November 1841 by Ambroise Rendu. With its more liberal orientation, Rendu's Cercle Catholique also offered conferences and gatherings that would rival those of the Institut's, and it counted figures such as Montalembert and Ozanam among its participants (23). The governing bylaws of the Institut Catholique were established during several meetings in January 1842, Cauchy being secretary (24). According to a notice that was issued in February 1842, the Institut Catholique was to be 'a union that is open to both the youths who come to the capital each year in search of instruction and to mature persons who have retained a desire for learning'. Weekly meetings and lectures were held in the rue de Verneuil where the Institut maintained a library and reading room; the lectures were under the direction of 'special committees consisting of persons who are veritable authorities in science' (25). The Committee of Law and Letters was directed by Pardessus, a member of the Académie des Inscriptions et Belles-Lettres, and Cauchy presided over the Committee of Sciences.

A short time later, the Arts Committee was added to the two initial ones, and it was directed by Raoul Rochette, the Permanent Secretary of the Académie des Beaux-Arts. Heading the Committee of Sciences, Cauchy gathered a group of legitimist scientists around himself: Coriolis, Binet, Freycinet, Beudant (all from the Académie des Sciences), Leroy, Cayol, Gautier de Claubry, Auguste de Sainte-Hilaire, and the doctors Cruveilher, Récamier, and Tessier (16).

Cauchy was also a member of the commission responsible for managing the business affair of the Institut. A general meeting was held once each month, and during these sessions public lectures were given. These lectures were published in the *Bulletin de l'Institut Catholique*, a publication that regularly appeared from 1842 until 1844 (27).

Along with Auguste de Sainte-Hilaire, Dr. Tessier, and Gautier de Claubry, who were in charge of the conferences on botany, medicine, and physics, Cauchy gave lectures on mathematics each Wednesday evening at 8 o'clock. There were also regular poetry readings, as well as conferences that took place during the course of the general meetings. The function of the scholars and scientists associated with the Institut Catholique was, in a word:

> to enlighten the youth and instruct their minds, to direct them in the ways and habits of study and work, to instill in them a taste and love for truth and beauty, and to act as models that they should pattern themselves on, to share in their works and efforts whenever possible, and to confirm them in the goodwill and affection that will encourage them to hold fast to the path of virtue and not be influenced or tempted by perfidious suggestions (28).

With regard to this statement of purpose, Cauchy himself wrote:

> A youth who is studious and eager to learn will after a few years devote himself to the cultivation of science and letters without losing the point of view of religion, which comes before science. Such a youth would want the benefit of the experience of those who have preceded him and would hope that the true teachers and masters of science, men who are distinguished, with proven abilities, and who have a sincere attachment to the Catholic faith, would serve as his guides. Such a hope cannot be denied. The members of the two committees vie in their zeal for working toward the success of so beneficial an undertaking. They all pray, and indeed, have long since prayed to God that He Himself will bless their works, which cannot fail to echo his glory since such labors as these have the search for truth as their real and final end (29).

In spite of the efforts of its associated scholars, the Institut Catholique had only limited success. This was no doubt due, in large part, to the competing efforts of the Cercle Catholique, an organization that held very prestigious conferences and gatherings under the direction of Ozanam.

Paralleling his efforts on behalf on Catholic education, Cauchy resumed his work with charities, the 'good works' to which he had devoted so much time and effort prior to 1830. The Congrégation was now disbanded, and a whole new generation of charitable organizations had since come into being, organizations in which there was an uneasy coexistence between conservative and liberal Catholics. The most important of these new charities was the Société de Saint-Vincent-de-Paul. This organization was founded on the ashes of the old Société des Bonnes Oeuvres in 1833 by a group of students and followers of Ozanam and Bailly. Upon his return from exile, Cauchy joined it and actively participated in its charitable works by creating the Conférence of Sceaux (30).

In any event, the modernity of Ozanam's works and efforts, which had issued from the more realistic charitable practices of the Restoration Era but

Augustin-Louis Cauchy, medallion by David d'Angers, 1843. Published by permission of the École Polytechnique.

were free of the reactionary bigotry of the old Congrégation, could hardly satisfy a person like Cauchy. This was so because Cauchy's hopes, dreams, and ambitions, like those of other legitimists, centered on a return to the counterrevolutionary ideology of the Société des Bonnes Oeuvres, which had flourished before 1830 (31).

In April 1839, an organization, Catholicisme en Europe, was founded under the inspiration of Monseigneur Gillis, the Bishop of Edinburgh. The purpose of this organization was to 'aid and comfort Catholics in the Protestant countries of Europe', and Cauchy actively participated in the group from the start. This group was led by Ferdinand Bertier de Sauvigny and had Cauchy as one of its most active propagandists. Entitled *Annales du Catholicisme en Europe*, the group's publication enjoyed a real success for a time, having the approval of some 40 bishops and counting some 1200 subscribers in France as of March 1841. Nevertheless, this group met with the resistance of the Commission of Lyon of the Association pour la Propagation de la Foi dans les Pays Infidèles (Association for the Propagation of the Faith in Heathen Lands), a missionary group that was headed by Ozanam and of which Cauchy was a member. Well-established in Paris and possessing a certain credibility, which was, no doubt, attributable to Monseigneur Affre, the Archbishop of Paris, Catholicisme en Europe seemed to be a formidable competitor with suspect political intentions.

Two letters that Ozanam wrote to Meynis in March 1841 throw light on the basis of the conflict. Denouncing the 'scheme of infiltration' that legitimists such as Cauchy were engaging in as members of the charitable associations, he wrote:

> M. Bailly thinks that the operations of Catholicisme are the business of certain religious legitimists who are now sorry that they let all the charitable associations slip through their fingers, and who, in particular, are now attempting to monopolize the Société de Saint-Vincent-de-Paul (32).

This remark was later followed up in a letter dated March 17, 1841, in which it was noted:

> The business of Catholicisme is being monopolized by a legitimist faction, which, using all means imaginable, seeks to worm its way into everything... (33).

During the same month, March 1841, the Archbishop of Lyon, Monseigneur Bonald, came to Paris, where he had a meeting with Cauchy. At this meeting, the archbishop sought to persuade Cauchy to renounce the undertaking, and at the same time, he spoke with Monseigneur Affre about the association (34). Paralleling the archbishops's initiatives, the Propagation de la Foi addressed itself directly to Rome and, as a result, secured the Pope's support. In the meantime, Cauchy continued to struggle vigorously. In order to counter the offensive now being launched by the Propagation de la Foi, he wrote a letter to Father Roothaan, the Vicar General of the Jesuits, in May 1841; he defended Catholicisme en Europe against its critics, explaining that this organization's purpose was complementary to and not in competition with that of the Propagation de la Foi. The Jesuits had, after all, initially supported Monseigneur Gillis' work. In any event, Cauchy's letter arrived too late; the Pope's decision in favor of the Propagation de la Foi movement was irrevocable, and Father Roothaan refused to intervene. Thus, this enterprise to which Cauchy had, as usual, given so much of himself, ended in defeat (35).

It did not take long for Cauchy's involvement on the side of the Jesuits to hurt his scientific career. His views and persuasions not only put him on a bad footing with the government but with his colleagues at the Académie as well. The scientists at the Académie were generally very much attached to the Université. In this regard, Cauchy suffered a cruel experience in 1843, just as the agitation of the Catholics in favor of freedom of education reached a pitch. This agitation was but one episode in the long educational struggle that pitted the partisans and adversaries of the Université's educational monopoly against each other and extended over the entire 19th century. Catholic secondary education grew considerably and, as matters stood, illegally. In fact, by the end of 1843, about half of the students attending the colleges were enrolled in those operated by Catholics, so that in departments everywhere in France the church-operated schools vied effectively with the state-sponsored colleges. The formation, in July 1843, of the Comité de Défense de la Liberté d'Enseignement revealed the ambitions of the Catholics who, at this time, had Guizot's ear, who was favorable to the notion of freedom of education, at least under certain conditions. One particular aspect of the dispute, the aspect that concerns us here, had to do with higher education. Some fairly limited experiments at this level, such as the École Normale Écclésiastique of the rue des Postes in Paris and the conferences of the Institut Catholique and the Cercle Catholique, were grounds for a belief that Catholic university education would also be created.

Although characterized by many conflicting tendencies, the Catholic campaign became increasingly violent and, rightly or wrongly, the Jesuits were accused of having inspired some of the more harmful pamphlets. One of their main targets was the Collège de France. Characteristically, the students of the Latin Quarter, often republican and always anticlerical, came to hear and applaud the words of the three oracles of liberal thought, Mickiewicz, Quinet, and Michelet. In much the same way, reactionary professors were booed and hissed and, indeed, in some cases, were kept from speaking altogether. A number of rightwing journals decried the closing of courses thought to be seditious. As if to crown this campaign of disparagement, a particularly venomous work, entitled the *Monopole Universitaire*, suddenly appeared on the scene in Lyon early in 1843. Although it was attributed to the Canon Desgarets, it was, in fact, written by the Jesuits. At this point, Michelet and Quinet decided to take the offensive. Both offered lessons and discussions on the Jesuits, and in these presentations they each denounced the system of

Ignatius of Loyola. Michelet's course on the Jesuits was held from April 26 through the first of June, while Quinet's was held from May 10 until June 14. On the whole, these two courses of workshops created quite a stir, and despite a few incidents, professor Michelet and Quinet scored a triumph (36).

It was precisely at this juncture that Cauchy decided to present himself as a candidate for a chair in mathematics at the Collège de France. The vacancy had been created by Lacroix's death on May 25, 1843. By law, an assembly of the professors of the Collège would meet and agree on a first candidate and this choice would be presented; similarly, the Académie des Sciences would present a second candidate. These presentations would then be submitted to the Minister of Public Instruction, who would then appoint the new professor. Generally, the Académie would confirm the choice made by the assembly of professors, and the minister would simply ratify the decision made by the professors and confirmed by the Académie.

On June 11, 1843, the assembly of the professors of the Collège de France met, with three candidates to be considered for the vacant chair in mathematics: Libri, Cauchy, and Liouville. That Cauchy would win the election seemed a foregone conclusion, because not only had several voting scholars assured him of their support, but Liouville had also declared in his letter of candidacy that should the assembly of professors choose Cauchy, he would be the 'first to applaud that choice'. Moreover, Cauchy seems to have been convinced—quite wrongly, as it actually turned out—of Libri's good will (37).

The assumption that Cauchy would win the election may have been reasonable, but it was made without taking political considerations into account, because the dispute about academic freedom and, more precisely, the quarrel about the role of the Jesuits could not but influence the election at the Collège de France. By offering himself as a candidate for the chair, Cauchy seems to have been determined to ignore the events that were taking place. Nevertheless, he was aware of the problems posed by his refusal to take a political oath should he be elected, and, with this in mind, he initially abandoned his goal. However, for some reason, which we have been unable to determine (but which, quite likely, was that he did not want to give grounds for the belief that he had withdrawn out of fear of being beaten), he finally decided to confirm his candidacy (38).

The assembly of professors took place on June 11, and during the meeting, there was an open discussion at which a number of professors spoke. The vote, however, was postponed to the following Sunday, June 18, 1843. From a standpoint of scientific scholarship, the matter was clear: Liouville, himself an able mathematician, was disposed to defer to Cauchy, whose merits he acknowledged. Libri had already given clear and certain proof of incompetence as a mathematician when he had replaced Lacroix. Moreover, certain persons—the historian Michelet, in particular—were already aware of Libri's embezzlements (39). But, the matter was essentially political: it was absolutely necessary that Cauchy be defeated because he was the Jesuits' candidate. Libri, of course, had once been Cauchy's protégé, as well as a friend of the Jesuit priest Moigno. Now, however, he presented himself as the determined enemy of the Society of Jesus. Accordingly, at election time, he published two articles in the *Revue des Deux Mondes*, that were very harsh on the Jesuits (40). Furthermore, according to Michelet's journal, Libri never ceased pestering and worrying him about the election, swearing that his defeat would mean a victory for the Society of Jesus, pure and simple.

On June 17, 1843, the eve of the election, Libri wrote a letter to Letronne in which he declared that nothing could stop him from 'keeping up his war' against the Jesuits and if, indeed, he should be rejected for the chair in mathematics, then 'the Jesuits would sing out their victory in their newspapers' (41). Meanwhile, in order to allay any suspicions that might occur to his competitor, Libri, on June 13, 1843, took part in the commission of the Académie that was to evaluate a paper by Jacques Binet. Cauchy and Sturm were also on this commission, and moreover, Binet, the author of the study to be evaluated, was Cauchy's close friend and a known supporter of the Jesuits.

On June 18, 1843, the assembly of professors proceeded to nominate a candidate after having rejected a motion by a member of the assembly who asked that the discussion of the candidates' qualifications be reopened in light of the fact that Cauchy had written a letter to Letronne in which he gave assurances that if he should be elected to the chair 'the government would have no cause to fear any serious obstacles' and that his lectures would not be the subject of any disruptions (42). On the first ballot, with 24 voting professors present, Cauchy received 3 votes against 12 for Libri and 9 for Liouville; on the second ballot, Liouville received 12 votes with 11 for Libri and 1 for Cauchy; on the third ballot, Libri carried 13 votes as opposed to 10 for Liouville and 1 for Cauchy.

That the professors at the Collège had chosen Libri caused a scandal in the mathematics community. The following day, June 19, 1843, Liouville sent a letter to Letronne, the Administrator of the Collège de France, in which he resigned his position as an adjunct professor, declaring that he was 'deeply humiliated as a man and as a mathematician by what took place yesterday at the Collège de France' (43).

Cauchy attended the Académie's meeting of June 19, 1843 and made a statement in which he asserted that he 'would never consent to the Académie's placing his name on any list of candidates for a chair in mathematics unless, on the one hand, no serious obstacles to his candidacy were raised if such [obstacles] were not related to science; and, on the other hand, the candidates themselves, being desirous of giving a new indication of their esteem for their former teacher, should endorse that candidacy and fully consent to it' (44).

In any event, the matter was settled. The geometry section of the Académie, which received the candidacies, announced to the Académie that only Libri had been presented. Cauchy and Liouville had renounced their candidacy. So declared to the Académie that they were formally opposed to Libri's candidacy (45). On July 3, just before the election of the Académie's candidate, Cauchy made a final statement in which he recalled his statement of the preceding week and the 'report from the geometry section that alleges that I stood aside and did not offer myself as a candidate' (46). In spite of this final statement by Cauchy, Libri obtained only 13 votes out of the 45 that were cast, because 28 were blank, 3 were for Cauchy, and 1 was for Liouville. In any event, Libri had an absolute majority of the votes and, accordingly, became the Académie's official candidate. Shortly afterward, a royal ordinance ratified this choice and appointed Libri professor of mathematics at the Collège de France.

This affair left deep scars. For example, not long afterward, a bitter dispute arose between Libri and Liouville following a report made by the latter on Hermite's study on the division of abelian functions. At this time, September 4, 1843, Liouville announced that he planned to publish Galois' study 'Sur les conditions de résolubilité des équations par radicaux' (47). Cauchy, careful to do nothing that would further exacerbate the matter, kept silent throughout the dispute, which lasted from August 14 until September 18, through six meetings of the Académie. A few weeks later, during this same period of bitter political feelings, the matter of Cauchy's refusal to take the oath at the Bureau des Longitudes came to a head, with negative outcome for Cauchy. That there was no geometer on staff at the Bureau des Longitudes presented a situation that could no longer be tolerated. Accordingly, the Bureau voted on November 15, 1843, to send a letter to the Minister of Public Instruction in which it declared that the 'present state of affairs cannot continue without doing real harm to the development of the astronomical sciences' (48). The minister responded by inviting the Bureau to proceed to make a new appointment as 'the previous one could not be approved and was thus null and void because the candidate refused to fulfill an obligation imposed by law' (49).

Cauchy promptly reacted to this decision by writing an open letter to the president of the Bureau des Longitudes in which he rejected the minister's arguments and asserted that for four years he had never received an official letter enjoining him to take an oath (50). At the same time, he asked that he be allowed to attend the Bureau's meeting on December 6, 1843, in order to give an oral explanation of his position. However, the Bureau, no doubt angered by Cauchy's publication of his letter to the president, refused to receive him and, accordingly, proceeded to elect a new mathematician. The outcome of this election was that Poinsot, Cauchy's enemy, took his place at the Bureau.

Cauchy's double defeat, at the Collège de France and at the Bureau des Longitudes, was but one sudden and unfortunate turn in the ongoing confrontation between the partisans and adversaries of the Université's monopoly. Since his return from exile, Cauchy had very unwisely taken part in the Catholic offensive. Now, deprived of all official positions except his seat in the Académie, he gave up active scientific research in order to throw himself into the political battle.

The question of freedom of education had, in effect, been brought before the Chamber of Deputies where those in favor of the Université's monopoly were in a majority. The stakes were high, because the real issue was whether the state or the church would control the intellectual development of the leading social classes. To the extent that this was so, the Jesuits could not fail to have great concern, because, within the span of a few years, they had created some 74 academic establishments across the country, even though they were not authorized to do so. Accordingly, those who were against the Jesuits demanded that the government expel them once and for all.

The defenders of the Jesuits mobilized, and Cauchy was among the first to do so. He took part in a kind of public relations campaign that was organized by Father Ravignan, a Jesuit priest who in 1837 had replaced Lacordaire as preacher at Notre-Dame and whose conferences and retreats at the Abbaye-aux-Bois Cauchy had assiduously attended since 1841 (51).

All in all, Cauchy's contribution to the Jesuit cause at this juncture was rather modest. He addressed himself to those scholars with whom he was well acquainted in a pamphlet entitled *Considérations sur les Ordres Religieux Adressées aux Amis des Sciences*. Published in March 1844, this work touched on the main arguments that were likely to appeal to men of science: the general spirit of sacrifice that motivated the religious orders, the Jesuits civilizing mission in Paraguay, the role that their colleges and students had played in the development of science, the value of Jesuit scholars, and the blindness of their enemies, etc. Shortly afterward, he distributed a short work entitled *Mémoire à Consulter Adressé aux Membres des Deux Chambres*. In this work of only a few pages, he summarized the arguments that he had explored in the preceding pamphlet. In 1844, Cauchy published a third tract entitled *Quelques Réflexions sur la Liberté d'Enseignement*. Here, after having recalled the promises made in the Charter of 1830, he affirmed the necessity for religious instruction within the framework of a good education, and then proceeded to give a comparison between scientific and religious instruction. He condemned Lamennais' neocatholicism and in conclusion exposed the inconveniences and shortcomings of the Université's monopoly.

The discussion of freedom of education that took place in the Chamber of Deputies ended inconclusively. However, Guizot had to take account of the opinions that had been expressed during the discussion. The Jesuit activities had been denounced, and there followed a long series of negotiations with the Pope with the outcome that a number of Jesuit institutions were closed and the Society of Jesus was disbanded in France. None of this, however, kept the Jesuits from discretely pursuing their activities until the Revolution of 1848.

The measures that were taken, with the approval of Pope Gregory XVI, as well as the support of Monseigneur Affre, the Archbishop of Paris, against the Jesuits were a stunning blow to the their supporters, who now had to keep quiet. At the same time, however, Montalembert, operating under the broad cover of freedom of education, was trying to gather various Catholics into a Catholic party, in which stalwarts and traditionalists such as Cauchy would be on the fringes.

Cauchy continued to pursue his charitable works after the eventful years of 1843 and 1844. However, he did so without any pretense of steering them in the direction of his personal persuasions. While actively working with the Société de Saint-Vincent-de-Paul, he tried to use his fame as a person of science to advance pious designs. As Valson stated:

> When it happened that he had to overcome a difficulty, an obstacle, that would have deterred a less courageous person, it was to his fellow scholars and scientists at the Institut that he would first turn for help. There, he would invariably find the means of exerting his influence, and it must be said to the honor of that illustrious body, that they never failed or let him down. With tremendous warmth and perseverance he went about his task. By using diplomacy, he was able—sometimes through mere sympathy for the cause that he was so ably defending; sometimes through his own friendship with an individual—to win over this one, to convince that one, and so obtain first a few signatures alongside his own. As time passed, the number of signatures would increase, and in the end, everyone would have signed. In this way, he associated the Académie—and often the entire Institut—with his own worthy efforts and good works (52).

Thus, he supported the Société de Saint-Régis, an organization to which he belonged. Founded in 1826 by Jules Gossin, this group worked to obtain civil and religious marriage for couples who were living out of wedlock and to legitimize the offsprings of such unions. In 1844, Gossin had entered a competition in statistics in which he had presented statistical data on persons who had recourse to the Société de Saint-Régis; and he had received an honorable mention from the commission in May 1846 (53). This initial success prompted Gossin and Cauchy to seek the Académie's help in obtaining an amelioration of the fiscal laws relating to the acts required for the celebration of marriage. Cauchy drafted a petition that he circulated among his colleagues and that was designed to stir 'the concern of the government and public authorities about a charitable organization known as the Saint-Régis' (54). There were 174 members who signed the petition, and a committee that was headed by Portalis and on which Cauchy served was appointed. This committee, which included men such as Tocqueville, Villeneuve, Villermé, and Pardessus, proposed to the Chamber of Deputies that the tax laws be amended. This proposal was supported by Villeneuve, and on June 19, 1846, it was enacted. Under the terms of this measure, distressed people were exempted from the stamp and registration fees that were required for any and all kinds of certificates pertaining to the celebration of marriages.

Encouraged by the success of this undertaking, Cauchy once again stirred the Institut to take a position, this time regarding the Oeuvre de l'Agonie

Irlandaise (work on the Irish agony). Here, too, the initiative had come from Jules Gossin, the president of the general board of the Société de Saint-Vincent-de-Paul. The question now centered on aid for the Irish, who were then suffering from the terrible famine of 1846. Writing in 1847, Henri de Riancey recalled:

> Early one morning, before dawn, Cauchy hastened to my home. A new idea, a new approach had occurred to him, and he wanted to act on it right away. A supplication would be addressed to the Father of the Faithful. The Sovereign Pontiff would be incited to directly appeal to the Catholics to snatch a God-fearing faithful people from the jaws of hunger and death. Members of the Institut would be urged, without regard to beliefs or [political] persuasions, or even to religion itself, to sign this supplication. Once the Institut had been canvassed, attention would be focused on the chambers, the salons, and indeed on any persons of standing and repute in France. Once the letter had been drafted, Cauchy would carry it from door to door, certain of success, and nothing was going to stop him. He was not mistaken; he had not placed too much faith in the generosity of the French people. Thus, how could anyone fail to lend himself to such a work of spirit and charity? Within a few days, Cauchy, weakened and exhausted with fatigue, had collected hundreds of signatures. These he submitted to the papal nuncio who, delighted and amazed, speedily sent them off to Rome (55).

Another issue that compelled Cauchy's attention during this time was the problem of prisons; this question had been debated in the Chamber of Deputies. Cauchy had served on a jury in the Court of Assizes of the Seine. In the name of all the jurors, he drafted a study that he published under the title *Considérations sur les Moyens de Prévenir les Crimes et de Réformer les Criminels.* He proposed in this study that the conditions of prison detention be reformed (56).

As we have seen, Cauchy's exclusion from all scientific institutions and his inability to get a professorship were direct consequences of his political views. Not only was he opposed to the regime that came to power on the heels of the July Revolution, but he also struggled against the Université's monopoly. This latter fact, of course, meant that he soon became all but isolated in the French scientific community; for, by and large, the members of that community were deeply attached to their alma mater.

It will be seen that Cauchy's virtual quarantining had an altogether negative effect on the evolution of his scientific work. Here, we note the particularly negative effects that his exile had on mathematics in France. By the mere force of the situation, an entire generation of French youths were deprived of instruction by Cauchy. Moreover, if the eight years of exile are taken into account, it is possible to take some measure of the harm that political events inflicted on the mathematical tradition in France.

The politicizing of faculty appointments was a consequence of the fact that the French scientific community of that time was overly concentrated in Paris, where it was under the discrete but ongoing surveillance of the authorities, as well as in the midst of the lively debates that stirred the political classes. It can even be said that it was the organization of French scientific life itself that permitted the ostracism of the country's most important mathematician.

Chapter 12
Scientific Works from 1838 to 1848

Between October 1838 and February 1848, Cauchy's scientific accomplishments were as abundant as during the Restoration Era. However, the conditions under which he worked were very different from those that had existed before 1830. Most important, Cauchy was now no longer teaching; and, by the very nature of things as they stood at this time, he found himself working on the fringes of the French scientific community. He had been away from that community for more than eight years, and during that time, considerable changes had taken place. He was still a member of the Académie, des Sciences, attending the weekly meetings during which he was able to communicate the results of his research. In 1836, the Académie began to publish its *Comptes Rendus Hebdomadaires des Séances*, which allowed members to publish their works quickly. This stimulated interest in Cauchy's communications on his research. In fact, Cauchy, more than any other member of the Académie, took advantage of this journal, publishing a note or memoir each week. Moreover, starting in September 1839, he resumed publication of *Exercices* (which had been interrupted since 1836) under the new title *Exercices d'Analyse et de Physique Mathématique*. Altogether, Cauchy presented about 240 notes and studies to the Académie during this period, of which most were published in the Académie's *Comptes Rendus*. Aside from these works, Cauchy published 27 reports during this period. Furthermore, there were some 40 installments of *Exercices* containing about 50 studies, along with several articles that appeared in reviews and two earlier studies that appeared in the *Mémoires de l'Académie*. The articles appearing in the reviews, like the material contained in *Exercices*, were generally based on—albeit sometimes in modified form—Cauchy's communications with the Académie.

Did Cauchy abuse the publication privileges that were now available to him? It is clear that he wanted to recover the position that was opening for him in spite of the ostracism that he was being subjected to for political reasons. This desire could well have prompted him to increase his activities at the

Académie, even at the risk of repeating himself (in certain papers). After all, the Académie was now the only forum, the sole tribunal, in which he could publicly and regularly participate. However, his exclusion from scientific institutions and the resulting lack of an established position rather aggravated a tendency that had been perceptible quite early on, namely, to extend his efforts in many directions, according to the circumstances. That Cauchy was so prolific in his research soon aroused criticism and sarcastic comments, first expressed in the autumn of 1842. The scientific writer of the *National*, a republican publication whose hostility to Cauchy probably sprang more from political grounds than from scientific ones, first broke the silence by publishing several articles during October 1842. In particular, on October 19, 1842, the *National* remarked:

> Might there really be a sickness that goes with the study and cultivation of geometry? Should the noble kind of intelligence that is devoted to it be subjected to chronic outbursts of a strange fever, which is nothing more than a sort of algebraical spurt in the form of a new research paper? In truth, though we may be little inclined to treat serious things lightly and though M. Cauchy's very name inspires deep respect in us all, we must ask the question, seeing that an honorable academician continues to pile one study on top of another, and, in this way, swells the number of mathematical works beyond all need or accounting. Thus, we repeat today that algebra, at least as we have grown accustomed to it, was so different that our surprise should be understandable. Instead of this carefully nurtured source exuding its precious liquids in a careful and measured way, we are now confronted with an inexhaustible urn that overflows its sides, always gushing new tides on which we find there are floating, in a pell-mell fashion, all kinds of bizarre signs and unknown symbols. It is perhaps quite natural that we should ask if his fervor does not alter the purity of all that he does; and if by thus increasing the quantity of his works so prodigiously, he does not diminish the very quality of geometry (1).

Convinced that time would vindicate him, Cauchy, as he wrote in a letter to Moigno, refused to reply to this attack (2).

However, a short time later, Jean-Baptiste Biot published an article on the Académie's *Comptes Rendus* in the November issue of the *Journal des Savants*. In this article, Biot reproached Cauchy, without mentioning him specifically. He claimed that Cauchy was threatening the very existence of the Académie's new publication policy through his excessive prolixity. After having praised the *Comptes Rendus*, he continued:

> But, it is said, these detailed extracts, written at the whim of the author, may very well be too verbose and are, in fact, no more than disguised publications of entire memoirs, whole extensive works. This is very

deplorable, but I shall not pretend that this habit prevails, however, in a single instance, with one particular author, it has happened. This author, a geometer and most assuredly a very cleaver and able one at that, has used the publication privilege to publish whole memoirs in almost every issue of the *Comptes Rendus.* These extensive works that he has published fairly bristle with symbols that have little or no connection with each other. Frequently, they simply restate the same result several different times, or they may just present the same idea in several different forms, so that today, if the author is indeed able to understand their interrelationships and all the agreements between them—and I have no doubts that he can indeed do so—then he is probably the only person who can make any use of them, or can follow the lines of thought that they follow. This is surely a highly unfortunate situation, a situation that is regrettable even for the author himself. This said, we come to the central issues: by what means and by what authority can this grievous situation be set aright? These essays—and the overhasty manner in which they were published justify the use of the term "essay" in referring to them—do indeed contain, despite their capricious diversity, much that is worthwhile: some very beautiful proofs and arguments, some very powerful computational procedures, some methods that appear to be very fruitful in terms of applications... If only the author would take the time to follow up on them. Would you deny him the liberty of making these things known? We should hope that the advice of his friends—and he does not lack friends; nor are they in disagreement on this point—will persuade him that he ought to be more mindful of his own interests as well as those of science and that he might very well lose this useful means of publishing by continuing to use it in this immoderate way. If he should agree to devote several months to the completion of the calculation of planetary perturbations, let us say, then there is no one who would regret the change in direction by which he will have come to focus himself fully on this important subject (3).

One particular circumstance not mentioned by Biot in the article probably prompted him to publicly criticize his colleague. In 1839, Biot had asked Cauchy if he could explain the phenomenon of rotatory polarization and give the laws governing this phenomenon by means of his theory of light. Cauchy accepted the challenge. However, in an article that appeared in the *Comptes Rendus* of November 14, 1842, he had to admit that the problem presented difficulties that he was unable to overcome (4). Although Cauchy still claimed that he could explain this phenomenon by a new method of analysis, it can hardly be doubted that this defeat served to convince Biot, who remained a strong supporter of the emission theory of light, of the sterility of the works in mathematical optics that Cauchy presented in the *Comptes Rendus* during the succeeding years.

Cauchy felt that he had to reply to Biot's criticisms; and, thus, without going into the background of the dispute with his colleague, he expressed a wish that he presented before the Académie in a note on December 19 and that, he said, was 'made in the interest of science':

> This wish is that, if in the future, the author of this article believes himself qualified, either on the basis of his experience or by dent of being a friend, to give advice to a colleague, then he should kindly confine his observation specifically to the Académie (5).

This response was typical of the method Cauchy always used in dealing with such attacks: simply remove polemic from the realm of personalities and place it in a purely scientific domain. Having replied to Biot's criticisms, Cauchy did not slacken the pace at which he published his research during the following months (6). Indeed, until 1848—and even beyond that, until his death in 1857—he continued to publish in this way. Thus, much later, in 1869, we find that Bertrand leveled criticisms against him that were quite similar to those that Biot had made. In fact, Bertrand observed:

> The dangerous tendency to rush into publication was an irresistible temptation to Cauchy, and quite frequently, it was a stumbling block. His mind was always active. Every week he would bring to the Académie his scarcely completed works, as well as plans for memoirs and studies, and notions that were sometimes downright unfruitful. But, then, out of all this, a brilliant discovery would crown his efforts. He obliged his readers to follow, trick by trick, along paths that were often sterile, meandering, and careless (7).

In order to fully appreciate Cauchy's approach to scholarly publications, it is necessary to take into account his relations with his fellow scientists. Abstract research is an endeavor that requires peace and solitary work, but the many contacts that come by way of teaching and participation in the scientific community are also beneficial. Now, Cauchy found himself increasingly in 'splendid isolation', which he was rather fond of simply as a matter of personal temperament. From this point of view, it is interesting to note that he published very little in the scientific reviews, such as Liouville's *Journal de Mathématiques Pures et Appliquées* or any of the other publications in which the other mathematicians of that era usually published. He preferred the Académie's *Comptes Rendus* or his own *Exercices.*

Learned society in Paris had undergone profound changes between 1830 and 1838. The great scientists of the Restoration Era had died one by one: Fourier in 1830, Legendre in 1833, Ampère and Navier in 1836, and after Cauchy's return from exile in 1838, these were followed by Prony in 1839, Poisson in 1840, and Lacroix and Coriolis in 1843. Imperceptibly, in spite of some brilliant individuals, the center of scientific life shifted from France to

Germany. In Paris, Cauchy, now approaching 50, represented the past. Generally not much appreciated by his contemporaries, men such as Poinsot (8), Biot, and Arago—and that more on the basis of his political ideas and behavior than on scientific grounds—he was regarded, if not as the leader of a school, then at least as the leader of a line by a new generation of French mathematicians. An analysis of the commissions of the Académie on which he served shows that he collaborated with members of all the various areas of mathematics, with the exception of those in geography and navigation. However, five mathematicians stand out in terms of the frequency with which they collaborated with Cauchy: Sturm, Liouville, Poncelet, Binet, and Lamé (9).

Sturm and Liouville were born in 1803 and 1809, respectively (10). A native of Geneva, Sturm had arrived in Paris toward the end of the Restoration Era and soon became associated with Fourier and his group. Liouville had attended the École Polytechnique, where he had studied under Ampère but not under Cauchy. During that period of time, Cauchy had become acquainted with these two youths, who attended his courses at the Faculté des Sciences. Later on, in 1829, when these two young scholars simultaneously presented studies to the Académie, Cauchy acted as reporter of the evaluation commissions for both papers. Unfortunately, because of his departure in 1830 he never submitted a report on either work. During the years when Cauchy was in exile, research on the number of roots of an algebraic equation contained within a given region tended to bring the three men together in spite of the differences that separated them, and it can hardly be doubted that the two young men hailed Cauchy's return to Paris in October 1838 as a major event for the future of science in France.

Scientific collaboration was especially close between Cauchy and Liouville, who appears to have been his most frequent associate on evaluation commissions. Cauchy and Liouville dominated French mathematics during this era, and they remained close up until the Revolution of 1848 and the election at the Collège de France in 1850, which was a consequence of the revolution. Liouville, a man of democratic leanings and a republican, was more open than Cauchy and less concerned with questions of priority of publication (11). Moreover, within the framework of the mathematical sciences, he was basically an organizer and leader (12).

We should also mention the scientific collaboration between Cauchy and Poncelet. These two mathematicians were very different and hardly cared for each other. We have already seen the unpleasant memories that Poncelet had of Cauchy's reception of his work on projective geometry. Nevertheless, they frequently worked together evaluating studies on mechanics. In particular, both participated in the evaluation of the numerous studies that Saint-Venant presented to the Académie between 1843 and 1847. As for Lamé and Binet, they had long been acquaintances of Cauchy. Although they were late in becoming members of the Académie—the one first taking his place on the Académie on March 6, 1843, and the other on July 10, following—they soon

Joseph Liouville (1809–1882), the best French mathematician of the 1840s, founder of the *Journal de Mathématiques Pures et Appliquées.*

began to collaborate regularly with Cauchy. In particular, once Binet had became a member of the Académie, he became Cauchy's closest and most faithful colleague on the evaluation commissions.

Cauchy's scientific pursuits enabled him to maintain contacts with a new generation of mathematicians whose studies he had evaluated. Thus, during

Jacques Binet (1786–1856), French mathematician and Cauchy's friend since 1806. Photograph by J. L. Charmet, permission of the Académie des Sciences.

this era, he was able to evaluate and report on the works of the younger men who represented the best hopes of French scientific life, men such as Bertrand, Leverrier, Bonnet, Saint-Venant, and Laurent. Some of these men, while firmly acknowledging Cauchy's influence, inspired him while he was charged with the responsibility of evaluating their works. Leverrier, for example, was able to persuade Cauchy to apply his analytical talents and powers to the study of the perturbations and to planetary inequalities. Meanwhile, Saint-Venant, who was one of Cauchy's most gifted and closest disciples, perfected and simplified his theory of elasticity; on the other hand, Cauchy used all his authority in support of Saint-Venant's election to the Académie. Saint-Venant was competing for election to the seat in the mechanics section that had been left vacant by Coriolis' death in 1843. Aside from Saint-Venant, there were other candidates for the seat: Morin, Fourneyron, and Combes; and Cauchy was responsible for presenting the section's report to the Académie. The mechanics section had jointly placed Saint-Venant and Morin at the top of the list of candidates for the seat. However, Cauchy intervened in support of his favorite by pointing out that only Saint-Venant had been given the unanimous vote of the mechanics section. But, this little finishing touch came to nothing, because Saint-Venant, who made only a very mediocre showing on the final vote, which was held on December 18, 1843, was defeated by Morin (13).

In spite of his scholarly activities and his friendly relationship with several young mathematicians of great talent, Cauchy's position in France's scientific community was marginal during these years. As a mathematician, he was admired and respected while, at the same time, his exclusion from the Université coupled with his proverbial lack of personal warmth kept him isolated from young students. Cauchy was completely aware of his isolation, and it pained him. In fact, he used to define himself—of course, with a certain bitterness—as an 'old professor to whom youth has listened to for such a long time and with such goodwill and who still works for the young in the quiet of his study, still seeking to be of use' (14). He deeply left the loss of his professorial chairs and regarded it as a loss of personal standing. For him personally, teaching mathematics was a veritable joy and, in a sense, his own element and his life (15). He had placed a great deal of hope in the École Normale Écclésiastique in 1839, just as he had done with regard to the Institut Catholique in 1842. This had happened again, and particularly, so, in 1843, when he attempted to gain a chair at the Collège de France. After his defeat, he became very dejected. Political requirements had definitely thrown up an insurmountable barrier between him and those he desired to have as students.

The sheer number and importance of the papers Cauchy published between 1838 and 1848 prohibit giving a detailed critical analysis here; this body of work could easily be the subject of one or even several studies in its own right. Hence, an analysis of Cauchy's methods of work and of the interplay of the influences to which he was subjected between 1838 and 1848 will be given. By using this approach, a certain order into this rather considerable body of written material will be introduced.

The first period lasted from October 1838 until December 1843. When he returned to France, Cauchy expected to resume his previous place in the

French scientific community, and his entire scientific production was oriented toward that goal. On the one hand, he sincerely wanted to publicize the contents of the works he had undertaken during the time he was in Turin and Prague. This work included investigations on series expansions of functions and the calculus of limits and its applications to the theory of light, as well as other topics in analysis that were developed during that time. On the other hand, he hoped to show by his studies in higher analysis and particularly in mathematical astronomy that he deserved a position at a scientific institution. The problem first arose in 1839 with his nomination to the Bureau des Longitudes. However, as the years passed, Cauchy permitted himself to be increasingly guided in his choice of research topics by the events of the day; this is especially true as concerns the subjects that were examined in studies by the young scholars that he evaluated.

Cauchy's double rejection, first at the Collège de France and then at the Bureau des Longitudes, was followed by an entire year in which his scientific production was reduced. From December 1843 until December 1844, Cauchy submitted relatively few studies to the Académie, publishing just three issues of *Exercices* in which he examined certain questions from classical analysis, infinitesimal analysis, and the calculus of variations. Moreover, he rarely participated on academic committees until November 1844 (16). He was obviously greatly affected by the quarantine that had been imposed on him. Beginning in December 1844, he resumed his normal rhythm of publishing. Now less preoccupied by questions relating to his career, he allowed academic life to dictate his choice of research to a large extent. Significantly enough, he became interested in probing his earlier works more deeply; thus, in 1845–1846, he examined the theory of permutations, while in 1846–1848, he worked on the theory of functions of a complex variable. This latter undertaking quickly became a very important area of interest to young mathematicians of the time.

During the first few months after his return to Paris, Cauchy publicized the work on light that he had undertaken while in exile and, as we have seen, had remained unpublished except for one part. Thus, until August 1839, he communicated to the Académie studies almost exclusively on mathematical optics and on the methods used in mathematical physics. In the spring of 1839, he published a collection of these papers at the publishing house Bachelier (17). Although he continued to submit papers on these subjects to the Académie during the following years, he did so at an increasingly reduced pace. Paralleling this development, he devoted the greater part of the first volume of the *Exercices d'Analyse et de Physique Mathématique* between September 1839 and June 1841 to problems that had to do with the theory of light (18).

In several studies, Cauchy resumed his research work that he had taken up in Prague on infinitely small motions in a homogeneous system of molecules, especially on the wave propagation initiated by a perturbation at a given point of the system. He considered the simple displacements, of molecules corresponding to the real parts of the plane-wave solutions,

$$c_i e^{ux+vy+wz-st} \qquad (c_i, u, v, w \text{ and } s \text{ are complex constants}),$$

of the fundamental equation of molecular physics:

$$\partial_0^2 u_i = \sum_{\mathrm{p}} m_{\mathrm{p}} \left(f(r_{\mathrm{p}}) \frac{\Delta_{\mathrm{p}} u_i}{r_{\mathrm{p}}} + (r_{\mathrm{p}} f'(r_{\mathrm{p}}) - f(r_{\mathrm{p}})) \mu_{\mathrm{p}i} \mu_{\mathrm{p}j} \frac{\Delta_{\mathrm{p}} u_j}{r_{\mathrm{p}}} \right). \qquad (6.13)$$

All the small motions of the molecules can be obtained by either finite or infinite superpositions of these simple displacements. Cauchy also investigated the motions in two mutually penetrating systems of molecules, in the hope of explaining the interactions between the light and the matter (19).

Cauchy focused special attention on the theory of reflection and refraction that he had developed during 1836 and 1837, while he was still in exile. He applied a general method for determining the conditions on the boundaries of bodies to the special case of light by attempting to deduce these conditions from Fresnel's and Brewster's laws. Cauchy maintained his confidence in the theory that he had developed in Prague. He assumed the existence of relations between the elasticities of the ethereal substance such that the vibration of the molecules were perpendicular to the plane of polarization. However, at the end of 1839, he proposed a new theory: he assumed that the speed of propagation of the longitudinal vibrations was reduced to zero (20). In this way, he managed to free optics from a constraining longitudinal wave that he, for want of a better idea, had momentarily tried to identify with a heat wave (21). But, this simplification could be carried out only by introducing some additional new relations between the elasticities of the ether, a substance that had a negative compressibility.

Cauchy did not probe his theory closely. Moreover, he increasingly abandoned optics and focused on general methods in mathematical physics. Still, toward the end of 1842, he once again took up the problem of diffraction, which he had already studied in 1837. Biot's very harsh article in the *Journal des Savants* made Cauchy drop his communications. It is true that he had to admit defeat over his research on rotary polarization. Cauchy presented no more communication on optics until the 1848 Revolution (22). The theory of light, which, since 1838, had occupied first place in his research interests and efforts, was now abandoned. Thus, the difficulties inherent in Cauchy's approach to the nature of light sprang from his own stubbornness and zeal. The biting comments from a colleague put an end to it (23).

Meanwhile, Cauchy continued to improve his developments on the applications of analysis to mathematical physics. Specifically, the question that had to be dealt with here was that of solving the linear ordinary differential equations and linear partial differential equations with constants coefficients, which occur so often and in such important contexts in mathematical physics.

In May 1839, Cauchy presented his great study 'Sur l'intégration des équations linéaires' to the Académie (24). In this work, he developed the method for integrating the linear ordinary and partial differential systems with constant coefficients that he had created during 1820. He considered first the first-order homogeneous differential system

$$\frac{dx_i}{dt} = a_{ij}x_j \qquad (1 \leqslant i \leqslant n, 1 \leqslant j \leqslant n). \tag{12.1}$$

Let $S(s)$ be the characteristic polynomial det $((a_{ij}) - s)$. If $\boldsymbol{\alpha}$ is the arbitrary vector (α_i) $(1 \leqslant i \leqslant n)$ and $\mathbf{V}(s, \boldsymbol{\alpha}) = (V_i(s, \boldsymbol{\alpha}))$ $(1 \leqslant i \leqslant n)$ the linear mapping $S(s)\ ((a_{ij}) - s)^{-1}\ (\boldsymbol{\alpha})$, Cauchy showed that the functions

$$x_i(t) = \mathcal{E}\frac{V_i(s, \boldsymbol{\alpha})e^{st}}{((S(s)))} \qquad (1 \leqslant i \leqslant n) \tag{12.2}$$

are the solutions of Eq. (12.1) satisfying the initial conditions $x_i(0) = \alpha_i$ $(1 \leqslant i \leqslant n)$. Then, he introduced the function

$$\Theta(t) = \mathcal{E}\frac{e^{st}}{((S(s)))},$$

which he called the principal function of the system. The principal function is the solution of the differential equation of order n,

$$S\left(\frac{d}{dt}\right)\Theta = 0,$$

satisfying the initial conditions

$$\Theta(0) = 0, \frac{d}{dt}\Theta(0) = 0, \ldots, \frac{d^{n-2}}{dt^{n-2}}\Theta(0) = 0, \frac{d^{n-1}}{dt^{n-1}}\Theta(0) = 1.$$

With the principal function, Eq. (12.2) can be written in the following form:

$$x_i(t) = V_i\left(\frac{d}{dt}, \boldsymbol{\alpha}\right)\Theta(t) \qquad (1 \leqslant i \leqslant n).$$

Cauchy extended the use of the principal function to linear differential systems with second members of any order. Then, he applied this method to the integration of certain linear partial differential systems with constant coefficients. For instance, let the system be

$$F_i\left(\frac{\partial}{\partial x}, \frac{\partial}{\partial y}, \frac{\partial}{\partial z}, \ldots, \frac{\partial}{\partial z}\right)(\xi_1, \xi_2, \ldots, \xi_p) = f_i(x, y, z, \ldots, t), \tag{12.3}$$

of n_jth order relative to t for ξ_j and satisfying the initial conditions

$$\xi_j(x, y, z, \ldots, 0) = \phi_j(x, y, z, \ldots), \qquad \frac{\partial}{\partial t}\xi_j(x, y, z, \ldots, 0) = \chi_j(x, y, z, \ldots), \ldots,$$

$$\frac{\partial^{n_j - 1}}{\partial t^{n_j - 1}}\xi_j(x, y, z, \ldots, 0) = \psi_j(x, y, z, \ldots) \qquad (1 \leqslant j \leqslant p).$$

Cauchy transformed Eq. (12.3) by Fourier transforms into

$$\int_{-\infty}^{+\infty}\int_{-\infty}^{+\infty}\int_{-\infty}^{+\infty}\cdots\left[F_i\left(u, v, w, \ldots, \frac{\partial}{\partial t}\right)(\xi_1(\lambda, \mu, \nu, \ldots, t), \ldots, \xi_p(\lambda, \mu, \nu, \ldots, t)) - f_i(\lambda, \mu, \nu, \ldots, t)\right] \times e^{u(x-\lambda)+v(y-\mu)+w(z-\nu)+\cdots}\frac{d\lambda\, du}{2\pi}\frac{d\mu\, dv}{2\pi}\frac{d\nu\, dw}{2\pi}\cdots = 0,$$

where $u = \mathrm{u}\sqrt{-1}$, $v = \mathrm{v}\sqrt{-1}$, $w = \mathrm{w}\sqrt{-1}, \ldots$ and thus reduced the integration of Eq. (12.3) to evaluating the integral

$$\xi_j(x, y, z, \ldots, t) = \int_{-\infty}^{+\infty}\int_{-\infty}^{+\infty}\int_{-\infty}^{+\infty}\cdots\bar{\xi}_j(\lambda, \mu, \nu, \ldots, t)e^{u(x-\lambda)+v(y-\mu)+w(z-\nu)+\cdots}\frac{d\lambda\, du}{2\pi}\frac{d\mu\, dv}{2\pi}\frac{d\nu\, dw}{2\pi}\cdots \qquad (1 \leqslant j \leqslant n)$$

where $\bar{\xi}_j$ are the solutions of the linear ordinary differential system

$$F_i\left(u, v, w, \ldots, \frac{\partial}{\partial t}\right)(\bar{\xi}_1(\lambda, \mu, \nu, \ldots, t), \ldots, \bar{\xi}_p(\lambda, \mu, \nu, \ldots, t)) - f_i(\lambda, \mu, \nu, \ldots, t) = 0 \tag{12.4}$$

satisfying the initial conditions

$$\bar{\xi}_j(\lambda, \mu, \nu, \ldots, 0) = \phi_j(\lambda, \mu, \nu, \ldots), \qquad \frac{\partial}{\partial t}\bar{\xi}_j(\lambda, \mu, \nu, \ldots, 0) = x_j(\lambda, \mu, \nu, \ldots), \ldots,$$

$$\frac{\partial^{n_j-1}}{\partial t^{n_j-1}}\bar{\xi}_j(\lambda, \mu, \nu, \ldots, 0) = \psi_j(\lambda, \mu, \nu, \ldots) \qquad (1 \leqslant j \leqslant p).$$

In the case of an homogeneous system, Eq. (12.3) with $f_i = 0$, of the first order relative to t ($n_j = 1$, for all j), where the coefficients of $\partial_t \xi_j$ are constants, Eq. (12.4) is a first-order linear system that can be solved by using its principal function $\Theta(t)$. Cauchy skillfully used this method and reduced the investigation of the linear partial differential system to the determination of the function

$$\Omega(x, y, z, \ldots, t) = \mathcal{E}\int_{-\infty}^{+\infty}\int_{-\infty}^{+\infty}\int_{-\infty}^{+\infty}\cdots\frac{e^{u(x-\lambda)+v(y-\mu)+w(z-\nu)+\cdots}}{((S(s)))}\omega(\lambda, \mu, \nu, \ldots) \times \frac{d\lambda\, du}{2\pi}\frac{d\mu\, dv}{2\pi}\frac{d\nu\, dw}{2\pi}\cdots,$$

which satisfies for $t = 0$ the conditions

$$\Omega = 0, \frac{\partial}{\partial t}\Omega = 0, \ldots, \frac{\partial^{n-1}}{\partial t^{n-1}}\Omega = 0, \frac{\partial^n}{\partial t^n}\Omega = \omega(x, y, z, \ldots).$$

The function $\Omega(x, y, z, \ldots, t)$, which generalizes the function $\Theta(t)$, was called by

Cauchy the principal function of the homogeneous linear partial differential system.

Following this study, Cauchy devoted several studies to the case in which it is possible either to simplify the principal function or reduce the degree of the characteristic equation $S(s) = 0$. In particular, he considered the case in which the characteristic equation is homogeneous. In this situation which he had considered in 1830, the principal function would reduce to a quadruple integral; and, if the characteristic equation should be of order two, then it would reduce to a double integral (25).

This last type of equation, the homogeneous equation, is of great importance in mathematical physics. In fact, in 1830, Cauchy showed this type of equation allowed representation of the propagation of a wave in a system of molecules. The mathematician P.-H. Blanchet, whom Cauchy had cited in his 1830 paper, 'Sur la théorie de la lumière', and who in 1838 had presented two studies on the propagation and polarization of motion in elastic media to the Académie (26), had, on Liouville's urgings, undertaken new investigations on the equation of the propagation of light.

On June 21, 1841, and on July 5, 1841, he presented two studies that were to be evaluated by a committee consisting of Cauchy, Sturm, Liouville, and Duhamel. In these two works, Blanchet determined the external boundaries of the wave fronts corresponding to solutions of a given homogenous equation by using the calculus of residues.

Cauchy immediately set about publishing weekly in the *Comptes Rendus* studies and papers on principal functions satisfying homogeneous equations and on the reduction of such functions. He also investigated characteristic surfaces, as well as the corresponding wave surfaces (27). He reproved certain results that had been obtained by Blanchet and published them right away in the *Comptes Rendus*, while all the evaluations of Blanchet's work have not even today been published. However, on August 9, 1841, Blanchet presented a note in which he showed that the results that Cauchy had obtained were equivalent to the results that he had obtained, and he was able to publish this note in the *Comptes Rendus* (28).

Even afterward Cauchy continued to publish papers on the same subject at a steady pace, and Blanchet once again requested permission from the Académie to publish a note in the *Comptes Rendus* on November 15, 1841. This time he spoke out against an incorrect statement that Cauchy had made (29). During the course of the meeting at the Académie, Cauchy recognized that he had, in effect, encountered a difficulty in his research. Thus, in a note published shortly thereafter, he abandoned his earlier statement without really disowning it. The roles of Cauchy and Blanchet were now completely reversed; Blanchet, now evaluating his former examiner, put an end to the debate by publishing a letter in the *Comptes Rendus* on December 20, 1841 (30). Furthermore, on March 14, 1842, Cauchy, speaking on behalf of the evaluating committee, issued a report that praised Blanchet's studies and to which he appended a note on his own regard.

On the whole, while recognizing the merits of the studies that had been submitted to him, Cauchy once more showed himself to be manifestly clumsy and lacking in tact; for he had, in effect, submerged Blanchet's works—before he had even evaluated and reported on them—under a mass of his own studies that dealt with the same questions as the works that had been submitted to him for evaluation. Cauchy habitually went about things in this way; later, we will see some more instances. The point of the story is that this time it put him in a difficult, uncomfortable position.

Following his election to the Bureau des Longitudes in 1839, Cauchy took an interest in theoretical astronomy, an interest that paralleled his research in mathematical physics.

We have already seen that in 1831 Cauchy had presented his Turin theorem on series in the context of celestial mechanics. It is therefore not surprising that he had quickly published this result in the *Comptes Rendus* in August 1839, while he was seeking election to the Bureau des Longitudes. This first proof, which was, in fact, identical to that of 1831, was followed by a second in 1840; the 1840 proof was based on the consideration of the mean value of a function on a circle (31).

The mean value relative to r of a function $\omega(z)$, which is finite and continuous with its derivative in the ring $r_0 \leqslant r \leqslant R$ is the value of the expression

$$\lim_{n \to +\infty} \frac{\omega(r) + \omega(\theta r) + \cdots + \omega(\theta^{n-1} r)}{n}, \tag{12.5}$$

where θ is an nth root of unity. By using the mean-value theorem, Cauchy proved that this expression remain constant inside the ring $r_0 \leqslant r \leqslant R$. Then, he stated that any finite and continuous function $f(u)$, with its derivative in the disc $0 \leqslant r \leqslant R$, can be represented by the mean value of $\omega(z) = (z/z - u)f(z)$; relative to r if $|u| < r < R$. Finally, by substituting the series expansion $\Sigma f(z)(u/z)^n$ for $\omega(z)$ in Eq. (12.5), he obtained a series expansion of $f(u)$ in the disc $0 \leqslant |u| < R$.

Over the following years, Cauchy often had occasion to recall the theorem of Turin. He had some difficulties in stating conditions for the applications of his result, because he had not really clarified the concept of an analytic function of a complex variable. In 1831, he had merely assumed such a function to be finite and continuous but, from 1839 on, he added the additional hypothesis of a continuous derivative, which he used in the second proof. Then, in December 1844, on Liouville's advice, he dropped this supplementary condition, without clearly justifying this new choice (32).

The Turin theorem could obviously be put to good use in mathematical astronomy in the study of the series expansions of certain functions, such as the perturbation function. In 1839, thanks to his calculus of limits, Cauchy was able to deduce the first really rigorous proof of the implicit function theorem, in the case of analytic functions (33). However, it was the problem of applications to the theory of differential equations that now absorbed

Cauchy's attention. In December 1840, he published his Prague study 'Sur l'intégration des équations différentielles' in *Exercices d'Analyse et de Physique Mathématique.* At various times, he would return to his method of solution by series. In particular, during the summer of 1842, he presented several important studies in which he simplified his method of finding series solutions, and for the first time, he applied his calculus of limits to systems of linear partial differential equations. Thus, he obtained the first rigorously proved existence theorem for such equations (Cauchy-Kovalevskaya theorem) (34).

Laurent's paper of August 21, 1843, 'Extension du théorème de M. Cauchy relatif à la convergence du développement d'une fonction suivant les puissances ascendantes de la variable', on which Cauchy wrote a favorable report on October 30, 1843, and to which he added a note with his own ideas—extended the applications of the Turin theorem (35). Laurent showed that it is possible to expand a function $f(z)$ in a series

$$\sum_{-\infty}^{+\infty} a_n z^n,$$

which today is known as a Laurent series, in the neighborhood of an isolated singular point. This was the first really important result that had been developed in the theory of the functions of a complex variable by any mathematician except Cauchy.

Paralleling the development of these very general mathematical theories, which were applicable to celestial mechanics, Cauchy developed new techniques for the calculations then used in mathematical astronomy. He focused special attention on the problem of determining the series expansion of the perturbation function, a topic he had earlier examined in his Turin paper of October 11, 1831. In several studies that he presented to the Académie in September and October 1840, he improved Liouville's method of 1836, which consisted of substituting simple approximate integrals for the double integrals representing the coefficients in the series expansion of the perturbation function (36).

Several weeks later, Cauchy sat on the committee that was charged with the responsibility of evaluating Leverrier's study 'Sur le développement de la fonction perturbatrice.' No doubt, on this occasion, he established a rapport with the young astronomer who was trying to calculate a large inequality in the Pallas motion and who had sought Cauchy's advice. Cauchy wrote:

> Responding to the requests made by this young scholar and scientist, I wrote my research results on a problem with a solution that might very well spare astronomers a great deal of tiresome labor and an equal amount of vexations' (37)

On August 9, 1841, Cauchy presented his method of calculating inequalities to the Académie. This was the same day on which Leverrier presented his study on the Pallas motion to the Académie. One year later, Cauchy came back to examine the problem of calculating the perturbation function, this time using the calculus of residues. This investigation resulted in a new theory of planetary motion that used an (unpublished) expansion of the perturbation function, in which the coefficients could easier be calculated (38).

The cancellation of his election to the Bureau des Longitudes did not dampen Cauchy's efforts. In April, July, and December of 1844, he presented new studies on the perturbation function and the calculation of planetary inequalities, and in March 1845, he submitted a report on a study by Leverrier, 'Sur l'inégalité du moyen mouvement de Pallas', which was essentially a follow-up on the 1841 study mentioned earlier. Here, the young astronomer calculated the great inequality describing the influence of Jupiter on the minor planet. Leverrier used interpolation formulas that required certain very long calculations. By using methods that he had developed during the preceding months, Cauchy easily simplified the calculational problems that had confronted Leverrier. Thus, he confirmed the correctness of Leverrier's calculations in two different ways. Cauchy's report and the notes that he added, and in which Cauchy, as was his habit, literally smothered the work that he was supposed to evaluate, made as much of an impression as Leverrier's study (39). Leverrier, encouraged by the success of his work on Pallas motion, immediately set about investigating Uranus's unexplained perturbations. This research effort ended in the discovery of a new planet, Neptune, first by calculations and then by observations in September 1846. As for Cauchy, he continued to present studies on celestial mechanics at regular intervals during the following years.

For a while, following his defeat at the Collège de France and his ousting from the Bureau des Longitudes, Cauchy slowed the pace of his mathematical production. He was exhausted by the part he had played on the Jesuits' side in the struggle against the Université's monopoly. After 1845, when he resumed publishing at his usual rate, his research was more oriented toward abstract areas and less toward the applied fields that had so long been the focus of his attention. Thus, he began to follow his own inclinations and, allowing himself to move with the tide of academic influences, he returned to investigate problems and questions that he had long neglected.

Cauchy resumed his research on the theory of permutations, an area that, after his first investigations in 1812, he had all but abandoned. In September 1845 and April 1846, he presented a long series of notes and studies on permutations to the Académie; these writings were soon assembled into a major work, 'Sur les arrangements...,' which was published in December 1845 and in March and April 1846 in three issues of *Exercices d'Analyse et de Physique Mathématique* (40). It was clearly the evaluation of Bertrand's study 'Sur le nombre des valeurs que peut prendre une fonction quand on y permute les lettres qu'elle renferme', which was presented to the Académie on March 17, 1845, and for which Cauchy was to serve as reporter, that prompted research in this area. In his study, Bertrand used a certain postulate on prime numbers to prove that any function of n letters assume at least n values (41). Cauchy sought to establish this same result without using Bertrand's postulate.

As usual, Cauchy did not wait to submit his evaluation of Bertrand's study until he was able to communicate his research in this area. He gave his report on Bertrand's paper on November 10, 1845, but went considerably beyond the framework of Bertrand's theorem in his own work. With great virtuosity, he set forth a subtle calculus on 'systems of conjugate permutations,' that is, on the subgroups of permutations that were as yet imperfectly defined. He obtained some powerful results, such as the theorem known today as Cauchy's theorem: if p is a prime divisor of the order of a finite group, there is an element of the group whose order is p (42).

It might be asked whether Bertrand's work by itself could have motivated Cauchy's important research on permutations during 1845–1846. About the same time, on September 4, 1843, Liouville announced that he was about to publish Galois' papers. It can hardly be assumed that Cauchy was not aware of the difficulties that were confronting Liouville relative to his publisher's tasks on Galois' works (43). Liouville and Cauchy were on good terms at the Académie and had frequent chances to exchange ideas. Moreover, Cauchy referred twice to Charles Hermite, a young protégé of Liouville who at this time knew of Galois' works (44).

Nevertheless, Cauchy's approach to the theory of permutations was quite different from Galois' approach. Cauchy, who was not at all interested in the problem of the solvability of equations by radicals, developed a formal calculus in which structural properties were scarcely clarified. Unlike Galois, Cauchy remained a prisoner of his calculation techniques. If, indeed, Galois' papers had prompted Cauchy to reexamine the theory of permutations, an area that he had not bothered with for 33 years, then it would certainly seem that Cauchy did not have a precise knowledge of their contents (45).

Starting in mid-1846, Cauchy's work was dominated by the construction of a theory of the functions of a complex variable. He made fundamental progress in the development of the theory of imaginary integrals, which he had created in 1825 and which he used in 1831 to provide his theorem on the infinite series expansions of functions and in determining the number and the nature of the roots of an equation within a given contour. Cauchy had neglected this theory for his work on applications, the calculus of residues, and the calculus of limits; this neglect, as we have seen, had led to certain difficulties in defining the class of functions to which these theorems applied. As was frequently the case with Cauchy, an outside force was needed to prompt him to reexamine the theoretical framework in which he was operating. It seems that an article by Anatole Lamarle, which appeared in the April 1846 issue of Liouville's *Journal de Mathématiques Pures et Appliquées*, actually triggered these new research efforts. In this article, Lamarle discussed the hypotheses of the Turin theorem; like Liouville, he affirmed that the continuity of the derivative of the function to be expanded was indeed a superfluous condition and that it was only necessary to assume 'a certain periodicity for the function', that is, that the function takes on the same value each time its argument is increased by 2π (46).

Cauchy replied with a long article in the August 1846 issue, in which he took up the question of multiform functions (47). Stressing the principle that 'it is obviously useful to adopt conventions that preserve the property of continuity for the functions used in this calculus for the longest possible time' (48), he extended the definition of imaginary logarithms—which had been defined in 1821 in his *Cours d'Analyse* only in the plane $R > 0$—for the whole complex plane with a slit, along the negative real axis: if $z = re^{(p+2k\pi)\sqrt{-1}}$ $(-\pi < p < +\pi)$, then $\log z = \log r + p\sqrt{-1}$. He thus took up an idea of Lamarle (who, in his article, had used the notion of a logarithm defined in the complex plane with a slit along the positive real axis to prove the Turin theorem) by modifying the domain of the definition in such a way as to keep the logarithmic function continuous in the neighborhood of positive real values.

At the same time that he was replying to Lamarle, Cauchy began to publish a series of papers in the *Comptes Rendus* in which he moved matters much further along. In these papers, he arrived at the theory of imaginary integrals by using a completely geometric point of view. In a first note, presented on August 3, 1846, to the Académie, he gave some general theorems on curvilinear integrals (S) taken along the oriented border s of a surface S (49). First, Cauchy supposed that S is changing from S_1 to S_2 by a continuous deformation. If the differential $X(x, y, z, \dots)\,dx + Y(x, y, z, \dots)\,dy + Z(x, y, z, \dots)\,dz + \dots$ is exact, he stated that the integral (S) of the function $k = X(x, y, z, \dots)\dfrac{\partial x}{\partial s} + Y(x, y, z, \dots)\dfrac{\partial z}{\partial s}$ $+ Z(x, y, z, \dots)\dfrac{\partial z}{\partial s} + \dots$ remains constant as long as k remains finite and continuous when s is moving continuously from the border of S_1 to the border of S_2 (i.e., when the borders are homotopic). If one of the functions, X, Y, Z, etc., fails to be finite and continuous at some points P', P'', P''' of S, then we have the formula

$$(S) = (a) + (b) + (c) + \dots \tag{12.6}$$

where $a, b, c, \dots$ are elementary surfaces around P', P'', $P''' \dots$, and (a), (b), $(c) \dots$ are singular integrals taken along the border of these surfaces. If these functions are finite and continuous at each point of S, then

$$(S) = 0. \tag{12.7}$$

In the case of the exact differential $X(x, y)\,dx + Y(x, y)\,dy$, Cauchy found this last theorem again by means of Green's formula:

$$\int_s X\frac{\partial x}{\partial s} + Y\frac{\partial y}{\partial s}\,ds = \iint_S \frac{\partial X}{\partial y} - \frac{\partial y}{\partial x}\,dx\,dy.$$

Cauchy applied Eqs. (12.6) and (12.7) to the theory of complex functions in a study of September 21, 1846. In this paper, he associated the complex variable $z = x + y\sqrt{-1}$ to the mobile point P in the plane whose coordinates are x and y. In order to prove the residue theorem, he used the same procedure that he had used in 1825 and 1826 (50). First, he established that

$$(S) = 2\pi\sqrt{-1} \tag{12.8}$$

for $k = \frac{1}{z - z_0} \cdot \frac{\partial z}{\partial s}$, if the mobile point Q relative to z_0 is inside the surface S. On the other hand, if $f(z)$ has the single pole Q inside S, $(S) = 0$ for

$$k = \left(f(z) - \mathcal{E} \frac{((f(u)))}{u - z} \right) \frac{\partial z}{\partial s}.$$

Consequently, this result combined with Eq. (12.8) gives

$$(S) = 2\pi\sqrt{-1}\,\mathcal{E}((f(z))) \qquad \text{for} \qquad k = f(z)\frac{\partial z}{\partial s}.$$

This last formula, which can be extended to functions with several poles, is the residue theorem.

Cauchy took an interest in other 'multiform functions' as well as in abelian integrals and, in particular, in hyperelliptic and elliptic integrals. In 1843, he had presented a series of notes on elliptic functions in which his point of departure was not the inversion of elliptic integrals but simple infinite products, which he called 'geometric factorials' (51). He also knew that Liouville was developing a theory of doubly periodic functions. On December 9, 1844, Liouville had stated, but did not prove, a very general theorem according to which a doubly periodic function that is bounded in the entire complex plane is constant. At the very next meeting, Cauchy gave the first proof of this result, which is known today as Liouville's theorem. Using the calculus of residues, Cauchy extended this theorem to any bounded continuous function of a complex variable (52).

In 1846, Cauchy's idea for dealing with multivalued functions was very simple (53): he considered the different ways in which the imaginary variable could pass continuously from one value to another. The value of an integral thus depended not only on the endpoints but also on the path selected to connect them. If the function $f(z)$ to be integrated is uniform, that is, if it takes the same value for each value of the variable, then the integral along a closed path depends on the number of revolutions of the path around the poles of the function. If $f(z)$ has only one pole and if the closed path turns n times around this pole, the value of the integral along this path is nI, where I is the residue of the function at the pole multiplied by $2\pi\sqrt{-1}$ (the residue theorem). Let the differential equation

$$dt = f(z)\,dz \tag{12.9}$$

satisfy the initial conditions $t = 0, z = a$ and suppose that $f(z)$ is uniform, finite, and continuous except in some poles C, C', C'', etc. Then, $t = \int_a^z f(u)\,du$ is a multivalued function—the complete integral of the differential equation, said Cauchy—that depends on the number of revolutions that makes the path of integration around the poles. For instance, for $z = a$, the paths of integration are closed paths and the values of t are contained in the formula

$$nI + nI' + n''I'' + \ldots, \tag{12.10}$$

where n, n', n'', etc. are any integers and I, I', I'', etc. are the residues of f at the poles C, C', C'', etc., multiplied by $2\pi\sqrt{-1}$. The constants $I, I', I'', \ldots$ are called by Cauchy the indices of periodicity of the path. By means of Eq. (12.10), it is possible to reduce the determination of the complete integral of the differential equation, Eq. (12.9), to rectilinear integrations, that is, integrations along rectilinear paths.

Thereafter, Cauchy began to study the case of a function $u(z)$ that is multivalued, for instance, if u is an algebraic function defined by $F(u, z) = 0$. The decisive idea, which was inspired by his research on logarithms, was to derive the value of the function on one point from its value on another point by joining the two points by a continuous path. So, he showed that the periodicity of the inverse function of $F(z) = \int_0^z f(u)\,du$ can be explained if one supposes that $f(z)$ takes the same value after a certain number of revolutions along a closed path. But, Cauchy restricted himself to general considerations. He did not investigate how the values of $F(z)$ interchange, a task that he left to his disciple Victor Puiseux (for the case of algebraic functions) (54).

Curiously enough, the geometric interpretation that Cauchy systematically used in these notes in 1846 did not mean that he had abandoned his symbolic theory of imaginaries. It was, in his opinion, just a matter of his simplifying and clarifying the exposition of the theory, of describing how this calculus actually goes (55).

During this period, he presented a paper entitled 'Mémoire sur les fonctions de variable imaginaire' in which he continued to adhere to his symbolic theory of imaginary expressions as presented in his 1821 publication *Analyse Algébrique*, while still allowing the possibility of representing an imaginary number as a variable point in the plane (56). This timorous attitude could at best be a temporary, provisional position, and Cauchy soon replaced his symbolic theory by two new theories of imaginaries, one in 1847 and the other in 1849.

The first of these two theories, the theory of algebraic equivalencies, was presented to the Académie in June 1847 (57). In this development, imaginary numbers were regarded as equivalence classes of polynomials with real coefficients modulo $(x^2 + 1)$ (58). Cauchy replaced the symbolic sign $\sqrt{-1}$ by the letter i (already used by Euler and Gauss), which he introduced into the polynomials as an unknown. Without giving any theoretical justification, he extended his theory of algebraic equivalencies modulo $(x^2 + 1)$ to convergent series in order to define imaginary exponentials (59).

Cauchy presented this theory following his research on Fermat's last theorem. This research was begun in March 1847. At the Académie's meeting of March 1, 1847, Lamé announced that he had found a proof of this theorem based on a complete factorization of $x^n + y^n$ into n relatively prime linear factors containing complex numbers (60). Liouville immediately expressed his scepticism by observing, in particular, that Lamé's proof assumed, but did not show, the uniqueness of the decomposition of $x^n + y^n$ into n relatively prime complex factors. Cauchy, however, seemed to have been persuaded that Lamé had achieved a fundamental result and quickly began his own investigations. The problem he faced was that of putting Lamé's ideas into better form and clearing up certain technical difficulties. On March 15, Wantzel stated that he had established the uniqueness of the factorization of $x^n + y^n$ into relatively prime complex factors (61). There followed a veritable race between Cauchy and Lamé, each trying to be the first to obtain a complete proof of Fermat's theorem. On March 17, both mathematicians registered sealed envelopes containing the main features of their respective proofs, the customary procedure in case a question of priority should arise later (62). The following meetings of the Académie were taken up by their discussions and communications with each other (63). In this connection, Cauchy investigated the theory of 'radical polynomials', that is, cyclotomic polynomials $a_0 + a_1\rho + a_2\rho_2 + \cdots + a_{n-1}\rho^{n-1}$ (where ρ is an *nth* root of unity and the coefficients are integers), without achieving any precise results.

At the next meeting, on May 17, Liouville read a letter from Kummer in which he stated that three years ago he had shown that factorization into relative prime complex factors was not unique (64). This development dealt a death blow to the hopes that had been generated by Lamé's paper of March 1. In any event, while Lamé recognized his error and kept quiet about the matter, Cauchy continued to present regularly notes on radical polynomials and their

Sealed enveloped containing the main features of the proof of Fermat's theorem by Cauchy, March 17, 1847. The text is written in Italian with Greek letters. Académie des Sciences de Paris. Published by permission of the Académie de Sciences.

decomposition until August (65). Relative to Kummer's investigations, Cauchy noted in a paper that he presented on May 24, that:

> At the last meeting, M. Liouville spoke of M. Kummer's works on complex polynomials. From the little bit that Liouville said on the matter, I was persuaded that the conclusions reached by M. Kummer are, at least in part, the same as the ones to which I have been led by the preceding considerations. If M. Kummer had investigated the question a bit further, even if he had been able to raise all the questions and difficulties inherent in the problem, I should have been the first to congratulate him on his success, because what we all really want more than anything else is that the works of all friends of science should join in determining the truth and making it known (66).

During the following meetings of the Académie, Cauchy expanded his work on the theory of radical polynomials of ideal complex numbers, without, however, bothering with Fermat's theorem at all (67). In June 1847, he applied Kummer's theory of algebraic equivalencies to the theory of complex numbers (68). This return to the theory of complex numbers clearly showed that Cauchy was no longer directly interested in Fermat's theorem.

Spread in all directions and presented in a multitude of notes, papers, studies, and reports, Cauchy's work between 1838 and 1848 has suffered from its very mode of publication. In 1857, Biot, expressing the view that Cauchy would have been one of the 'leading lights of the mathematical sciences' had his life, like the lives of Euler and Lagrange, somehow been able to flow along without trouble, observed:

> By dint of the instability and disorder that events stamped on his personality and thinking, the influence that he exerted on them [that is, on the mathematical sciences] will not be completely felt until sufficient time has passed to evaluate all the consequences [of his scientific works]. Young geometers who have the courage to read his works in detail and with care will find them to be a mine of ideas, with rich veins of discoveries and insights to follow through on and to bring up to date (69).

Today, Cauchy's natural intellectual prolixity, encouraged by circumstances, allows the historian of the sciences to follow the development of mathematical ideas on an almost week-by-week basis.

Chapter 13

Practices and Principles in Cauchy's Works

In this chapter, we examine how, relative to the theme of rigor, principles and practices were articulated in Cauchy's works. The age of Cauchy was a historical period in which scientific practices asserted an increasing degree of autonomy vis à vis philosophical reflections. If it is true, on the one hand, that the philosophers of the first half of the 19th century were less and less interested in the development of the sciences, with Auguste Comte as a notable exception, then it is also true that scholars—particularly mathematicians—increasingly insisted that they and they alone had the right to determine the content of their discipline and its standards of scientificality. The divorce between a particular scientific practice and general philosophical reflection, in a discipline whose foundations were quite uncertain at the time, gave rise to a blossoming of what might be termed local and spontaneous philosophies among the scholars and scientists who had dedicated themselves to this discipline. While these scholars' philosophies were incomplete, approximate, and often anomalous, they nevertheless worked to the extent that they could justify current practices and thus mitigate existing theoretical shortcomings. Thus, the principle of rigor in mathematics seems to be an excellent example of what we earlier referred to as local and spontaneous philosophies.

Although Cauchy was not the only mathematician at the beginning of the 19th century to call for rigor in mathematics, he is justly regarded as one of the main initiators and supporters of the movement to attain that goal. When Cauchy embarked on his career, France's scientific institutions, which had been developed, for the most part, during the era of the Great Revolution and First Empire, were beginning to assume a definitive form, a shape that they would retain almost unchanged throughout the century. In particular, the professionalization of mathematical activity, together with the gradual appearance of a standard curriculum within this framework, could not remain without consequences: little by little both the state and the content of the mathematical sciences changed.

Certainly, mathematics was a great deal more than a means of making a living for Cauchy: it was a reason for living, as well as a vocation. Working, he resembled a monk cloistered away in his study:

> His tastes and ways were simple and peaceable in the extreme. His study was small and modestly furnished. Most often, he did his writing at a small desk—a table, really, without pigeon holes—which, in the evenings and at night, was lighted by two simple wax candles with shades. Everything, whether in this small room or in his library, was in perfect order. He never went out without raising the wood in the fireplace and extinguishing the fire. He never wanted anything for himself, being oblivious to himself, as it were, and having no desires to satisfy (1).

From time to time, in frustration, he would utter a complaint:

> There is no let up! No end to it! Accursed problems! Innumerable calculations. Endless fighting. Signs. Formulas. Theorems besetting me from dawn to dusk (2)!

Far from living from his work as a mathematician, Cauchy used a considerable amount of money—especially during the 16 years that he was kept on the fringes of the scientific community—to guarantee the publication of his works. Institutional practices, both pedagogical and theoretical, had been rapidly evolving since the end of the 18th century, and it would be wrong to overlook the influence that these practices had on a mathematician such as Cauchy. The advancement and expansion of technological education, a development that civil and military engineers pushed to new heights, as well as the development of mathematical physics to a stage that was infinitely more ambitious than the rational mechanics of previous ages, revived the old theoretical and practical problem of the relationship between mathematics and the applications of mathematics. The creation and early success of the École Polytechnique, an institution at which Cauchy was first a student and later a professor, reveal the will and desire on the part of the political authorities and the scientific community to solve this problem. Within the span of a few years, the École was a center producing the best engineers for the public services, as well as exceptional scholars and scientists who quickly blazed a path of progress in mathematical physics.

But, even at the École, the problem of defining the relationship between mathematics and the applications of mathematics was sharply debated. From the very beginning, the École Polytechnique had accorded a special place to analysis in its educational program, and during the opening years of the 19th century, this special position for analysis was strengthened, as is witnessed by the fact that between 1801 and 1806 the time allotted to analysis in the instructional program, in the first and second years, grew from 16 to 29% and from 11 to 18% of the total instructional time (3). In spite of a slight decrease in these percentages during the following years, the leading place of analysis in the École's educational program was never seriously questioned. Thus, it would appear that analysis was the most general and most widely taught subject at the École, required of all the engineering students, regardless of later areas of specialization. Placed at the very head of the official instructional program, analysis was the heart of the curriculum. Several courses in the area of applications, such as analysis applied to the geometry of three dimensions and mechanics depended on it very closely, and other areas, such as machines and physics, did so to a lesser degree, at least after Petit joined the faculty.

This leading role had at least two important implications for the content of the analysis courses. On the one hand, it largely determined the structure and organization of the courses to the extent that the basic idea was always to proceed from the general to the particular, from algebraic analysis to analysis applied to the geometry of three dimensions. On the other hand, it had a direct influence on methodology. Two notions come together at this juncture: some who were associated with the École—and indeed they were the majority—thought that analysis should be presented in the most simple and straightforward way possible, always keeping in mind the possible applications in the engineering sciences. This approach encouraged systematic appeals to intuition and to geometric representations. Others—and they were supported by Laplace and Lagrange—held the opposite view. They felt that it was necessary to present analysis in a way that was sufficiently abstract so as to permit its methods to be easily used in quite diverse situations. Cauchy, who, of course, shared this latter view, went so far as to propose in 1816 that instruction in mechanics, a subject for which he was responsible, be relegated to the second-year program in order that the entire first year could be devoted to infinitesimal calculus (4).

In his reasearch, Cauchy, despite his training as an engineer, rarely bothered with applications per se. Being little inclined to the physicists' view, his efforts were really focused, except, of course, for his fundamental contributions to the theory of elasticity, on the development of mathematical tools that would be applicable to physics and mathematical astronomy, as well as on the development of the calculus of residues, the calculus of limits, and on characteristics, i.e., on differential operators, etc.

These works illustrate the taste and flair for formalism that is characteristic of Cauchy's creative works. This is particularly so of his research in mathematical physics. Cauchy tried to develop various formal calculi with universal claims, not only in analysis but in algebra also, with his work on the calculus of permutations, his work in geometric calculus, and his theory of algebraic keys. On that basis, then, he was an 18th-century mathematician, a worthy successor of Euler and Lagrange.

Nevertheless, as far as the history of mathematics is concerned, Cauchy remains firmly connected with the development of rigor. This trend, as we have seen, legitimized the dominant position that analysis enjoyed at the École Polytechnique, as well as that generally enjoyed by mathematics at scientific institutions relative to the applied, empirical sciences. Cauchy went further in this respect than most. He justified the requirement of rigor in mathematics by a philosophical and religious theory of knowledge that did not change significantly during the 46 years of his scientific life.

To understand this, we will examine several texts of a philosophical and epistemological nature that he wrote at one time or another—introductions to books and papers, as well as nonscientific discussions, lectures, and publications. The fact that Cauchy's epistemological concepts did not really change will facilitate our work.

The philosophical and religious ideas to which Cauchy remained faithful all his life were well in line with traditional Catholicism and extreme conservatism as espoused by the Jesuits and, during the Restoration Era, by the Congrégation, to which he then belonged.

> Rightfully anointed kings have regained the throne [wrote Haller in 1816]. We will similarly enthrone a rightful, lawful science, a science which will serve the Sovereign Lord and whose truth is confirmed by the entire Universe (5).

Here we have it. Truth: truth was an essential term to Cauchy:

> Truth is a treasure beyond value, and no remorse or anguish in the soul ever stems from its acquisition. The mere contemplation of these heavenly wonders, of their divine beauty, suffices to compensate us for all that we may sacrifice in discovering it, and the joy of heaven itself is no more than the full and complete possession of immortal truth (6).

Cauchy, of course, did not define truth:

> Here, on earth, truth will never be complete, it will never be wholly revealed (7).

That could only be encompassed by God. Nevertheless, he did distinguish two orders of truth.

The first order consisted of philosophical and moral truths. These were revealed truths, verities 'too sublime for our thoughts and understanding ever to attain them' (8). In 1811, he stated:

> The most submissive [obedient] persons are also the wisest; and by force of various sophistries, an individual might very well come to doubt the truths that have been taught to him, but he will not learn any new ones (9).

Accordingly, it would be far better to leave the teaching of these truths to those whom the creator of the universe has entrusted with that mission and responsibility. This mission, according to Cauchy, was entrusted to the priests, a group whom, Jullien wrote, Cauchy revered (10). All the doctrines other than the revealed truths as contained in the Holy Scriptures, and as interpreted and taught by the Church, were false and dangerous. For this reason, then, religious education was necessary. The Université's monopoly was culpable not because it was a monopoly, but because it planted 'chaos and anarchy' in beliefs by not accepting religious precepts as the foundation of education (11). This kind of reasoning, which was quite widespread among the supporters of the free school movement, provoked mockery and raillery from their opponents who accused the Catholics, and especially the Jesuits, of wanting to use the cover of religious education to reestablish the Church's control over the Université.

With Cauchy, as with all the counterrevolutionaries, religion was necessary for the maintenance of the established order, because it served to 'hold man's passions in check and make him practice virtues' (12). 'The vain and pernicious philosophy' of the last century, 'after having overrun the higher classes in society, then descended into the huts of the poor and there turned the lower classes into its toys, making these classes the authors of its misery and the instruments of its crimes' (13). From this stemmed all the evils that beset the 19th Century. Cauchy was thus being quite specific in 1844 when he declared:

> Unless it be accompanied by a good education, instruction can become more troublesome than useful... Of what use is it for the child of a poor man to learn how to write, if he only takes up the pen to snare the innocent, to deceive and undermine the good faith of others... (14).

The second order of truth was composed of scientific truths; these were 'conquered' verities as opposed to 'revealed' truths, which constituted the first order. 'The pursuit of truth should be the sole aim of any science' (15). In 1811, Cauchy paid homage to the efforts that generations had made in increasing the scope of human knowledge (16), and he regarded the times he lived in as 'an extraordinary era in which a renascent, ceaseless activity devours all thought' (17). In spite of remarks such as these, Cauchy did not set forth any distinctive defining criteria for scientific truths. Indeed, if he regarded exactness as an 'essential and necessary feature of any true science', then he also saw exactness as a crucial feature of 'the most beautiful creations of the human mind, even in literature, even in poetry' (18).

It is necessary to contrast Cauchy's religious dogmatism with his epistemological relativism. Moreover, the classification scheme that he used in the sciences is imprecise. He regarded the sciences only from the point of view that was important to him, namely, in light of his own religious faith and his practices and perceptions as a scholar and scientist. In the introduction to his *Analyse Algébrique* of 1821, he contemplated what he referred to as the sciences of reason which, aside from mathematics, included 'those sciences that are called "natural" sciences in which the only method that can be successfully used consists of observing the fact and then subjecting these facts to [mathematical] calculations' (19). In more modern words, he said nearly the same in 1842 but replaced the term sciences of reason by exact sciences, a term that included the physical and mathematical sciences (20). In 1833, he set forth 'the [sequence of] steps that one should follow in order to arrive at the knowledge of truth' in the sciences. The method he proposed was a positive one:

> First, it is necessary to study the facts, to multiply the number of observations made, and then later to search for formulas that connect them so as thus to discern the particular laws governing a certain class of phenomena. In general, it is not until after these particular laws have been established that one can expect to discover and articulate the more general laws that complete theories by bringing a multitude of apparently very diverse phenomena together under a single governing principle (21).

In 1842, the natural sciences founded on observation only were not regarded as exact sciences; the term no longer carried the same meaning as in 1821. Mathematics was of no help in these cases.

Scientific truths, being of second order, were inferior to first-order truths. In fact, a knowledge of the basic essence of things or an understanding of what created objects were actually composed of was, in Cauchy's view, a divine privilege. Confined as they are to the here and now, human beings would never be able 'to penetrate the innermost nature of beings, of things, and would never clearly discern the secret mechanisms that underly the motion of things' (22). Being of an inferior nature, second-order truths were to be subjected to first-order truths; and a scholar or scientist 'ought not hesitate to reject any hypothesis that contradicts revealed truth. This point is of crucial importance', Cauchy emphasized (23). Second-order truths simply lay the groundwork for first-order truths. For this very reason, then, the Christian religion is so highly favorable to the advancement of the sciences and to the development of the most noble faculties of our intelligence' (24). Thus, for example, astronomy brings us closer to God. In fact, at the end of his 'Leçon d'astronomie', Cauchy gave a description of this 'spiritual elevation', which proceeds from the observation of the stars rising to the contemplation of God:

> We go on. Always keeping a close eye on that brilliantly lighted path, that great belt that is painted on the heavens even on a dark and cloudy night. From the purest milk, a whiteness shines forth. In this dazzling trance, we can see still other suns. We perceive suns in each bank of clouds and haze. My very soul is smitten by such spectacles, and I am hushed with awe, adoring that Being whose glorious name is read by such wonderful traits in the fires of the aurora and on the pavillon of the skies (25).

But, a danger always threatens scholars and scientists: the desire to extend second-order verities at the expense of those of the first order, as the philosophy of the Enlightenment had sought to do. On this, Cauchy declared:

> The great crime of the last century was that of wanting to raise nature itself up against its very Author, of desiring to set creatures in a state of

> permanent revolt against the Creator and even to arm the sciences against God himself, the sciences whose only real aim must be the search for truth (26).

Mathematicians were not spared the risks, because they might be tempted to place the propositions of their particular science over first-order truths. Cauchy not only rejected this platonic interpretation of mathematics, but, in a famous passage in the introduction to his *Analyse Algébrique* of 1821, he asserted that mathematics cannot be applied to historical, political, or moral questions, insofar as these relate to first-order truths:

> Let us cultivate the sciences with true ardor, but without falling prey to the desire to extend them beyond their proper domains. Let us not think that we can come to grips with history by means of formulas, nor should we attempt to base morality on theorems from algebra or from integral calculus (27).

On the basis of the above remarks, it is clear that Cauchy was taking a stand in opposition to the use of statistics and the calculus of probabilities in the human sciences, which had been developing quite rapidly since Laplace's pioneering works on probability.

Slightly modifying his views, Cauchy later came back to this indictment. In January 1827, he used Dupin's statistical studies to vindicate the works of the Brothers of the Christian Schools (28). In particular, in 1845, he presented the paper 'Mémoire sur les secours que les sciences du calcul peuvent fournir aux sciences physiques et même aux sciences morales et sur l'accord des théories mathématiques et physiques avec la philosophie' in which he rehabilitated statistics (29), declaring:

> Geometers [mathematicians] and physicists have sometimes been accused of wanting to apply the methods and procedures of calculus and mathematical analysis to the search for all truths. It is beyond question that there have been exaggerations, just as it is not to be doubted that there have been abuses of other things, even of numbers. However, it is only fair to state that in many instances the science of numbers and analytical methods cap help us to discover the truth or, at the very least, how to recognize it (30).

According to Cauchy, statistics, in effect, presented 'a rather secure means of determining whether a given doctrine is true or false, safe or depraved, and whether a given institution advances or hinders a people and their welfare', This is so because statistics enables rigorous estimates to be made of facts and effects.

The diagram below gives a schematic representation of the types of human knowledge relative to revealed truth. From this representation, it is clear that in Cauchy's view the mathematical sciences were to be placed at the opposite end of the 'table of knowledge' from revealed truth.

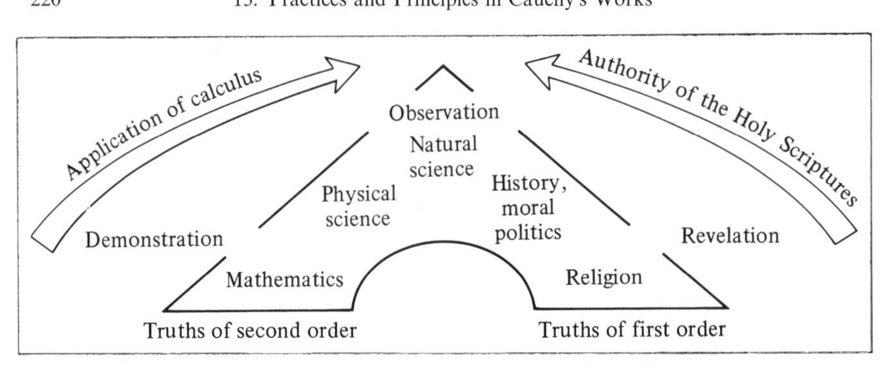

Cauchy's table of knowledge (interpretation scheme).

Thus, science reigns supreme in the domain of second-order truths, and mathematics could not claim that it could be extended to areas beyond the exact sciences. Yet, on the other hand, the position of mathematics at the other extreme of the table assured this discipline of its autonomy with regard to both religion and metaphysics. Moreover, the autonomy enjoyed by mathematics was greater than that accorded any other science. Strangely enough, on the basis of his religious philosophy, Cauchy completely isolated the mathematical sciences in the sense that, by his philosophy, mathematics could not find justification for its propositions and methods in any appeal to metaphysics. So, after having taken the position in 1811 that mathematics had now reached its 'highest stage of development and could thus be regarded as complete' (31), he did not hesitate to change this point of view and to subject it to serious internal analysis and criticism.

Mathematicians of the preceding era had developed the procedures of analysis, the method of infinitesimals, the expansions of series, the passage from the reals to the imaginaries by appeal to reasons based on the generality of algebra. Only a belief in the indefinite progress of human knowledge and the unusual outpouring of fundamental results in mathematics could possibly have justified the use of so metaphysical a principle. Mathematicians at the end of the 18th century and the beginning of the 19th century were acutely aware of the fact that this principle was highly questionable from the standpoint of mathematical rigor, although they did not doubt the validity of the various theories that had been obtained by appeal to that dubious principle. Lagrange continued to use the method of infinitesimals in mechanics, even though he would have replaced it by the method of derivatives (of functions), because he regarded the latter method as being more satisfactory from the standpoint of rigor. Similarly, Laplace, though searching for more orthodox methods of calculating integrals whose values he had determined by passing from the reals to the imaginaries, did not doubt the correctness of the basic results that he had obtained by means of the latter method. Cauchy's position, however, was quite different. It was not enough simply to replace old methods that were based on a kind of principle of continuity by new methods that might be more mathematically rigorous. It was, in his view, necessary to determine precisely the domain of validity for methods and results that the mathematicians of the Age of Enlightenment had improperly (and, quite frequently, erroneously) extended. Cauchy, a man whose philosophical ideas were formed in reaction to those held by 18th-century thinkers, the philosophes, essentially had no reason to place any confidence in metaphysics, an area from which he completely detached the mathematical sciences. He similarly had good reason for restricting the domain of validity of mathematical propositions, which he regarded as being second-order truths, and he struggled forcefully against any tendency to extend mathematics into sciences where it was inapplicable.

Thus, it was that the confident age of indefinite extension was followed, in analysis, by a more rigorous age of voluntary restriction. It was, then, in this spirit that Cauchy undertook a work on the foundations of analysis. This proved to be very fruitful, leading not only to the articulation of many existence theorems on series, integrals, differential equations, etc., but also to the creation of completely new concepts and methods. By strictly defining the boundaries beyond which the traditional mathematical methods and procedures could not be applied, he opened up a vast and new domain for research, which he explored by inventing his theory of functions. Starting in 1814, he approached the theory of integration from this perspective. We recall that in connection with his investigation of the rigorous passage from the reals to the imaginaries in integral calculus he studied the behavior of a simple integral when the function (to be integrated) was discontinuous at a point within the interval of integration. In such a case, the theorem that states that the value of the definite integral is equal to the difference between the values of the primitive at the endpoints of the interval no longer holds. For this simple reason, from 1815 on, Cauchy preferred to use Leibniz's definition of the integral as the sum of the elements that correspond to different values of the variable rather than the definition that was derived from the primitive function and had been generally adopted by mathematicians of his day. In the same way, the restrictions on the theorem concerning the substitution of the order of integration in a double integral led to the definition of a completely new type of integral, the singular integrals, and later to the development of the calculus of residues.

But, it was in *Analyse Algébrique* of 1821 that Cauchy first elaborated on rigorous methods, which he defined in the following terms:

> As to the methods [used here], I have sought to endow them with all the rigor that is required in geometry and in such a way that I have not had to have recourse to the generality of algebra. Reasons of this kind, although commonly accepted—particularly in the passage from convergent to divergent series, and from real quantities to imaginary expressions—cannot be considered, it seems to me, as anything other than proper inductions to be used sometimes in guessing the truth. Such reasons, however, ill agree with the mathematical sciences' much-

> vaunted claims of exactitude. It should also be observed that they tend to attribute an indefinite extent to algebraic formulas when, in fact, most of these formulas hold only under certain conditions and for only certain values of the variables involved. In determining these conditions and these values and in settling in a precise manner the sense of the notation and symbols I use, I eliminate all uncertainty. In this way, the different formulas only describe certain relationships that are always easy to verify by merely substituting specific numbers for the quantities themselves. It is true that in order to remain faithful to these principles, I sometimes find myself forced to depend on several propositions that perhaps seem a little hard on the first encounter.... But, those who will read them will find, I hope, that such propositions, implying the pleasant necessity of endowing the theorems with a greater degree of precision and of restricting statements that have become too broadly extended, will actually benefit analysis and will also provide a number of topics for research, which are surely not without importance (32).

Thus, for the first time, Cauchy gave a foreword on the requirement of rigor in analysis. If he was, in fact, unable to do all for analysis what Euclid had done for geometry, then it would be well to keep in mind the stubborn opposition that he encountered from the conseils at the École Polytechnique, as well as the difficulties involved in the tasks he had undertaken. This aside, he was able to make the theme of rigor a fruitful one, one whose effects it is easy to note in the sequence of works that he would produce during the coming years. From this standpoint, then, Cauchy, along with Gauss, should be regarded as the leading mathematician of the 19th century.

Chapter 14

The Final Years 1848–1857

Cauchy was not unhappy about the Revolution of February 1848. The fall of Charles X in July 1830 had left him dejected, but that of Louis-Philippe revived his hopes. Would not his former student, the Duc de Bordeaux, now Comte de Chambord, and for the legitimist party, Henri V, soon take his place on the throne that had been vacated by the Orleans? That hope was quickly dashed. However, Cauchy had other reasons, more personal ones, for being happy about the change in government. His scientific career would now be free and unfettered; it was no longer necessary that he take a loyalty oath, because it had been eliminated by the provisionary government of the Republic during its first days in power.

He felt a kind of jubilation, a deeply felt happiness, during the weeks following the fall of the July monarchy. Certain of his rights, and without informing anyone of what he was about to do, he appeared at the very first meeting that the Bureau des Longitudes held after the Revolution. Now, he could claim the place that had been denied him on political grounds. This act was, of course, unprecedented and was highly displeasing to the members of the Bureau.

> The session had not yet begun when M. Cauchy came into the meeting room, without having informed anyone of his intentions, and signed the attendance roster. Since M. Cauchy was not a member of the Bureau, the president, having the unanimous support of the Bureau, requested that he withdraw and struck his name from the attendance roster. After M. Cauchy departed, the meeting began (1).

He also played an active part in the preparation for the elections to the Constituent Assembly, which the provisionary government had announced on February 24 and were to take place on April 23. Discounting the special case of the Convention, this would be the first time that there would be elections by universal sufferage, and this raised the problem of recording the vote. Cauchy agreed to serve on a commission that was to examine various proposals that were made to the Académie des Sciences regarding methods of counting the vote; during April and May 1848, he gave several reports on studies concerning this question (2).

The Revolution of February 1848 was not an anticlerical uprising as that of 1830 had been. This point was obviously of paramount importance to Cauchy (3). It should be observed, however, that Cauchy's family—his father and brothers—did not take the same sympathetic view of these political events; for this revolution had cost them their positions at the Palais du Luxembourg, which they had held since the beginning of the century (4). It seems that Louis-François was unable to adjust to this turn of events and withdrew to Arcueil, where he died a few months later, on December 28, 1848. He was 88 years old.

Several weeks after his father's death, Cauchy was able to get himself reinstated in higher education and at the Université. The oath of loyalty no longer presented an obstacle to his appointment, and it was, then, merely a matter of waiting until there was a position available. Leverrier, who occupied a chair for mathematical astronomy that had been specifically created for him, switched on October 23, 1848, to the vacant chair for physical astronomy. Was this done merely to create a position for Cauchy? That would seem to have been the case, for everything appears to have been well prepared in advance. Even before the vacancy was declared, the Faculté des Sciences submitted a list of candidates, which the Minister of Public Instruction rejected as irregular. However, on February 11, 1849, a second list was submitted; Cauchy was first choice and Delaunay second. On February 17, 1849, the Academic Committee approved the choices that the Faculté des Sciences had made, and on March 1, 1849, Falloux, Minister of Public Instruction, signed the decree appointing Cauchy (5).

Cauchy's reinstatement at the Université, though something that he had heartily desired, did not mean that he had renounced his views in any way; with the enactment of Falloux's law, he took two opportunities to express his views publicly. In February 1850, he published two essays in *L'Ami de la Religion*, as well as in separate abstracts, entitled 'Lettres sur la compagnie de Jésus, adressées à un représentant du peuple à l'occasion de la discussion de la nouvelle loi sur l'enseignement'(6). This was a reply to the attacks that had been leveled against the Jesuits during the debates on the educational law. Enacted on March 15, 1850, this law established freedom of secondary education and teaching and instituted a Superior Committee of Public Instruction. Bishops would serve on this committee as a matter of law, as would three members of the Institut. Obviously quite pleased with these developments, Cauchy made a speech at the meeting of the general assembly of the five Académies, which met on May 29 and June 12 to select three persons who would represent the Institut on the Superior Committee of Public Instruction. Following a clumsy, ill-conceived attack on the principles and ideas that had characterized the preceding century and some passionate words in praise of truth, Cauchy concluded his speech with some remarks that encapsulated the spirit of Falloux's law:

> Everybody feels that we have now reached one of those solemn periods in which we see a society that, though severely shaken from top to bottom, can nevertheless be saved if good and decent citizens hasten to contribute to the well-being of their homeland and unite in their common efforts to attain their goal. It is especially by a wise policy based on public education and enlightenment that a social order that has been shaken to its very foundation shall be strengthened and consolidated (7).

Such a decisively political statement could hardly be expected to find much support in university circles, where, of course, there was a rather general animosity to Falloux's law, as this was seen as something that had placed the Université under the Church's thumb. Cauchy's position probably had a negative influence on the election at the Collège de France several months later, and we must take this into account if we are to understand the second defeat that he suffered in the elections for the Collège.

Libri, the encumbent in the mathematics chair at the Collège de France, gave up his position in the wake of the Revolution of February 1848, having occupied that post since his election in 1843, when he had been selected against both Cauchy and Liouville. Having renounced his professorial chair, Libri fled France, afraid of judicial proceedings for his theft of precious books from the public libraries. Until the outbreak of the Revolution of February 1848, Guizot, Libri's protector, had always intervened to suppress the proceedings. But, with the revolution, Guizot fell, and Libri was forced to flee. In April 1848, Hermite was appointed as special replacement to teach the courses that had been assigned to the hastily departed Libri.

In June 1850, following the judgement against Libri, who, from a safe distance abroad, continued to insist on his innocence, the Minister of Public Instruction called a meeting of the Assembly of Professors of the Collège de France in order that this body might express its opinion relative to the 'deplorable event' which had just taken place. The Assembly decided to allow a 'grace period' of several months during which time Libri could return to France and clear himself of the charges that had been brought against him. However, the decision would be sustained if he did not return by December 1, 1850. The Collège de France would regard him as having officially resigned his chair and then seek a replacement. The Minister of Public Instruction, however, did not consider himself as being bound by the Collège's resolution, and the vacancy was announced. On November 15, 1850, Cauchy and Liouville offered themselves as candidates for the chair. Thus, it was that the two most outstanding French mathematicians of that era found themselves competing with each other. It should be recalled that earlier, in June 1843, the same situation had arisen and that at that time Liouville had publicly declared his intention to withdraw his own candidacy in favor of Cauchy. But, it was Libri who, in fact, had won the appointment, thus creating a great scandal in the mathematical community (8). It seems that Cauchy declared his intentions of seeking this position quite a while before Liouville, who only announced his candidacy shortly before the election following certain changes that were made in the program of the chair in question (9). The professors met on November 18, and, following a reading of a letter of candidacy that Cauchy had submitted, a letter in which he spoke of some of his works, the assembled professors discussed the two candidates' qualifications (10). At this meeting, it was decided that the ballot would be held at the next meeting, which took place on November 25. Of the 23 votes cast, Cauchy received 11 as opposed to 10 for Liouville, with two abstentions (11). Accordingly, a second ballot was held. This time, Liouville took 12 votes against Cauchy's 11, with a single abstention. Thus, without any objections, the Collège selected Liouville as its candidate.

Cauchy and his supporters were quick to react. The fact that there had been two abstentions on the first balloting raised some questions about the validity of the election. Should these abstentions count as having established a majority, as it was in the view of the faculty assembly? If that was the case, then clearly Cauchy had failed to get a majority of the votes, since he had only managed to get 11. However, if the blank votes were disregarded, then obviously Cauchy had won on the first ballot, since he had received 11 of the 21 nonabstaining votes that were cast. At the meeting of December 1, 1850, one of the professors arose and demanded that there be a clarification as regards the conditions under which the results of the first ballot had been set aside. Once this had been placed in the minutes of the meeting, Cauchy's friends were able to start their offensive.

In their search for material that would support their position that Cauchy had indeed won the election on the first ballot, his friends were able to find, with the help of Sédillot, the Secretary of the Collège de France an example that dated from 1837. In that case, Letronne had been elected to a position on the first ballot, and no attention had been paid to the abstentions (12). On December 6, Desgranges, Quatremère, and Binet, acting with Cauchy's full consent, sent a letter of protest to Parieu, the Minister of Public Instruction; Quatremère, the most determined of the three, took the floor in the Assembly of Professors and demanded 'that the Collège settle once and for all the jurisdictional question of whether or not the abstentions should be counted in the voting'. The real point was, of course, to open a debate on the election of November 25. This discussion, which set Quatremère and Barthélémy Saint-Hilaire, the Administrator of the Collège, against each other, ended inconclusively. The meeting, on Duvernoy's motion, decided to adjourn without making a decision (13).

Cauchy's three supporters had not mentioned their letter of protest during the discussions; however, Barthélémy Saint-Hilaire had heard of it, and he promptly wrote a letter to the minister in which he set forth his own views on the matter. In the meantime, the minister had received a third letter, this one sent by Cauchy himself on December 14. Embarrassed, the minister replied to the administrator that 'it seemed to him necessary that a formal decision was now called for and that the Collège de France should apprise him definitively of which candidate had been properly chosen by the professors. Although Barthélémy Saint-Hilaire confirmed on December 23 that Liouville was the candidate that had been properly elected, Parieu requested on December 30 that another election be held. He easily justified his decision by requiring an absolute majority of 13 votes. This number was not based on the number of professors present at the balloting but on the number of incumbent professors. That much so, he asserted, neither Cauchy nor Liouville could really claim to have been properly elected (14). That same day the geometry section submitted the names of two candidates to the Académie: Cauchy and Liouville.

The administrator now called a meeting of the professors in order that a new election could be organized. However, Cauchy's attitude remained firm, and he refused to accept what he called the minister's 'Solomonic judgment'. He was utterly convinced that he had carried the election of November 25 on the first ballot. Accordingly, as he explained in a letter to the professors dated January 4, he refused to go through the election process again (15). In any event, the election took place as scheduled, and of the 23 votes cast, Liouville received 16 to Cauchy's 7. Cauchy had, of course, declared that he would withdraw his candidacy should an election be held. Depressed and humiliated by what had taken place at the Collège de France, Cauchy now read a note to the Académie. Entitled 'Sur l'influence souvent exercée par des circonstances étrangères à la science dans la solution des questions qui paraissaient purement scientifiques, et sur le pouvoir attribué, dans une élection récente, à un bulletin blanc,' this note also appeared in a lithographic version (16). The Académie did not even have to proceed with the selection of a candidate, since Liouville was the only person to present his candidacy.

In the wake of this affair, relations between Cauchy and Liouville deteriorated. On March 31, 1851, when Cauchy presented a paper by Hermite on doubly periodic functions, an incident flared up between the two mathematicians. In his report, Cauchy presented the theory of doubly periodic functions within the general framework of his own theory of functions. At this juncture, Liouville referred to the fact that quite a while ago he had given a general theory on this topic as could be substantiated by notes taken by Joachimsthal and Borchardt, and the manuscripts of these notes, Liouville went on, had been registered with the Secretariat of the Académie. Cauchy replied by citing his own research on the topic, which dated back to 1843 and by observing that Liouville's theorem, a result that provided the foundations (for the general theory now under discussion) could be easily derived from his calculus of residues (17). Eight years had now passed, and Hermite found himself placed, by pure chance, in the same situation that he had been in 1843. As we have already seen, his first study had given rise to a lively quarrel between Liouville and Libri shortly after the latter's election to a position at the Collège de France (18). After this incident, Liouville and Cauchy continued to work on the same evaluation committees at the Académie, but there was now little if any real scholarly collaboration between them. In fact, relations between these two mathematicians became so strained that in July 1856 Liouville complained to Cauchy about his rather cavalier attitude. This complaint stemmed from the fact that Cauchy submitted a report on a study entitled 'Sur l'intégration des équations différentielles au moyen des fonctions elliptiques' by Briot and Bouquet, even though Liouville, who was also a member of the committee responsible for evaluating this work had neither seen nor read either the study itself or the report (19).

Cauchy's uncompromising views in the political sphere caused him one last problem shortly after the coup d'état of December 2. The new regime, like the one that had been in power before February 1848, quickly came around to requiring that a loyalty oath be taken by all state functionaries, including professors at the Faculté des Sciences. However, on May 14, 1852, Cauchy sent a letter to the dean of his faculty in which he recalled his loyalty to the Bourbons and declared that he would give up his chair in mathematical astronomy if he were required to take on oath (20).

Cauchy, in effect, gave up teaching in 1853 and was replaced by Faye. In December 1853, Leverrier was appointed to replace Cauchy in 1854 (21). But, as in 1839, there were attempts to find an arrangement. Toward the end of 1853, Marshall Vaillant, who had only recently been made a member of the Académie des Sciences, wrote a letter favoring Cauchy to Minister Fortoul:

> You are aware, just as I am, of M. Cauchy's reputation as a scholar. It is commonly said by even the most learned persons that he is the greatest mathematician in Europe. I should also point out that he is a man of tremendous integrity and exemplary character. He is a very religious man, alert, learned, and sincere. Everybody admires and respects him. The reserve that M. Cauchy has been said to show should not be regarded as hostility toward the Emperor's government. Indeed, anyone who considers it as such would only be fooling himself. Working alongside M. Cauchy at the Académie as I do, I can assure you that he holds His Majesty in the highest esteem and is deeply respectful of the protection that the Emperor has accorded the sciences and the arts (22).

Finally, Fortoul was able to get Emperor Napoleon III to exempt Cauchy from the loyalty oath requirement. In this affair, at least, the Emperor showed himself to be a more generous and intelligent man than his predecessors had been. This was the first time that Cauchy had ever seen his political intransigence pay off, and he promptly resumed teaching his classes at the Faculté des Sciences, where he remained until his death.

After the Revolution of February 1848, Cauchy continued his old habit of regularly presenting notes and studies to the Académie; however, the pace at which he published slowly declined. At the end of 1853, he ceased publication of *Exercices*, though he continued teaching a course at the Faculté. His scientific interests, research, and production had always been very broad—mathematical physics, the theory of light, celestial mechanics, the theory of functions, and so on. Aware of the need to assemble his works, which were now scattered about in a multitude of studies, papers, and memoirs, into a coherent collection, Cauchy contemplated a synthesis, a kind of central theme, around which his works could be organized. The question, of course, was not merely one of making his works more accessible and better known but also of putting them on a more solid foundation. This project, an undertaking that would haunt Cauchy until he died, was never finished. Still, it did provide, at least in a certain sense, the starting point for the definitive *Oeuvres complètes*, which was decided on nine years after his death, but not actually begun until 1882.

In 1855 or 1856, Cauchy requested help from the Father Superior of the institution in the rue des Postes, asking that Michel Jullien, who was a young Jesuit student, be assigned to help him in putting his papers and scientific notes in order. Jullien later gave the following picturesque account:

> I spent several days with him actively engaged in this task. A good portion of my time was spent searching for the kindly old gentleman's eyeglasses, beneath the stacks of papers where he would always put them. His papers were written in such a scrawl that I do not know how the compositors at the printers will ever be able to decipher them (23).

This work concluded in the development of a manuscipt consisting of several large notebooks that were compiled by Jullien and that he later entrusted to Valson for use in his biography on Cauchy (which was written in 1868) and the *Oeuvres complètes* (24).

From 1848 to 1850, Cauchy presented a considerable number of studies on mathematical physics to the Académie. A particularly large portion of these works dealt with theory of light, a subject that he had neglected since the end of 1842. The research carried out by Laurent, Jamin, Bravais, La Provostaye, and Desains was praised by Cauchy, as was that of Saint-Venant. These works probably account for Cauchy's revived interest in that topic (25). As had been true of his investigations during the period 1829–1842, he approached the theory of light from two different points of view. On the one hand, during 1848, he resumed the study of the homogeneous equations of motion in an isotropic system of molecules. These investigations were the subject of many notes; however, only the titles of these papers were published in the *Comptes Rendus* (26). On the other hand, the preparation of a note on Jamin's study 'Sur la réflexion de la lumière à la surface des corps transparents' inspired Cauchy to write several papers on the simple motions in a system of molecules. These studies were presented to the Académie at the end of 1848. In these investigations, Cauchy showed that when the modulus of the simple motion was less than 1, then the ray of light grew progressively weaker; the ray was evanescent (27). In addition, Cauchy advanced the hypothesis that the obstructing longitudinal vibrations in the case of reflection of simple motion gave rise to an evanescent ray that moves along the boundary (28).

On January 2, 1849, a few days after Louis-François, death, Cauchy announced he was going to write a treatise on molecular mechanics. This work, he explained, would be written not only in response to the expressed wishes of a number of illustrious scholars and scientists but also as 'a kind of filial piety, since this had also been the wish of a loving father, who all during his life had combined a real love of knowledge and cultivation of letters with every virtue and who now is sleeping with the just, removing himself to a better world.'

In this treatise, Cauchy wanted to set forth the general principles governing molecular mechanics and then successively apply them to problems on the theory of light, sound, heat, elastic bodies, etc. (29). Unfortunately, he never realized this project, but rather contented himself with the publication of several articles in *Mémoires de l'Académie des Sciences* of 1850. As Valson noted:

> Hardly would he have taken his pen and begun to write before a new idea would come to the fore, and with an irresistible charm, overcome what he had been working on just moments before and lead him to pursue an important result that he, like a hunter in pursuit of a most exciting prize, would enjoy in advance (30).

Cauchy continued to work on calculation methods that would be applicable to celestial mechanics. The Revolution of February 1848 had interrupted a long series of works that he had begun in September 1847 that dealt with the determination of the orbits of the planets and comets (31). During the following years, indeed until his death, he would occasionally present studies on mathematical astronomy: in July 1849, he discussed the problem of calculating the errors in the study of planetary motion (32); in December 1851 and in January 1852 he took up the question of expansions of functions into convergent or even divergent series and their application to astronomy (33); in May and June 1854, he discussed the problem of transformations of implicit functions by 'isotropic means' and the use of the logarithmic spiral in determining the orbits of planets and comets (34); in August 1854, he discussed the solution of the differential equations governing the motion of the stars (35); and, finally, in April and May 1857, on the very eve of his death, he discussed the use of regulators in astronomy (36).

However interesting these studies are, they occupy a place of lesser importance in any ranking of the works he produced between 1848 and 1857. This might seem somewhat surprising because during this period Cauchy held a chair in mathematical astronomy at the Faculté des Sciences. However, in reality, he taught quite a few other things in addition to calculational methods in celestial mechanics.

In an article written in 1869, Bertrand spoke of how Cauchy's first lectures at the Sorbonne in 1849 progressed:

> His first lectures, it must be admitted, completely disappointed the expectations of his audience, who were more surprised than charmed by

> the slightly confused variety of subjects discussed. I distinctly remember the third lecture: it was completely devoted to the problem of extracting square roots, and the number 17 was taken as an example. Calculations were performed up to the tenth decimal place, calculational methods which were well known to the audience, but which Cauchy thought were new, no doubt because he had spontaneously thought them up the evening before.

This, of course, was typical of Cauchy's sometimes strange pedagogical style. However, Bertrand added:

> I did not attend any further lectures, and I was quite wrong in that decision, because, ten years later, these lectures could have provided me with an introduction to some of this illustrious teacher's most exciting discoveries (37).

Bertrand's remarks could only have been an allusion to the theory of complex functions, a theory that Bertrand characterized as having a 'main theme and appeal that was as subtle and as valuable as his most exciting discoveries'.

Finally, it should be observed that teaching gave Cauchy the opportunity to impose at least some order on the vast theoretical structure that was begun in 1814 and was finally ready in 1846.

In effect, Cauchy's lectures at the Sorbonne primarily consisted of an exposition of this theory of functions. The fact is confirmed by two partial transcripts: the first was written by G. Lespiault (38), probably in the academic year 1855–1856, and the second by the Norvegian C. R. Bjerknes in the academic year 1856–1857 (39). The venerable old mathematician could now exert an influence proportional to his ability: there was a small audience indeed, between three and five persons in 1856–1857 according to Bjerknes (40); among those who attended his lectures, however, were some talented young analysts, such as Hermite, Puiseux, Briot, Bouquet, and Méray, who would later develop the ideas and concepts that Cauchy had presented to them. Cauchy's courses at the Faculté des Sciences should be placed on the same level as those given by Liouville at the Collège de France, because these two teachers exerted a parallel influence on the new generation of French mathematicians. Briot and Bouquet, in their treatise *Théorie des Fonctions Doublement Périodiques, et en Particulier des Fonctions Elliptiques* synthesized the concepts that Cauchy and Liouville had developed in their courses (41). Thus, until 1857, Cauchy's work with the theory that he had initiated advanced in step with the works of his disciples. Valson noted:

> When he taught the methods he had developed, he presented them in a neater, more precise way, carefully coordinating method with basic principle. It seems that once he had taken chalk in hand, he could explain his formulas more easily on the board than he was able to do when he wrote them out in his notebooks. This was so because his mind was so alert that once it was in contact with his audience, a new liveliness would

take place. A beam of joy would light up his whole being when the proof that he was trying to get his listeners to understand suddenly became clear to them (42).

On the basis of his work as a teacher, Cauchy sensed that it was necessary that his complex variable theory be put on a clear and rigorous footing. A direct consequence of this realization was his adaption of the 'theory of geometric quantities' as a new theory of complex numbers. In the preceding chapter, we saw that Cauchy had been rather indecisive on this point since, until 1848, he had made use of geometric representations in order to develop his theory of curvilinear integrals while, at the same time, holding on to what he referred to as a 'symbolical theory of imaginaries,' which he replaced in 1847 by a 'theory of algebraic equivalences'. Although this latter theory may have been aesthetically appealing, it was inadequate as a basic principle for complex variable theory. The question had to be settled, because it was now no longer possible to continue in this indecisive way in an official course (43). In September 1849, he articulated his new theory of imaginaries (44). Aside from the works of Budé and Argand, he cited a study by Saint-Venant, 'Sur les sommes et les différences géométriques et sur leur usage pour simplifier la mécanique', which had been presented to the Académie on September 15, 1845. As a member of the commission charged with the responsibility of evaluating this study, Cauchy had read this work closely and discussed it with the author. In it, Saint-Venant had exhibited a kind of vector calculus that was very similar to the one that had been published a year earlier by the German mathematician Grassmann in his *Ausdehnungslehre*. Thus, Cauchy knew the principles on which his new theory rested a full two years before his paper 'Mémoire sur les quantités géométriques' appeared. We thus get an idea of Cauchy's reluctance to use a geometric approach, a reluctance that was so strong that it could only be overcome by the requirements of teaching (45). As he wrote, 'some new and mature thought was needed if advantage was to be obtained by replacing imaginary expressions by the geometrical quantities whose use gives algebra not only a clarity and new precision, but also a greater generality' (46).

The evaluation of Hermite's note 'Sur la théorie des fonctions elliptiques' in March and April of 1851, a work that had been presented to the Académie in November 1849, and even more importantly, two studies by Puiseux, 'Sur les fonctions algébriques', which were presented on January 13, 1851, and March 17, 1851, prompted Cauchy to focus his attention once again on the theoretical foundations of his theory.

In his study, Hermite applied Cauchy's methods to the study of doubly periodic functions for the very first time. It is possible—indeed, quite probable—that Hermite's study had been inspired by Cauchy's lectures at the Faculté des Sciences in 1849. In any event, Hermite acknowledged his debt to the great mathematician by presenting his study to the Académie as an application of the principles of Cauchy's calculus of residues (47). He integrated around the periphery of a period parallelogram and thus showed that the sum of the residues of a doubly periodic function reduces to 0. From this, he obtained a decomposition of these functions into a sum of terms that were proportional to transcendental functions of the form $\theta(z - z_i)$ and their derivatives. He then examined the function $\theta(x)$, which he expanded in a series, and its derivative $\varphi(z)$(48). Hermite discussed the contents of his paper in his courses at the Collège de France during the academic year of 1849–1850 (49). But, only on March 31, 1851, did Cauchy finally present a report on this work to the Académie; and, although the report was very favorable, it, as we have seen, gave rise to a quarrel between Cauchy and Liouville

Thus, it took Cauchy more than a year to submit his report on Hermite's paper. Cauchy decided to act in this manner once he had made up his mind to resubmit the report on a study that Puiseux had just presented and that he obviously liked. He could not bring himself to favor one of these young mathematicians over the other insofar as they both had applied his theory to the study of 'multiform functions.' One week after he had submitted the report on Hermite's paper, he submitted the report on Puiseux's study, 'Sur les fonctions algébriques'; the paper itself had been presented to the Académie on January 13.

Puiseux, who had been a young maître de conférences at the École Normale Supérieure since 1849, had already published several important works by the time he met Cauchy. He took courses under Cauchy at the Faculté des Sciences and became his best disciple. Moreover, these two men also quickly became good friends. No one better fits Valson's description of the 'young people at the schools' of whom Cauchy was so fond, 'young scholars who were invited into his study as well as into his parlor, and with whom he would talk and chat amiably, more like a friend than like a teacher' (50). Puiseux professed to holding the same religious views as Cauchy and, according to Jullien, was an excellent and fervent Christian who assisted Cauchy with various undertakings on behalf of the faith (51).

Using the still somewhat confused ideas of Cauchy's 1846 papers, which were, perhaps, discussed in lectures given at the Sorbonne, Puiseux showed how, in the neighborhood of a branch point, the values of an algebraic function change according to the closed path along which the function is evaluated, thus forming several circular systems. He then used this result to calculate the value of the different periods of elliptic integrals. Puiseux's study was written in the careful and rigorous style that characterizes Cauchy's best works (52).

Cauchy knew the essentials of what was contained in this study even before it had been submitted to the Académie. On January 20, he presented a study in which he investigated questions that dated back to 1846 and had now been raised again by Puiseux, and he introduced what he termed 'lignes d'arrêt' (lines of stoppage), these being straight lines that cut the complex plane in such a way that the values of an algebraic function could be separated (53). However, Cauchy, though tempted, as usual, to append his own work to the work that he was evaluating, directed his efforts in another direction. Instead of following the study of elliptic functions along the lines that Puiseux and Hermite had taken, he tried, in effect, to make his theory of functions more precise.

In a short paper bearing the very general and nondescript title 'sur les fonctions de variable imaginaire,' he first set forth the definition of the derivative or differential coefficient of a function of a complex variable and then followed up on this definition by giving the conditions for the differentiability of a function $u = v + wi$ of a complex variable $z = x + yi$: $D_u v = D_y w$, $D_y v = -D_x w$ (54).

Later, on April 7, 1851, he defined what he termed a monotypical function, which is continuous and uniform, and a monogenic function, which has a derivative at each point. Cauchy's intention here was clear: he was setting concepts and vocabulary that were needed to define the class of functions to which his calculus of residues and calculus of limits could be applied (55). In a new paper of May 12, he showed in substance that a monotypical and monogenic function can be expanded into a Laurent series, and its integrals can be evaluated by the theorem of residues if its singular points are poles (56).

In a February 1852 paper in which he sought to give the appropriate hypothesis for implicit functions and especially for those defined by differential equations in order that they should have series expansions, he also clarified, connected, and completed his terminology. He replaced monotypical by monodromic, he further asserted that the derivative of a monogenic function is continuous and proposed to refer to a monodromic function that is monogenic and finite as a synectical function. Today, these functions are called holomorphic (57). He then showed that functions that can be expanded in Taylor series must be synectical.

As we have seen, Cauchy stopped teaching a few weeks after this last paper was published, because he would not take a loyalty oath. At the end of 1853, when he resumed teaching, he published several articles in *Exercices* in which he explored elements of the theory of functions that probably constituted the basic material for his lectures. We specifically mention here the articles 'Sur les fonction des quantités géométriques', in which Cauchy gave a very general definition of the notion of function (58); 'Sur les fonctions continues de quantités algébriques ou géométriques', in which he restated the definition of lines of stoppage and monodromic functions (59); and, finally, 'Sur les différentielles des quantités algébriques ou géométriques et sur les dérivées des fonctions de ces quantités', in which he introduced the concept of directional derivative (60).

During the following years, Cauchy closely followed the work of young mathematicians as they went about developing his theory, because he was frequently charged with the responsibility of examining their works at the Académie. The most important contributions were undoubtedly those by Briot and Bouquet on the integration of differential equations, works that prompted Cauchy to undertake new research on differential equations as a branch of the theory of functions (61). Accordingly, on April 2, 1855, he

(344)

Si, d'ailleurs, on pose

$$\frac{dy}{dx} = \operatorname{tang} \varpi, \quad (5)$$

ou, ce qui revient au même,

$$\frac{dx}{\cos \varpi} = \frac{dy}{\sin \varpi}, \quad (6)$$

l'équation (4) donnera

$$D_z Z = \frac{D_x Z \cos \varpi + D_y Z \sin \varpi}{\cos \varpi + i \sin \varpi}; \quad (7)$$

puis, en ayant égard aux formules

$$1_\varpi = \cos \varpi + i \sin \varpi, \quad 1_\varpi 1_{-\varpi} = 1,$$

on tirera de l'équation (7),

$$D_z Z = 1_{-\varpi}(D_x Z \cos \varpi + D_y Z \sin \varpi). \quad (8)$$

Il est bon d'observer que, dans les formules (4), (7), (8), les dérivées partielles $D_x Z$, $D_y Z$ varient généralement avec la position du point mobile A, et se réduisent à des fonctions des deux coordonnées rectangulaires x, y de ce même point, ou, ce qui revient au même, à des fonctions de l'affixe z. D'autre part, tandis que les coordonnées x, y varieront par degrés insensibles, le point A décrira généralement une courbe continue; et si par ce point on mène une droite qui forme, avec l'axe polaire, un angle ϖ propre à vérifier la formule (5), la direction de cette droite, appelée *tangente*, sera ce qu'on peut nommer la *direction* de la courbe au point dont il s'agit. Si la ligne décrite par le point A se réduit à une droite, la tangente ne différera pas de cette même droite. Cela posé, il suit immédiatement de la formule (4), (7) ou (8) que la dérivée de Z, considérée comme fonction de z, dépend en général, non-seulement de la position du point A sur la ligne qu'il décrit, mais encore de la direction de cette ligne. Si cette direction devient parallèle à l'axe des x ou à l'axe des y, on aura

$$dy = 0 \quad \text{ou} \quad dx = 0,$$

et la dérivée de Z, prise par rapport à z, sera, dans la première hypothèse,

$$D_x Z; \quad (9)$$

dans la seconde hypothèse,

$$\frac{D_y Z}{i} = -i D_y Z; \quad (10)$$

Differentiability of a complex function. 'Sur les différentielles des quantités algébriques ou géométriques, et sur les dérivées des fonctions de ces quantités', *Exercise d'Analyse et de Physique Mathématique*, **4**, p. 336–347. Published by permission of the École Polytechnique.

presented Méray's paper 'Sur les fonctions périodiques monogènes et monodromes' to the Académie (62). Méray took Cauchy's courses at the Faculté des Sciences; and, later on, at the request of Saint-Venant and the Cauchy family, he would do the editorial work on this material for publication in *Cours d'Analyse* (63).

Parallel to the development of his theory of functions, which had now come to be based on geometric quantities, he developed a calculus of keys, which he thought would have a certain universality, because, in relation to this calculus, he noted:

> Their use not only allows us to easily introduce certain quantities that come as a natural result [of the study of this calculus], and to which they 'open the door', so to speak, but they also enable us to resolve efficiently a considerable number of other questions and problems of a rather diverse nature (64).

He made particular use of this calculus of keys in presenting the theory of determinants. He also applied, without any great success, this formalism to the calculus of geometric quantities. Thus, it was that Cauchy returned to a formalistic approach at the very time that he was increasingly geometrizing his theory of functions. This attitude, which is ostensibly so contradictory, is reminiscent of the attitude that, beginning in 1825, led to the simultaneous development of the formal calculus of residues and the geometric theory of imaginary integrals.

When, with the help of Möbius, Grassman discovered Cauchy's works on algebraic keys and those of Saint-Venant on the same topic, he felt that his own research had been plagiarized. At the beginning of the 1840s, he had, in fact, created a certain geometrical analysis involving vector concepts, and it strongly resembled Saint-Venant's geometric calculus, as well as the calculus of algebraical keys. Accordingly, after reading a paper by Saint-Venant that was published in an 1845 issue of the *Comptes Rendus*, Grassman sent copies of his *Ausdehnungslehre* to Saint-Venant and Cauchy in 1847. This was followed by an exchange of letters between Grassmann and Saint-Venant at the end of 1847 and the beginning of 1848 (65). Grassmann thus had good grounds for believing that Saint-Venant, and perhaps even Cauchy, knew of his geometric analysis. He also sent a protest to the Académie des Sciences, which was received on April 17, 1854 (66). Accordingly, a commission consisting of Cauchy, Lamé, and Binet was appointed to investigate this matter; however, it never submitted its findings. Nevertheless, Cauchy was occupied by this affair until the end of 1856, just a few months before his death. On December 8, 1856, he asked Saint-Venant to write a note for him on Grassmann's works and on the correspondence that had taken place during 1847–1848. Saint-Venant submitted this note on December 15 following, and at the same time, he wrote a letter to Grassmann as a token of his good faith (67). Based on Saint-Venant's note and letter, it seems that the copies of the *Ausdehnungslehre*, which Grassmann sent to the two French mathematicians in 1847, were mislaid by the postal service and that, even if Saint-Venant knew anything about the studies and papers that Grassmann had sent in 1848, Cauchy himself had read nothing by Grassmann. The great similarity between Cauchy's theory as based on algebraic keys and the theory that Grassmann had developed thus cannot be explained as a gross plagiarism, an act that would have been out of character for a person like Cauchy. However, Cauchy, through Saint-Venant, who did have a vague idea about Grassmann's work, may have been indirectly influenced at the beginning of his research by Grassmann's research.

We have seen that Cauchy exercised a great influence on the young generation of scholars and that, therefore, many of them, Puiseux, Hermite, Briot, and Bouquet, shared his ideas. Furthermore, he was very active at the Académie where, until his death, he was busy evaluating a considerable number of studies, which the secretariat had to demand that the Cauchy family return after his death (68). He often spoke out at meetings of the Académie, sometimes doing so to support an opinion of a colleague, to support a scientific project, or to criticize a study. Thus, in July and August of 1853, he engaged in a long argument with Bienaymé on his interpolation method of 1835 as compared with the method of least squares (69).

He also engaged in quite a bit of active proselytizing among his colleagues at the Académie, at least among those who seemed to him to be capable of being brought back to the faith of their childhood. It seems that he won Duhamel over. However, as Jullien observed, Duhamel was content to merely perform the obligatory religious practices, such as attending mass and taking the sacraments. We will later see that at the Académie relations between these two scholars grew steadily worse. Cauchy also tried to lead Lamé, who had become 'disillusioned with the world and saddened by his declining state of health', back to religion. In order to accomplish this, Cauchy asked Jesuit students to attend Lamé's courses in mathematical physics, which were very poorly attended (70).

On another level, right up until the end of his life, Cauchy continued to devote himself to activities and projects that were nonscientific in nature and had now come to absorb his attention to a greater degree than science. He worked tirelessly, visiting people at all hours of the day or night, seeking to advance his charitable works (71).

In particular, he was the guiding light behind the work of the Écoles d'Orient, which was created in 1856 (72). The Crimean War was concluded by the Treaty of Paris in March 1856. In exchange for a French–British guarantee of the independence and integrity of his empire, the sultan had to grant equal rights to his Christian subjects. The new perspectives that were opened up by the Hatti-i-Humayoun lighted fires of hope in Cauchy's breast; he saw a possibility for a new crusade (73). The problem was less one of converting Muslems to Christianity than of bringing Orthodox Christians, who were now under the sultan's rule and to whom the Tsar of Russia had sought to extend his protection, back to the Catholic faith. Tsar Nicholas I's amibition had been to pressure the sultan into granting him the right to protect the holy places for the Greek monks that lay within the sultan's domains. These places were under the protection of the Latin faiths and the tsar's scheme had made the French keenly aware of the whole question since 1853.

Cauchy's first idea on this point was to work toward creating an educational institution along the lines of the Frères des Écoles Chrétiennes to win the people over to Catholicism. Accordingly, he collected a mass of documents and papers on this question and, after having discussed it with his friends, sought basic support 'from persons belonging to the highest levels of society, and particularly from among the ranks of scholars and scientists as well as from his colleagues at the Institut' (74). A provisional committee held meetings at his home in the de Bure townhouse in Paris. Cauchy, along with the archaeologist Charles Lenormant and Father Gagarin, a Jesuit priest, played a leading role at these meetings. The organizational meeting for the project was held on April 4, 1856, at the home of Mandaroux-Vertamy (75), and Marshall Bosquet was selected as honorary president of this project, with Mathieu from the Bureau des Longitudes as acting president. Cauchy and Lenormant became vice presidents. Of the many persons who claimed membership in this group, we mention Armand de Melun, President of the Administrative Committee; Montalembert; and Falloux; and certain academicians, such as Wallon, a historian who served as Secretary General; de Wailly, a paleontologist; de Rouge, an Egyptologist, and the mathematicians, Hermite and Binet; Abbé Lavigerie was director of the project (76). Of all the charitable works that Cauchy created, the work of the Écoles d'Orient, which still exists today, was by far the most successful.

In spite of the time and effort devoted to the sciences and to the Écoles d'Orient project, Cauchy continued to work actively with the Society of Saint-Vincent-de-Paul, an organization from which he had founded the Conférence of Sceaux. He used the income from his chair in mathematical astronomy to support charitable works in the community, particularly the activities of the Sisters of Saint-André and a young boys' orphanage. Shortly before his death, he was also concerned about the difficulties that were encountered in setting up a school of the Frères des Écoles Chrétiennes at Sceaux, and in order to assure its success, he made some rather 'considerable pecuniary sacrifices' (77).

The last few months of Cauchy's life were painful for him. A rather silly and utterly useless quarrel took place between him and Duhamel, one of his so-called converts. Centering on a question of priority, this disagreement clouded his final scientific activity. On December 22, 1856, in the wake of certain observations that Bertrand had made with regard to a study that Ostrogradski had done in 1854, Cauchy claimed that in 1818 he had generalized Carnot's theorem on inelastic shocks in arbitrary bodies. At the next meeting of the Académie, Duhamel questioned Cauchy's conclusion by referring to a note that had been presented in 1832. The quarrel continued well into the following weeks, until finally Poncelet, supported by Morin, exhibited the general principles governing the shock of inelastic bodies and recalled studies of his own that had been written in 1826 and the objections he had raised in 1829 to Cauchy's results (78). Cauchy, of course, lost the argument; in fact, this quarrel did nothing to increase his standing, because during the discussion, he stubbornly refused to concede priority, even though he admitted the inadequacy of his study of 1828. The attacks mounted against him during these discussions and 'gave the final days of his life a basic sadness and bitterness that only his friends were aware of' (79).

According to Valson, during this time, he experienced a feeling of emptiness (80). In May 1856, his old friend Jacques Binet passed away. Speaking for the Académie des Sciences at the burial, Cauchy declared:

> More fortunate than we are, Binet has now gone to the source of all Light, to learn the secrets that we ourselves shall one day know by traveling the road that he has already treaded. Lost in thoughts such as these, thoughts of higher things, I know I will be forgiven, Messieurs, for abridging these remarks (81).

Less than a year later, on March 30, 1857, his younger brother Alexandre died, and Cauchy was profoundly affected by his death. Now afflicted with what he called 'great rheumatism', he left Paris for Sceaux on May 12, 1857, where, on his doctor's advice, he was to spend the summer. Little by little, the malady seems to have weakened its hold, and then suddenly, on Tuesday, May 21, 1857, the symptoms became sharply aggravated. On the following day, he

Cauchy's country house, at Sceaux, near Paris. Cauchy died in this house on May 23, 1857. Published by permission of the École Polytechnique.

received the Last Sacraments, and at about 4 o'clock a.m. during the night of May 23, 1857, Cauchy died.

> Our fears have now come to pass [his daughter Alicia announced later that same day]. Having remained fully alert, in complete control of his mental powers, until 3:30 a.m., my father suddenly uttered the blessed names of Jesus, Mary, and Joseph. For the first time, he seemed to be aware of the gravity of his condition. At about 4 o'clock, his soul went to God. He met death with such a calm that made us ashamed of our unhappiness (82).

Thus ended the life of the greatest French mathematician of his times—scarcely two years had passed since Gauss had died in Germany. A new age was now opening in the long history of mathematics, an age in which the leading figures in the mathematical sciences would be Germans. Between 1854 and 1859, Riemann, Weierstrass, and Kronecker came onto the scene on the other side of the Rhine. Meanwhile, however, in France, there was a blossoming of works on Cauchy's theory.

Concerned about his works, Cauchy had requested that someone be appointed to have his unpublished papers edited and published. To the greatest extent possible, his family followed this request and, accordingly, entrusted the project to Jullien, and later to Méray, who had taken Cauchy's courses at the Faculté des Sciences, and then finally to Valson, who in 1876 prepared Cauchy's *Oeuvres Complètes* under the auspices of the Académie des Sciences.

Notes

Notes to Chapter 1

1. From the marriage contract between Louis-François Cauchy and Marie-Madeleine Desestre. Arch. Nat., *Minutier central des notaires*, Étude CXVIII, 640, October 13, 1787.
2. Th. Lebreton, *Biographie Normande*, **1**, Rouen, 1857, article on Louis-François Cauchy.
3. C. A. Valson, *La Vie et les Travaux du Baron Cauchy*, **1**, Paris, 1868, p. 3.
4. See. Dr. Delaunay, 'Le Parnasse du temps de Napoléon: L. F. Cauchy, correspondant de la Société des Arts du Mans'. *Bulletin de la Société d'Agriculture, Sciences, et Arts de la Sarthe*, 1949–1950, 2nd fascicule, p. 140.
5. *Almanach Royal*, 1788.
6. From the marriage contract, Arch. Nat., *Minutier central des ntoaires*, Étude CXVIII, 640.
7. The property included a country house, a farm house, and an enclosed area of 12 acres (Archives de la Seine, *Mutations après décès*, Ivry-Villejuif, DQ 14858, December 25, 1848, inheritance of Louis-François Cauchy).
8. C. A. Valson, *La Vie et les Travaux du Baron Cauchy*, **1**, p. 3.
9. Undated letter from Louis-François Cauchy to his mother residing at Rouen, cited in C. A. Valson, *La Vie et les Travaux du Baron Cauchy*, **1**, p. 13.
10. Archives E. P., VI, 2, a, Cauchy file, medical certificate.
11. Arch. Nat. F1b 1* 531. *Organisation des Bureaux du Ministère de l'Intérieur par ordre chronologique*, 1792–1811.
12. The expression is from C. A. Valson, *La Vie et les Travaux du Baron Cauchy*, **1**, p. 14.
13. Archives de la Société d'Agriculture, Sciences, et Arts de la Sarthe, XIV, B8, quoted by Dr. Delaunay, 'Le Parnasse du temps de Napoléon...' p. 143.
14. During the Consulate, Louis-François Cauchy published *Ode latine adressée au Premier Consul de la République Française Napoléon Bonaparte* in 1802. During the period of the First Empire, Louis-François published at least nine pieces of poetry in honor of Napoleon (see the bibliography in Dr. Delaunay, 'Le Parnasse du temps de Napoléon...').

15. See. C. A. Valson, *La Vie et les Travaux du Baron Cauchy*, **1**, p. 10. It should be noted that Vitry, one of Louis-François' colleagues at the Ministry of the Interior and Chief of the Bureau of Agriculture, was the uncle of the Marquis de Fontanes.
16. Ibid., **1**, p. 17.
17. Ibid., **1**, pp. 16–18, and J. B. Biot, 'Lettre à Monsieur de Falloux', *Mélanges Scientifiques et Littéraires*, **3**, p. 144.
18. C. A. Valson, *La Vie et les Travaux du Baron Cauchy*, **1**, p. 18.
19. See J. B. Biot, *Mélanges...*, **3**, p. 144, and C. A. Valson, *La Vie et les Travaux du Baron Cauchy*, **1**, p. 18. According to Valson, ibid., **1**, p. 18, one day, Lagrange told Louis-François:

    Unless you hasten to give Augustin-Louis a solid literary education, his tastes (for it) will get swept aside; he will be a great mathematician, for sure; but, he won't be able to write his native language.

20. Presently the Lycée Henri IV.
21. See the *Almanach National*, 1802.
22. C. A. Valson, *La Vie et les Travaux du Baron Cauchy*, **1**, pp. 20–21.
23. Ibid., **1**, p. 21.
24. Ibid., **1**, p. 22.
25. Following is the letter that L. F. Cauchy wrote to Coulomb, Secretary General of the Ministry of the Interior, on that occasion [Arch. Nat. AA 63 (167)].

    Paris, 14 Fructidor, Year XI of the Republic.

    I should be very much obliged to you, Citizen Secretary General, if you would kindly obtain for me two or three tickets for the general awarding of prizes, which is to take place at the Institut the day after tomorrow, Sunday, 16 Fructidor. My eldest son, who successfully participated in the competition between the students of the three Écoles centrales, has received a letter from the director of his school summoning him to the [general] awards ceremonies, and I should like his mother and his brother, by their presence, to be able to partake in a small part of this honorable ceremony.

    If it pleases you, Citizen Secretary General, I assure you of my sincerest and utmost devotion.

    L. F. Cauchy

    P. S. Ten tickets have just been sent to the Senate; but *guid hoc inter tantos*? You can well imagine that there will not be one left that I can get.

26. Letter from Marie-Madeleine Cauchy to her granddaughter, cited by C. A. Valson, *La Vie et les Travaux du Baron Cauchy*, **1**, p. 21.
27. Alexandre-Laurent Cauchy was born on March 12, 1792, in Paris. He became assistant to his father at the Senate in 1810 and, after 1815, at the Chamber of Peers. In 1816, he entered the Congrégation. After the retirement of his father, he was named by the king, Keeper of the Archives of the Chamber of Peers in 1825. Still, his father, Louis-François, remained at the Palais du Luxembourg as Honorary Keeper. Meanwhile, Alexandre-Laurent was raised to the Bench: first judge–auditor in 1815, he was soon appointed judge at the Court of Paris and, in 1825, president of the division (président de chambre). He successively became president of the division at the Court of Appeals of Paris in 1847 and judge at the Cour de Cassation in 1849 and at the High Court in 1851. He died on March 3, 1857, a few

weeks before his brother Augustin-Louis. Alexandre-Laurent had married Clémentine Blanchet de la Salière in 1825 and had three sons and four daughters.

Eugène-François Cauchy was born on October 16, 1802, in Paris. He entered the Congrégation in 1822. In 1835, he replaced his brother Alexandre-Laurent, whose assistant he had been since 1815, as Keeper of the Archives of the Chamber of Peers. At the same time, he became maître des requêtes at the Conseil d'État. He lost both of these offices in 1848. Eugène Cauchy was a famed jurist, who was an expert in international law. He was elected to the Académie des Sciences Morales et Politiques in 1866. He died on April 2, 1877. Eugène had married Marie Anthelmine Richerand in 1833 and had one son and two daughters.

Louis-François Cauchy's daughter Marie-Madeleine was married to a relative, Pierre-Louis Guignon, an official of the Cour des Comptes in Paris. Thus, connections were not formed outside of the central administration in Paris. His second daughter, Marie-Thérèse, was born in 1796. She died at an early age.

28. An expression by A. L. Cauchy, taken from a letter dated August 17, 1835, and cited by C. A. Valson *La Vie et les Travaux du Baron Cauchy*, **1**, p. 90.
29. *Histoire de l'Académie Royale des Inscriptions et Belles-Lettres*, 2, **14**, p. 85.
30. On this occasion, Dinet became the friend of the Cauchy family, his neighbors at Arcueil (C. A. Valson, *La Vie et les Travaux du Baron Cauchy*, **1**, p. 22).
31. A. Fourcy, *Histoire de l'École Polytechnique*, Paris, 1828, pp. 376–379.
32. Ibid., pp. 255–257. In Arch. Ac. Sci., Ampère File, carton 5, folder 101, there is a detailed table of the analysis course taught to the second division by Lacroix in 1805–1806. This table shows that Lacroix closely followed his *Traité Élémentaire de Calcul Différentiel et Intégral* for his teaching.
33. According to the *Tableau présentant le mérite des élèves sur chaque partie de l'instruction d'après les examens intérieurs*, Cauchy earned the following merits in the first division: analysis, 20; statics, 18; geometry, 13; map drawing, 5; figure drawing, 12; general comments: hardworking and outstanding student.
34. *Correspondance sur l'École Polytechnique*, **1**, no 6, July 1806, p. 227, note. The geometrical solution of a problem proposed by Malus for the examination during this same period and published in the same issue of the *Correspondance*, p. 227, is simply signed C.; it was undoubtedly done by Cauchy.
35. Ibid., **1**, no. 6; July 1806, pp. 188–191. Cited by R. Taton, *L'Oeuvre Scientifique de Monge*, Paris, 1951, p. 262.
36. *Correspondance sur l'École Polytechnique*, **1**, no 6, pp. 193–195 (*O.C.*, 2, **2**, pp. 399–401).
37. Here is the description of A. L. Cauchy when he entered the École des Ponts et Chaussées: light brown hair and eyebrows, low forehead, Roman nose, grey eyes, middle mouth, round chin, long face, size: 1,63 m (Archives E. P., VI, 2, a, Cauchy file).
38. The Cauchy file at the École Polytechnique contains two letters from Louis-François Cauchy to the governor of the École regarding this matter. These letters are given here. The first letter, dated January 26, 1807, and addressed to Devernon, the Assistant Governor of the École, states:

Sir

My son has been helped in regaining his health by the special permission that you were so kind as to have granted him. However, in spite of the very best of care that we have given him, the illness continues, and I find that he cannot resume his work. I have had him remain in bed this morning, and I request that you allow him to remain here at home until he has recovered. I respectfully recognize this new obligation that I have to you, and I am, sir,

Your very humble and obedient servant,

Cauchy

The second letter, though undated was probably written early in October 1807 and was addressed to General Lacuée, governor of the school:

My Dear Governor:

My son has completed his examinations and thus his presence at the school is pointless. I would like to request that he be allowed to take advantage of these last days of the season to come to Arcueil and improve his health while awaiting news from the schools to which he applied. I therefore humbly request that you grant him a leave of absence, sir. This, I know, is a new request, and I shall not be less grateful to you for granting it than I have been for those that you have granted in the past. Please be assured of my deepest respects,

Most sincerely,

Cauchy

39. See A. Debauve, *Les Travaux Publics et les Ingénieurs des Ponts et Chaussées depuis le XVIIIème Siècle*, Paris, 1893.
40. C. A. Valson, *La Vie et les Travaux du Baron Cauchy*, **1**, p. 24. According to a letter that Louis-François wrote to Count Molé, Director-Général of the Ponts et Chaussées service, in 1811 (Arch. Nat. F^{14} 2187^2, Cauchy file) Cauchy won the competition for wooden bridge engineering.
41. ENPC library, Ms. 1845. The autographic manuscript, dated February 15, 1808, in Paris, and signed Aug. Louis Cauchy consists of 27 pages with 25 lines per page.
42. See Arch. Nat. F^{14} 2234^5, Girard file and Archives Ac. Sci. Girard file. P. S. Girard (1765–1836) entered the École des Ponts et Chaussées in 1784. He was appointed engineer-ordinary in 1789. In 1792, he won the Académie's double prize for his study on canal locks, and in 1798, he presented his *Traité Analytique de la Résistance des Solides et des Solides d'Égale Résistance, auquel on a joint une Suite de Nouvelles Expériences sur la Force et l'Élasticité Spécifiques des Bois de Chêne et de Sapin.* He was appointed member of the Institut d'Égypte in 1798 and was elected to the First Class of the Institut in June 1815. Because of a disagreement with Ponts et Chaussées inspector Cahouët, he was removed from his position directing the Ourcq canal in 1817.
43. In 1806, P. T. M. Égault (1777–1829) invented a new water level that was adopted by the Ponts et Chaussées service. Under his direction, Cauchy worked on the surveying of the circle aqueduct, which ran along the inner walls of Paris from la Villette to Mousseaux and from which large underground conducts or galleries ran to the center of the city. Following this, he worked on the raising of the land along which the Galerie de Saint-Laurent would pass. Located between the Faubourg Saint-Laurent and the Faubourg Saint-Denis, this gallery's length measured some 600 m.

44. Report from Girard to Count Molé, December 20, 1808, Arch. Nat. F^{14} 2187^2, Cauchy file. This report was published in A. Brunot and R. Coquand, *Le Corps des Ponts et Chaussées*, Paris, 1982, p. 95, note 2.
45. See C. A. Valson, *La Vie et les Travaux du Baron Cauchy*, **1**, 25. Cauchy's study of navigation on the Marne gives support to the idea that he worked also on the Ourcq Canal in 1809.
46. Arch. Nat. F^{14} 2148. An anonymous two-page note entitled 'Sur la solution de M. Cauchy relative au problème proposé pour le concours de 1809' is in the ENPC library (Ms 1982).
47. From a letter written by Louis-François Cauchy to Count Molé in May 1811 (Arch. Nat. F^{14} 21872, Cauchy file). Louis-François heard of the loss of these studies from Monge several days before the awarding of the prize for the competition of 1809. Concerning these studies, Louis-François observed that his son 'had spent considerable time on this work' and, according to his professors, was successful. C. A. Valson (*La Vie et les Travaux du Baron Cauchy*, **1**, p. 43) indicated that Cauchy's first work at Cherbourg was a study 'on the theory of stone bridges and of arches in general'. The question obviously had to do with one of the studies of 1809. Valson's errors came from a letter that Cauchy wrote his father. In it, he asked for 'this study and that you make an effort to find it and send it to me, if not completely, then at least the main formulas that I need in order to continue my research'. From what was said, the date of this letter from Cauchy to his father can reasonably be set as May 1811.
48. 'Mémoire sur les moyens de perfectionner la navigation des rivières en général et celle de la Marne en particulier', handwritten manuscript of 51 pages, ENPC library, Ms 1982.
49. 'Mémoire sur les ponts en pierre, par A. L. Cauchy, élève des Ponts et Chaussées', handwritten manuscript of 32 pages, ENPC library, Ms 1982.
50. 'Second mémoire sur les ponts en pierre, théorie des voûtes en berceau, par A. L. Cauchy, élève ingénieur des Ponts et Chaussées', handwritten manuscript of 52 pages, ENPC library, Ms. 1982.
51. C. A. Valson, *La Vie et les Travaux du Baron Cauchy*, **1**, p. 43.
52. See C. A. Geoffroy de Grandmaison, *La Congrégation (1801–1830)*, Paris, 1889.

Notes to Chapter 2

1. Arch. Nat. F^{14} 2187^2, Cauchy file. Letter from L. F. Cauchy to Count Molé, September 15, 1812.
2. On the history of the site of Cherbourg, see A. Demangeon and B. Fortier, *Les Vaisseaux et les Villes. L'Arsenal de Cherbourg*, Bruxelles, 1978.
3. C. A. Valson, *La Vie et les Travaux du Baron Cauchy*, **1**, p. 27.
4. Arch. Nat. DD. 2 144, 'Rapport sur l'établissement maritime projeté à Cherbourg' by Cachin, 23 Germinal year XI (April 14, 1803), published in A. Demangeon and B. Fortier, *Les Vaisseaux et les Villes...*, pp. 106–117
5. Arrêté of 25 Germinal Year XI (April 16, 1803), published by A. Demangeon and B. Fortier, *Les Vaisseaux et les Villes...*, pp. 44–45.
6. Ibid., p. 154.
7. It is estimated that there were at least 200 steam engines in France around 1810.
8. Cited by C. A. Valson, *La Vie et les Travaux du Baron Cauchy*, **1**, p. 30.
9. Arch. Nat. F^{14} 2187^2, Cauchy file. Letter from Louis-François dated July 9, 1810. On July 16, the director of personnel replied that he had just requested that the

Minister of the Marine agree that Augustin-Louis Cauchy retain his rank of aspirant while assigned to work at Cherbourg.

10. 'The enormous basins that ceaselessly resound under the blows of steel' is part of a verse that Augustin-Louis Cauchy included in a letter to his mother in August 1810, cited by C. A. Valson, *La Vie et les Travaux du Baron Cauchy*, **1**, p. 3.
11. Letter from A. L. Cauchy to his family, written on July 3, 1811; ibid., **1**, p. 28.
12. Letter from J. F. Cachin to A. L. Cauchy, undated; ibid., **1**, p. 32.
13. Letter from A. L. Cauchy to his mother, written in 1810; ibid., **1**, pp. 37–41. The people mentioned by Cauchy have not all been identified. Hippolyte Franqueville was the chief commissioner of the Port of Cherbourg. L. B. Fouques-Duparc (1772–1848), engineer of the Ponts et Chaussées, entered the Marine Service in 1803 and succeeded Cachin as the director of the site in 1813. S. Vallot (1772–1847), engineer–geographer and then engineer of the Ponts et Chaussées, taught construction at the École des Ponts et Chaussées. As to Monsieur L..., the reference is probably to Count of Latour-Maubourg.
14. Ibid., pp. 38–39.
15. Letter from A. L. Cauchy to his father, dated June 8, 1810; ibid., **1**, p. 31.
16. Letter from A. L. Cauchy to his family, undated; ibid., **1**, p. 31.
17. 'Sur les limites des connaissances humaines', Bibliothèque de l'Institut, Ms. 2038; F^0 83r–86r, (*O.C.*, 2, **15**, pp. 5–7).
18. Letter from Augustin-Louis Cauchy to his family, dated December 10, 1810, cited by C. A. Valson, *La Vie et les Travaux du Baron Cauchy*, **1**, p. 29.
19. Ibid., **1**, p. 45. No source is given as a basis for this statement. It seems quite likely, although Valson apparently confused the 1811 study with the study of January 20, 1812, entitled 'Sur les polygones et les polyèdres'.
20. *Procès-verbaux, Ac. Sci.*, **4**, p. 449. The manuscript, entitled 'Recherches sur les polyèdres', is in the meeting's packet of February 11, 1811. A résumé was published in the *Correspondance sur l'École Polytechnique*, **2**, n^0 3, January 1811, pp. 253–256, (*O.C.*, 2, **2**, pp. 402–405). The complete study appeared in the *J.E.P.*, **9**, 16th cahier, May 1813, pp. 68–86, *O.C.*, 2, **1**, pp. 7–25). On Cauchy's proof of the Euler formula, see J. C. Pont, *La Topologie Algébrique des Origines à Poincaré*, Paris, 1974, pp. 21–24, and N. Briggs, E. K. Lloyd, and R. J. Wilson, *Graph Theory (1736–1936)*, Oxford, 1976, pp. 74–83.
21. *Procès-verbaux, Ac. Sci.*, **4**, pp. 467–477, and *Correspondance sur l'École Polytechnique*, **2**, n^0 4, July 1812, p. 361 (*O.C.*, 2, **2**, pp. 406–408).
22. *Procès-verbaux, Ac. Sci.*, **5**, p. 6.
23. The manuscript is in the meeting's packet of January 20, 1812. A résumé was published in the *Journal des Mines*, **31**, n^0 184, April, 1812, pp. 314–318. (*O.C.*, 2, **15**, pp. 8–10). The complete study appeared in the *J.E.P.*, **9**, 16th cahier, May 1812, pp. 87–98. (*O.C.*, 2, **1**, pp. 26–38).
24. C. A. Valson, *La Vie et les Travaux du Baron Cauchy*, **1**, pp. 47–48.
25. Legendre report, *Procès-verbaux, Ac. Sci.*, **5**, pp. 17–18, and *Correspondance sur l'École Polytechnique*, **2**, n^0 4, July 1812 (*O.C.*, 2, **2**, pp. 408–413).
26. See. *O.C.*, 2, **15**, p. 6.
27. Relative to this point, see, for example, J. B. J. Delambre, *Rapport Historique sur les Progrès des Sciences Mathématiques depuis 1789 et sur leur État Actuel*, Paris, 1810. According to a letter mentioned by C. A. Valson, *La Vie et les Travaux du Baron Cauchy*, **1**, p. 43, it seems that in 1811 Cauchy contemplated pursuing the investigations on the theory of vaulted arches that he had undertaken at the École

des Ponts et Chaussées in 1809. This would have been an abandonment of pure mathematics.

28. From J. Mandelbaum, *La Société Philomatique de Paris de 1788 à 1835*, 3rd cycle, thesis, typewritten, Paris, EHESS, 1980, **1**, pp. 199–200.
29. C. A. Valson, *La Vie et les Travaux du Baron Cauchy*, **1**, p. 54.
30. Letter from Louis-François Cauchy to his son, undated; cited by C. A. Valson, *La Vie et les Travaux du Baron Cauchy*, **4**, p. 54.
31. Ibid., **1**, pp. 48–49.
32. Ibid., **1**, p. 29.
33. Ibid., **1**, p. 244 and p. 245.
34. Arch. Nat. F^{14} 2187^2, Cauchy file.
35. See the letters from Augustin-Louis, Louis-François, and Marie-Madeleine Cauchy, September and October 1812, in the private correspondence of the family Cauchy, kept by M. de Leudeville, Paris.

Notes to Chapter 3

1. *Procès-verbaux, Ac. Sci.*, **5**, p. 121. The manuscript is lost. The memoir was published in two articles, 'Sur le nombre des valeurs qu'une fonction peut acquérir lorsqu'on y permute de toutes les manières possibles les quantités qu'elle renferme' and 'Sur les fonctions qui ne peuvent obtenir que deux valeurs égales et de signes contraires par suite des transpositions opérées entre les variables qu'elles renferment', *J.E.P.*, **10**, 17th cahier, January 1815, pp. 1–28 and pp. 29–112, (*O.C.*, 2, **1**, pp. 64–90 and pp. 91–169). For a critical study of this paper, see A. Dahan, *Les Recherches Algébriques de Cauchy*, 3rd cycle thesis, typewritten, Paris, EHESS, 1979, pp. 11–34, and 'Les travaux de Cauchy sur les substitutions. Etude de son approche du concept de groupe', *Archive for History of Exact Sciences*, **23**, 1980, pp. 279–319.
2. On March 30 and April 3, Cauchy read a paper entitled 'Sur le nombre des polygones que l'on peut former en prenant pour sommets les points de division d'une circonférence divisée en plusieurs parties égales' at the Société Philomatique. The problem had been proposed in Gergonne's *Annales de Mathématiques*, **3**, pp. 231–232. This unpublished work was clearly an application of this study in which Cauchy represented a p-cycle as a polygon.
3. Cauchy obtained his theorem on the product of two determinants during the summer of 1812, while in Cherbourg. He was attempting 'to generalize the formulas of M. Gauss'. Jacques Binet, who was Cauchy's friend, independently obtained a similar set of formulas, but they were much more difficult to work with.
4. *Procès-verbaux des Séances* of the Committee of the École des Ponts et Chaussées, meeting of February 3, 1811, cited by André Lorion, 'L'École des Ponts et Chausées sous le Premier Empire (documents inédits)', *Revue de l'Institut Napoléon*, n⁰ 66, 1958, pp. 81–86.
5. Arch. Nat. F^{14} 2187^2, Cauchy file, letter from the personnel director to the Minister of the Marine and Colonies, dated February 16, 1813:

> Monsieur Cauchy, an engineer-ordinary who is assigned to the port of Cherbourg, has informed me that his health will not easily allow him to take his post and has asked to be assigned to service in the interior. Accordingly, I now ask your excellency to inform me by May 1 next whether or not, in the event that it is found that his presence is not essential, I can replace him at Cherbourg.

On March 18, 1813, Cauchy sent a letter of thanks to Count Molé, expressing gratitude for his (Cauchy's) reassignment to Paris. Five years later, under rather similar circumstances, the physicist Augustin Fresnel was appointed to a post with the Ourcq Canal Project.

6. Later, by a decree of December 24, 1816, Lehot became a répétiteur in physics at the École Polytechnique, where he and Cauchy once again crossed paths.
7. See Arch. Nat. F^{14} 2263^2, Lehot file and letter from Cauchy to Lehot, Paris, April 5, 1813, in Archives Ac. Sci., Bertrand autographs' collection, carton 1:

   My dear friend:

   I have just visited M. Picard, with whom I have an appointment for tomorrow at the Saint-Martin Canal. I will visit you at 11:30, and together we will be able to check the measurements and rule definitively on M. Picard's report. I am going to write to M. Prosper and to M. Delozanne, so that the former will assist us, while the latter will be able to support his interests. I assure you of my continuing sincere devotion.

   Picard, Prosper, and Delozanne were probably master builders. According to the report made by a commission of the Ponts et Chaussées on January 1, 1816 (Arch. Nat. F^{14} 7012), concerning the Ourcq Canal Project and related works, the work had not progressed very far. 'Meanwhile, earth had been removed from the slope at the bend of the Villette' and in continuation of the foundations of the Fontaine de l'Eléphant, a stretch of canal in stone, 90 m in length, a part of which is already vaulted, has been constructed. During his brief time with the Ourcq Canal and Paris Waterworks Project, Cauchy probably worked at these construction sites. Lehot did not leave until the autumn of 1813.
8. Arch. Nat. F^{14} 2263^2, Lehot file.
9. See G. Bigourdan, 'Le Bureau des Longitudes; son histoire et ses travaux de l'origine (1795) à ce jour', *Annuaire du Bureau des Longitudes*, 1928; A30–31 and A36.
10. *Procès-verbaux, Ac. Sci.*, **5**, p. 216, (*O.C.*, 2, **15**, p. 16). The commission was composed of Legendre, Carnot, and Poisson.
11. *Rapports sur divers mémoires lus à la première classe de l'Institut impérial par A. L. Cauchy, ingénieur des Ponts et Chaussées*; Paris, 1813, 15p. Copy in the Bibliothèque de l'Institut; 8^0 HR 25, I (10), (*O.C.*, 2, **15**, pp. 11–16).
12. 'Recherche sur les nombres', *J.E.P.*, **9**, 16th cahier; pp. 99–123 (*O.C.*, 2, **1**, pp. 39–63).
13. The meeting's packet of May 17, 1813, contains two manuscripts on this memoir:

   1. A 21-page manuscript entitled 'Théorie des équations: méthode pour déterminer à priori le nombre de racines réelles positives et le nombre des racines réelles négatives d'une équation d'un degré quelconque par A. L. Cauchy ingénieur des Ponts et Chaussées. Mémoire'.
   2. A 7-page manuscript entitled 'Théorie des équations: méthode pour déterminer à priori le nombre des racines réelles positives, et le nombre des racines réelles négatives d'une équation de degré quelconque par A. L. Cauchy, ingénieur des Ponts et Chaussées employé à Paris aux travaux du Canal de l'Qurcq. Exposé sommaire de cette méthode'.

   A note signed by Delambre on the first page of the latter memoir suggests that it was presented on May 13, 1813, and that Laplace, Biot, and Poisson were responsible for evaluating it. It is the manuscript that begins with 'Messieurs, . . . ,' which was to be presented at the session. It was printed by Veuve Courcier in 1813 (see *O. C.*, 2, **15**, pp. 11–16).

14. 'Mémoire sur la détermination du nombre des racines réelles dans les équations algébriques', *J.E.P.*, **10**, 17th cahier, pp. 457–558, (*O.C.*, 2, **1**, pp. 170–257). This memoir consists of extracts of the papers Cauchy had delivered at the Institut during 1813.
15. *Procès-verbaux, Ac. Sci.*, **5**, p. 217 and p. 218.
16. Arch. Nat. F^{14} 2187^2, Cauchy file, letter of thanks from A. L. Cauchy to Count Molé, dated June 8, 1813. See also Arch. Nat. F^{14} 2210^1, Denoël file.
17. Arch. Nat. F^{14} 2187^2, Cauchy file.
18. The meeting's packet of October 18, 1813, includes the manuscript of a study entitled 'Méthode pour déterminer à priori le nombre de solutions positives et le nombre de solutions négatives d'une équation de degré quelconque'. The file of the meeting of November 22, 1813, contains several other manuscripts:

    1. A small study, probably written shortly after October 18, entitled 'Théorème sur la différence entre le nombre de racines positives et le nombre de racines négatives d'une équation de degré *n*';
    2. A short note in which he explains how 'the general method, which would seem very complicated at first, actually has an unexpected degree of simplicity'.
    3. A long paper that was presented on November 22 and is entitled 'Sur un moyen d'éviter l'emploi des indéterminées dans la formation des équations auxiliaires qui servent à fixer le nombre des solutions positives et le nombre des solutions négatives d'une équation quelconque'.

    Each of these papers has remained unpublished, but the basic ideas contained in them are found in the long article in *J.E.P.*, cited in note 14.
19. Arch. Nat. F^{14} 2187^2, Cauchy file, letter from Louis-François Cauchy to Baron Costaz, dated January 12, 1814.
20. A note written in pencil in the margin of the letter of January 12, 1814, which is cited in the preceding note, points out [that] 'it is necessary to keep M. Cauchy on the Ourcq Canal Project'. Moreover, in the *Livre du Mouvement du Personnel des Ponts et Chaussées* for the year 1816, we find, that among the engineers assigned to the Ourcq Canal Project: 'Augustin-Louis Cauchy, engineer 2nd class, attached to the Institut, received no salary from Ponts et Chaussées'. No document, however, reveals when Cauchy effectively ceased working with the Ponts et Chaussées.
21. 'Sur le système des valeurs qu'il faut attribuer à divers éléments déterminés par un grand nombre d'observations pour que la plus grande des erreurs, abstraction faite du signe, devienne un minimum', *Bulletin Phil.*, June 1814, pp. 92–99, (*O.C.*, 2, **2**, pp. 312–322).
22. The Great Referendary, under whose order Louis-François worked, was the peer of France who was responsible for putting the seal of the assembly on enacted measures and bills and taking care of the archives.
23. Letter cited by C. A. Valson, *La Vie et les Travaux du Baron Cauchy*, **1**, pp. 56–57.
24. *Indication sommaire des mémoires présentés à la première classe de l'Institut par A. L. Cauchy, ingénieur des Ponts et Chaussées. Conclusions des rapports faits à la classe sur ces mémoires*, Paris, 1814. Copy in the Bibliothèque de l'Institut, Rec. H. R. 26 (t. I, n^0 9). The titles of the papers and studies presented to the Institut in 1811, 1812, 1813, and 1814 are formed here (**M1** to **M8** from the nomenclature of the *O. C.*'s bibliography, *O. C.*, 2, **15**, pp. 589–595).

25. C. A. Valson, *La Vie et les Travaux du Baron Cauchy*, **1**, pp. 55–56.
26. Letter from A. M. Ampère to Bredin, dated October 13, 1814, *Correspondance du Grand Ampère*, **2**, Paris, 1936, p. 486.
27. *Procès-verbaux, Ac. Sci.*, **5**, p. 431 and p. 434.
28. J. Mandelbaum, *La Société Philomatique de Paris de 1788 à 1835*, 3rd cycle thesis, typewritten, Paris, EHESS, 1980, **1**, pp. 199–200.
29. *Procès-verbaux, Ac. Sci.*, **5**, pp. 495–496 and p. 502.

Notes to Chapter 4

1. Louis-François got through this troubled period without undue difficulties, being Keeper of the Archives in the Chamber of Peers during the First Restoration; Secretary of the Archives of the Senate during the Hundred Days; and, again, Keeper of the Archives of the Chamber of Peers during the Second Restoration. Fontanes explained Louis-François' staying power in the following terms: 'He was appreciated at the Palais du Luxembourg for his Latin verses as well as for his perfect integrity and his capacity to work in a very useful and unobtrusive, inoffensive way'. (Cited by A. F. Villemain in his *Souvenirs Contemporains d'Histoire et de Littérature*, Paris, 1855, p. 42).
2. Letter from A. L. Cauchy to the Abbé Jean-Baptiste-Armand Auger, dated September 3, 1815, published by J. Pelseneer in the *Archives Internationales d'Histoire des Sciences*, **4**, 1951, pp. 631–633. Born in 1784, Auger was a professor of mathematics, a member of the Congrégation, and from 1814, the Vicar of the Saint-François Parish in Le Havre. (See Th. Lebreton, *Biographie Normande*, Rouen, 1857–1861, J. B. A. Auger, article.)
3. See P. Lafitte, 'Relations d'Auguste Comte avec Poinsot', *Revue occidentale*, March **1**, 1886, p. 147.
4. Archives E. P., Poinsot file, Letter of November 15, 1815, from Durivau to the governor, the Count Dejean:

   My General,

   Since it is urgent to temporarily replace M. Poinsot, in order to provide an unbroken progress of teaching, I beg to subject to your examination the names of two persons who can be equally chosen. These are MM. Cauchy and Lefebvre. Both of them are known by the favorable reviews the Conseil de Perfectionnement, and especially several first-class scientists, has given. Thus, in respect to ability, it is unnecessary to collect more information than you already have. But, another kind of consideration can guide your choice: M. Lefebvre is already busy at the École as répétiteur. We cannot take him from his position without replacing him temporarily. Such a temporary appointment is not attractive, and we cannot propose it, I believe, to M. Cauchy. Thus, we could not seize the opportunity to make this eminent person attached to the École. However, it is beyond doubt that he would gladly accept to replace a *professor*, even temporarily. Moreover, M. Lefebvre has just been promoted as répétiteur. This first advantage would be followed very closely by another, if he was chosen this time. As for M. Cauchy, he should have obtained nothing but promises. I think, therefore, General, we must appoint him to the temporary post. If you will engage him, I shall beg him to come to the École in order to be questioned for his new job and to open the analysis course tomorrow, November 16, if possible. We will see which decision M. Poinsot takes. By that time, the teaching will not be broken.

   Durivau

5. The Governor Dejean answered the treasurer of the École, Marielle, who proposed increasing Cauchy's pay (Archives E. P., Poinsot file, letter from Dejean to Marielle):

   M. Cauchy is a beginner. He has, therefore, to reduce his claims for his pay. Obtaining the post will compensate him later. I persist. December 8.

   Count Dejean

6. Joseph Bertrand wrongly stated in his *Éloges Académiques*, Paris, 1902, **2**, p. 109, that 'a few months after his return [from Cherbourg], Cauchy became a répétiteur at the École Polytechnique'.
7. In the draft copy of a letter to te minister of the Interior, dated December 22, 1815, the Governor of the École himself avowed that:

   In political terms, the committee's choices are not less satisfactory. Among the persons who have been nominated there are only true and faithful Frenchmen who are quite likely to inspire young people to be serious about their duties and obligations and to seek to perfect their education. I add that this situation excludes from our institution certain persons whose names became, unfortunately, too notorious during the Revolution and admits no one who, at this time, is not acceptable to the Court.

   This last phrase is crossed off (in the draft). Dejean added in the margin:

   A person might well state this by word of mouth—but it ought not be put in writing.

8. *P.V. Ac. Sci.*, **5**, December 26, 1815, p. 738.
9. *Procès-verbaux, Ac. Sci.*, **5**, November 13, 1815, p. 576. An abstract appeared in the *Bull. Phil.*, Dec. 1815, pp. 196–197, (*O.C.*, 2, **2**, pp. 204–206). Legendre, who was the reporter of the study, gives Cauchy's proof in a supplement to the second edition of his *Essai sur la Théorie des Nombres* in February 1816. Cauchy's paper was published in the *Mém. Institut*, 1, **14** (1813–1815), 1818, pp. 172–220, and in the *Ex. Math.*, **2**, Nov. 1826, pp. 265–296, (*O.C.*, 2, **6**, pp. 320–353).
10. Cauchy presented to the Société Philomatique on January 28, 1816, a solution to a similar but more general problem than the one he had dealt with in his paper on Fermat's theorem: to decompose a given integer into several squares whose roots shall form a given sum. His solution remains unpublished (see J. Mandelbaum, *La Socité Philomatique de Paris*... **1**, pp. 199–200).
11. *Procès-verbaux, Ac. Sci.*, **5**, December 26, 1815, p. 596.
12. J. Bertrand, *Éloges Académiques*, Paris, 1902, **2**, p. 112.
13. On Cauchy's teaching course of 1815–1816, we have the *Registre d'instruction* (See Appendix I) and the notes of the student Auguste Comte (See T. Guitard, 'La querelle des infiniment petits à l'École Polytechnique au XIXe siècle' *Historia scientarium*, n^0 30, pp. 1–61, especially pp. 28–31).
14. For a discussion of these events, see H. Gouhier, *La Jeunesse d'Auguste Comte et la Formation du Positivisme*, Paris, 1933–1941, **1**, pp. 116–122.
15. See C. Maréchal *La Dispute de l'Essai sur l'Indifference*, Paris, 1924, pp. 87–107.
16. Archives of the Collège de France, *Registre des délibérations prises aux Assemblées des lecteurs et professeurs du Roi au Collège Royal de France*, Vol. III; Meeting of November 10, 1816.
17. See Chapter 7, p. 113.
18. Archives of the Collège de France, vol. IV; meeting of November 21, 1824:

> A letter from M. Biot to the administrator was read. In this letter, M. Biot, who was on a trip that was to last for approximately one year, asked the professors to decide in favor of M. Cauchy teaching the course in mathematical physics during M. Biot's absence. M. Cauchy is a member of the Académie Royale des Sciences and a professor at the École Polytechnique. This proposition wàs supported by the assembly, and M. Cauchy's classes became part of the curriculum.

Cauchy's courses effectively began at the end of November.

19. Archives of the Collège de France, Biot file:

> Finding it impossible to attend the next meeting, I ask the honorable administrator to please request in my name from the professors that they agree that M. Cauchy (who had already substituted several times for me) should assume my duties and receive, in payment, 2000 francs of my annual pay, effective January 1, 1826 (in agreement with the guidelines submitted by our school to the Minister of the interior). I nevertheless desire to preserve the possibility and the right to reclaim my whole salary, if new circumstances in my present situation should come about so as to oblige me to reconsider and resume the duties of my chair.
>
> Paris, November 21, 1825
> J. B. Biot.

20. For a discussion on the method of solving linear differential equations, see *C.R. Ac. Sci.*, **8**, p. 829, (*O.C.*, 1, **7**, p. 369). In 1827–1828, according to a program of the courses, which is preserved in the Archives of the Collège de France, Cauchy expounded on 'The general methods by which it is possible to solve the main equations of mathematical physics'. Finally, in 1829–1830, Cauchy presented his theory of light (see Chapter 6, pp. 104–105).
21. Arch. Nat. AJ 16 25 (file 1822). The salary amounted to 2000 francs, to be deducted from the titular's salary.
22. Arch. Nat. AJ 16 5126.
23. According to a letter from Ampère to Bredin, dated March 16, 1821 (*La Correspondance du Grand Ampère*, **3,** Paris, 1943, p. 908):

> I suffer a good deal from chest and lung ailments, and it has been necessary for me to suspend my course at the Faculté. M. Cauchy will replace me for a month or six weeks.

24. There is little material bearing on these courses. From Cauchy, there is the note given in the appendix to his September 16, 1822, paper and the note given in *C.R. Ac. Sci.*, **16**, February 20, 1843, p. 413, (*O.C.*, 1, 7, p. 261):

> In December 1821, in my course on mechanics at the Faculté de Sciences, I presented a general method by means of which I have been able to obtain certain formulas pertaining to surfaces, volumes, masses, etc., relative to an arbitrary system of curvilinear coordinates...

In the preface of his *Leçons de Mécanique Analytique*, Paris, 1867, Moigno wrote:

> This material was first given quite some time ago, because it is essentially the course that my illustrious and venerated teacher. A. L. Cauchy, taught at the École Polytechnique during the years from 1820 to 1830. I faithfully recorded and edited this material; and, for his own part, Cauchy, in *Exercices de Mathématiques* and in *Nouveaux Exercices de Géométrie et de Physique Analytique* (sic), published the most complete theorems, such as those on linear moments and on the investigation of the general equations of equilibrium, which constituted the basic material for his courses.

> From 1838 to 1843, I used this handwritten material as the text for courses I taught at the École Normale Ecclésiastique de la rue des Postes. Moreover, if the unfortunate incident I discussed in the preface to my lessons on integral calculus had not taken place, they would have appeared a long time ago.

As Moigno stated later in his preface, Cauchy was the real author of *Leçons*, especially of the first 9 and the 12th. They provide an idea of the material (at least in part) that Cauchy presented in his courses at the Faculté des Sciences. Also, see the 10 articles contained in *Exercices de Mathématiques* (issues for the Ist and 2nd years) in which the principles of mechanics are discussed. In particular, there is a discussion of the theory of linear moments (see p. 259, note 56).

25. Arch. Nat. AJ 16 207. In 1824, Cauchy seems to have been somewhat hesitant to assume the vacant chair in astronomy. This is according to a letter Ampère wrote to Monseigneur Frayssinous (*Correspondance du Grand Ampère*, **2**, p. 667):

> I have learned that M. Cauchy, who has so brilliantly filled the chain in mechanics at the Faculté des Sciences, in fact, prefers the now vacant chair in astronomy. If your excellency has not already chosen another professor of mechanics, I should like very much to be considered for the vacancy in mechanics. I have taught mechanics for seven years at the École Royale Polytechnique. At the Faculté, M. Cauchy was teaching precisely the same material as that which constitutes the course at this school. Since I have been continuously engaged in teaching this material, perhaps it is I who might do the best of any of those who might be selected, in terms of presenting a course that assumes special study and in-depth investigation of this branch of the mathematical sciences.

26. Cauchy specified his criticisms during the following year. As to generating functions, see the paper that was presented to the Académie on December 27, 1824, and was published under the title 'Mémoire sur le calcul intégral' in *Mémoires Ac. Sci.*, **22**, 1850, pp. 39–130; (*O.C.*, 1, **2**, pp. 195–281). For Poisson's study, see Chapter 7, p. 122.
27. In a letter to Paolo Ruffini, dated September 20, 1821, which was published in P. Ruffini, *Opere Mathematiche*, **3**, Rome, 1954, pp. 88–89, Cauchy indicated the following on this point:

> I have long since been bound to the author, whom you have refuted and I owe him much, [still] I have never hidden my feelings nor my principles from him. In the introduction to my course on analysis, in which I otherwise gave him all possible credit, I formally stated—as you can read for yourself—that history ought not be investigated from a standpoint of formulas; nor should sanction be sought for ethics and morals in the theorems of algebra or integral calculus.

As to Cauchy's views on the calculus of probabilities, see Chapter 13, p. 219.

28. See the list of studies and papers for which Cauchy acted as reporter in *O.C.*, 2, **15**, pp. 518–526.
29. The study was not published and the manuscript disappeared. The title of this work is uncertain: 'Intégration des équations linéaires aux différences finies ou infiniment petit' or 'Recherche sur le calcul intégral aux différences partielles.' See Chapter 7, p. 122.
30. On the dispute of April 1829, see the minutes of the meetings of the Académie of April 6, 1829, and April 27, 1829. Cauchy's arguments were published in *Ex. Math.*, **4**, November 29, pp. 214–216. (*O.C.*, 2, **9**, pp. 254–258). Cauchy's observations on a question of priority were kept in the meetings's packet for June 14, 1830.

31. For example, in the introduction to *Analyse Algébrique* of 1821 and in the foreword of *Leçons sur le Calcul Différentiel.*
32. Poinsot said of his successor at the École Polytechnique: 'Cauchy is affected by a diarrhea of x'. He told Auguste Comte of his dislike for Cauchy. See P. Lafitte, 'Relations d'Auguste Comte avec Poinsot', and H. Gouhier, *La Formation du Positivisme: La Jeunesse d'Auguste Comte*, **1**, Paris, 1933, p. 129.
33. See Poinsot, note in *Bull. Fer.*, **7**, April 1827, pp. 224–226, and Cauchy's reply in the same journal, *Bull. Fer.*, **7**, May 1827, pp. 333–337 (*O.C.*, 2, **15**, pp. 138–140). The following remark by Cauchy should be noted:

    The time that scholars and scientists spend in 'making war' on each other I regard as nothing short of a loss for science. I believe that it is far better to solve problems and investigate questions than to get involved in disputes.

    It is unfortunate that Cauchy himself did not always adhere to such a wise attitude!
34. The bonds of friendship that existed between Cauchy and Binet were strong, durable, and old. Cauchy mentioned Binet, his close friend since 1812, in his paper on determinants. On May 1, 1821, he officiated at the reception for Binet when he (Binet) became a knight of the Legion d'Honneur, and he spoke for the Académie at Binet's funeral in 1856.
35. See, for example, the article on Cauchy in *Galerie Historique des Contemporains*, 2nd supplement, Mons, undated, signed D.M. (1828–1829?):

    He has a dry, rigid personality and his lack of tolerance for or indulgence with young people who would carve out careers for themselves in science has made him one of the least likeable—and certainly one of the least liked—scholar–scientists.
36. For these 32 papers that were examined outside the commission, Cauchy prepared 9 written evaluative reports and 17 verbal reports.
37. On this academic work, see the documentary appendix established by R. Taton in *O.C.*, 2, **15**, pp. 518–526 and pp. 579–580.
38. J. V. Poncelet, *Applications d'Analyse et de Géométrie*, **2**, Paris, 1864, p. 564.
39. Translated into French in *Niels-Henrik Abel, Mémorial Publié á l'Occasion du Centenaire de sa Naissance*, Christiania, 1902, 3rd pagination, p. 45.
40. Letter to Holmboe in Berlin, dated January 20, 1827; ibid., 3rd pagination, p. 57.
41. See *O.C.*, 2, **15**, pp. 572–573.
42. See R. Taton, 'Sur les relations scientifiques d'Augustin Cauchy et d'Evariste Galois', *Rev. Hist. Sci.*, **24**, 1971, pp. 123–148, esp. p. 138.
43. See A. Dahan, *Les Recherches algébriques de Cauchy*, 3rd cycle, Thesis, typewritten, 1979, pp. 80–83.
44. A. Iushkevich, *Michel Ostrogradski et le Progrès de la Science au XIXème Siècle, Conférence donnée au Palais de la Découverte*, Paris, 1967, p. 13.
45. See the chronological table of the studies presented to the Académie between 1816 and 1830 and the chronological table of publications for the same period. Arranged by R. Taton in the documentary appendix of *Oeuvres Complètes*, *O.C.*, 2, **15**, pp. 590–594 and pp. 598–601.

Notes to Chapter 5

1. See the introduction to *Analyse Algébrique* and the foreward to *Leçons sur les Applications du Calcul Infinitésimal à la Géométrie*. Cauchy also consulted Coriolis on his first investigations on the theory of light (see Chapter 6, p. 105).

2. Archives E. P., *Registre des procès-verbaux des séances du Conseil d'Instruction*, November 9, 1816.
3. Archives Ac. Sci., Ampère's papers, box 4, chapter 4, file 75, 'Programme du cours d'analyse par Augustin Cauchy'. See Appendix II pp. 303–305.
4. Falling under the traditional heading of algebraic analysis were basic notions of analysis, real and complex numbers, functions, infinitesimals, series, etc.
5. See especially *Mémoires Sav. Étr.*, **I**, 1827, p. 687, (*O.C.*, 1, **1**, pp. 402–403). In this passage, Cauchy asked how can the value of the definite integral

$$\int_b^{b'} \phi'(z)\,dz$$

be derived from the value of the primitive function $\phi(z)$ (see Chapter 7, p. 110):

> If the function $\phi(z)$ increases or decreases in a continuous manner between the limits $z=b$ and $z=b'$, the value of the integral will be represented, ordinarily, by $\phi(b'')-\phi(b')$. However, if for one value of z represented by Z and lying between the limits of integration, the function $\phi(z)$ passes suddenly from one fixed value to a value sensibly different from the first, in such a way that for a very small quantitiy ζ, we have $\phi(Z+\zeta)-\phi(Z-\zeta)=\Delta$, then the ordinary value of the definite integral given by $\phi(b'')-\phi(b')$ should be diminished by the quantity Δ, as can be easily shown.

The implicit notion of continuity ($\Delta=0$) and the explicit definition of a jump discontinuity ($\Delta\neq 0$) in this passage are consistent with the definition in *Analyse Algébrique*.

I. Grattan-Guinness assumed that Cauchy had plagiarized Bolzano's definition of continuity (see I. Grattan-Guinness, 'Bolzano, Cauchy, and the "New analysis" of the early nineteenth century', *Arch. Hist. Ex. Sci.*, **6**, pp. 372–400). In fact, this thesis is not convincing. By an internal study, H. Freudenthal has clarified the differences of the approach between the two mathematicians (see H. Freudenthal, 'Did Cauchy plagiarize Bolzano?', *Arch. Hist. Ex. Sci.*, **7**, 1971, pp. 375–392, and also H. Sinaceur, 'Cauchy et Bolzano', *Rev. Hist. Sci.*, **26**, 1973, pp. 97–112). The 1816 instructional plan and the *Registre d'instruction* of 1816–1817 (see note 6) corroborate the anteriority of Cauchy's definition of continuity.
6. Archives E. P., XII C7, *Registres d'instruction*. These registers, kept daily, give the title and occasionally a brief summary of these lessons, according to the indications of the teachers. See Appendix II, pp. 305–307, and C. Gilain 'Cauchy et le cours d'analyse de l'École Polytechnique', *Bulletin de la Société des Amis de la Bibliothèque de l'École Polytechnique*, **5**, pp. 2–145.

In 'La querelle des infiniment petits à l'École Polytechnique au XIXe siècle', *Historia Scientiarum*, **30**, 1986, pp. 1–61, Thierry Guitard inferred a chronology of the genesis of *Analyse Algébrique* from the *Registres d'instruction*. He defined the year 1817, the 'annus mirabilis', as the turning point in the elaboration of Cauchy's course, especially with the 'invention of the continuity' on March 1, 1817. I do not believe that this inference is correct, since the *Registres* give only a *terminus ad quem*: for example, according to the *Registres*, Cauchy knew the modern concept of continuity as far back as March 1817, but the 'invention' was anterior, as shown by the instructional program of December 1816.
7. 'Sur les racines imaginaires des équations', *Bull. Phil.*, January 1817, pp. 5–9 (*O.C.*, 2, **2**, pp. 210–216); this article resumes the memoir that Cauchy presented to the

Académie on December 23, 1816 (*Procès-verbaux, Ac. Sci.*, **6**, p. 132). Cauchy presented a new proof of the fundamental theorem of algebra to the Académie on October 13, 1817 ('Sur la décomposition des polynômes en facteurs réels du deuxième degré'; see *Procès-verbaux, Ac. Sci.*, **6**, p. 228) and published it immediately: 'Seconde note sur les racines imaginaires des équations', *Bull. Phil.*, October 1817, pp. 161–164 [*O.C.*, 2, **2**, pp. 217–222; see also *J.E.P.*, **11**, 18th cahier, January 1820, pp. 411–416 (*O.C.*, 2, **1**, pp. 258–263) and Chapter X, §1, of the *Analyse Algébrique*, pp. 331–339 (*O.C.*, 9, **3**, pp. 274–288)]. In the second proof of October 1817, Cauchy explicitly used the property of real continuous functions to be bounded and to reach their boundaries on any closed set; he gaves no proof of this statement, however.

8. *Procès-verbaux, Ac. Sci.*, **6**, p. 395. Cauchy published a resume in the *Bull. Phil*, January and February 1819, pp. 10–21 (*O.C.*, 2, **2**, pp. 238–252) (the manuscript is kept in the meeting's packet), and the whole paper only in 1842 [see the 'Mémoire sur l'intégration des équations aux dérivées partielles du premier ordre', § I, *Ex. An. Phys. Math.*, **2**, November 1842, pp. 241–260 (*O.C.*, 2, **12**, pp. 275–295)].
9. Quoted by C. A. Valson, *La Vie et les Travaux du Baron Cauchy*, **1**, pp. 62–63.
10. Archives E. P., *Registre des procès-verbaux des séances du Conseil d'Instruction*, March 4, 1819.
11. *Ibid.*
12. *Ibid.*
13. Archives E. P., *Registre des procès-verbaux des séances du Conseil d'Instruction*, June 15, 1820.
14. See Chapter 4, p. 55.
15. In the introduction to *Analyse algébrique*, Cauchy recalled, in a few words, how he was led to restructure this course:

    > Since several persons who were kind enough to guide me in the early stages of my scientific career—and in this connection, I am particularly pleased to mention MM. Laplace and Poisson—have expressed a desire to see me publish the *Cours d'Analyse de l'École Polytechnique*, I have decided to develop this course in such a way that would be useful to my students. Accordingly, I am now submitting the first part [of the course], which is called algebraic analysis. In this development, I have successively treated the various kinds of functions, both real and imaginary; divergent and convergent series; the solutions of equations, and the decomposition of rational fractions.

16. *Analyse algébrique*, Introduction, pp. ii–iii (*O.C.*, 2, **3**, pp. ii–iii).
17. *Ibid.*, Préliminaires, p. 4 (*O.C.*, 2, **3**, p. 19).
18. *Ibid.*, Préliminaires, p. 4 (*O.C.*, 2, **3**, p. 19).
19. *Ibid.*, Chapter II, § 2, pp. 34–35 (*O.C.*, 2, **3**, p. 43).
20. *Ibid.*, Chapter II, § 2, pp. 37–39 (*O.C.*, 2, **3**, pp. 45–47).
21. *Ibid.*, Chapter II, §2, pp. 43–44) (*O.C.*, 2, **3**, pp. 50–51). In note III, pp. 460–462 (*O.C.*, 2, **3**, pp. 378–380), Cauchy presented an analytic proof; in this proof, however, he missed the difficulty by implicitly assuming the existence of a limit for an upper-bounded (resp. lower-bounded) increasing (resp. decreasing) sequence of real numbers.
22. See P. E. B. Jourdain, 'The origins of Cauchy's conception of the definite integral and of the continuity of a function', *Isis*, **1**, 1913, pp. 661–703; J. V. Grabiner, *The Origins of Cauchy's Rigorous Calculus*, Cambridge: The MIT Press, 1981; and F.

Smithies, 'Cauchy's conception of rigor in analysis', *Arch. Hist. Ex. Sci.*, **36**, 1986, pp. 41–61.

23. By 'analytical expression', Euler and his successors extended algebraic and transcendental operations as well as infinite processes; that is, sums of series, infinite products, and infinite continued fractions (see C. Houzel, 'Euler et l'apparition due formalisme', in C. Houzel, J. L. Ovaert, P. Raymond, J. J. Sansuc, *Philosophie et Calcul de l'Infini*, Paris, Maspéro, 1976, pp. 130–135, and A. P. Iushkevich, 'The concept of function', *Arch. Hist. Ex. Sci.*, **16**, 1976, 37–85).
24. *Calcul Infinitésimal*, Avertissement, p. 1. (*O.C.*, 2, **4**, p. 1).
25. *Analyse Algébrique*, Chapter VI, §2. pp. 125–126 (*O.C.*, 2, **3**, pp. 115–116).
26. Cauchy's statement is true only if the series of continuous functions is uniformly convergent. Cauchy corrected his theorem and introduced the concept of uniform convergence in 1853 [see 'Note sur les séries convergentes dont les divers termes sont des fonctions continues d'une variable réelle ou imaginaire entre des limites données', *C. R. Ac. Sci.*, **26**, March 14, 1853, pp. 454–459 (*O.C.*, 1, **12**, pp. 30–36)].
27. Cauchy simply wrote $\log(z)$ to denote logarithm, defined in the half-plane $R(z) > 0$.
28. N. H. Abel, 'Recherche sur la série $1 + \frac{mx}{1} + \frac{m(m-1)x^2}{1\cdot 2} + \frac{m(m-1)(m-2)x^3}{1\cdot 2\cdot 3} + \cdots$', *Journal für die reine und angew. Math.*, **1**, pp. 311–339, especially, p. 313 (N. H. Abel, *Oeuvres Complètes*, **1**, 2nd ed., Christiania, 1881, pp. 221–250, especially, p. 222).
29. Archives E. P., Reports of April 13 and 14, 1821. A significant extract was published by I. Grattan-Guinness, *Annals of Science*, **38**, 1981, p. 680.
30. Archives E. P. *Registre des procès-verbaux des séances du Conseil d'instruction*, April 17, 1821.
31. Archives E. P., VI 2a2–1805, report of April 21, 1821. An important extract was published in the *Annals of Science*, **38**, 1981, p. 681.
32. Ch. Dupin, *Discours aux funérailles de M. Augustin Cauchy*, May 25, 1857, Paris, 1857.
33. J. Bertrand, 'Notice sur Louis Poinsot', in L. Poinsot, *Éléments de Statique*, 11th edition, Paris, 1873, pp. XXII–XXIV.
34. Archives E. P., *Registre d'instruction*, 1821–1822.
35. Archives E. P., *Registre des procès-verbaux des Séances du Canseil d' instruction*, January 30, 1823.
36. In the foreword of *Calcul Infinitésimal*, Cauchy wrote:

    This work, which was undertaken on the request of the Conseil de Perfectionnement de l'École Polytechnique, gives the basics of the lectures that were presented at that institution on the infinitesimal calculus. It will be composed of two volumes corresponding to the two-year course. Today, I am publishing the first volume, which consists of forty lessons, the first twenty of which have to do with the differential calculus, while the last twenty deal with a portion of the integral calculus.

37. See Lagrange, *Théorie des Fonctions Analytiques*, Paris, 1797, in which Lagrange uses the term derivative functions and employs the notation $f', f'', f''', \ldots, f^{(n)}$.
38. Nevertheless, Cauchy's treatment was confusing by the use of a notation borrowed from Lacroix:

$$du = \lim_{\alpha \to 0} \frac{\Delta u}{\alpha} = \frac{[u(x + \alpha h) - u(x)]}{\alpha},$$

where h is an arbitrary finite constant and α an infinitesimal (see Lacroix: *Traité Élémentaire de Calcul Différentiel et Intégral*, appendixe no 345).

39. On the theory of singular integrals and its use in the *Calcul Infinitésimal*, see Chapter 7, pp. 115–118.
40. Cauchy explicitly criticized Lagrange's point of view in the foreword to *Leçons sur le Calcul Différéntiel* of 1829 and in the first of the articles 'Sui metodi analitici', which was published in December 1830 (see p. 274, note 19).
41. See also *Ex. Math.*, **1**, May 1826, pp. 25–28 (*O.C.*, 2, **6**, pp. 38–43).
42. Archives E. P., *Registre des procès-verbaux des séances du Conseil de Perfectionnement*, November 29, 1823,

> One member [Laplace?] stated these pamphlets are too complicated and are beyond the comprehension of the students. The lectures, which these pamphlets should clarify, are helped but little; only a very few students can benefit by a study of this material. If this serious problem is not quickly remedied, it is to be feared that in the following classes the training in mathematics, which has contributed so much in elevating the École to the eminent rank it now occupies, will be weakened.

43. A. M. Ampère, *Précis du Calcul Différentiel et Intégral*, an unpublished work that was never completed.
44. The study is the memoir 'Sur la Théorie des Ondes' of 1815. Cauchy deposited it, along with some supplementary notes, with the secretariat of the Académie on May 17, 1824, for final publication.
45. These lectures were published in 1981 by C. Gilain under the title *Équations Différentielles Ordinaires*. They were taken from copies that Cauchy himself had deposited at the Bibliothèque de l'Institut.
46. For more on the lectures, see the introduction by C. Gilain, pp. XIII–XX.
47. Archives E. P., *Registre des procès-verbaux des séances du Conseil d'Instruction*, November 24, 1825. Remarks reported by the governor of the École and confirmed by Baron d'Hautpoul, member of the Conseil de Perfectionnement. See A. d'Hautpoul, *Quatre Mois à la Cour de Prague*, Paris, 1912, p. 242.
48. Archives E. P., *Registre des procès-verbaux des séances du Conseil d'Instruction*, March 10, 1825, and November 24, 1825. He stated, for example, that:

> It is quite necessary to say it: the students come into the various schools of applications with no knowledge of the integral calculus—or they very soon forget the little that they may have learned. At Metz, they say only one student could do integrations. The École Polytechnique was not founded for the training of mathematicians, but rather it was founded for the purpose of training students to enter certain public services. It is therefore necessary that their education should be in those fields that are applicable to the areas that fall within the scope of the [public] services concerned.

49. Archives E. P., *Registre des procès-verbaux des séances du Conseil d'Instruction*, March 10, 1825.
50. Archives E. P., *Registre des procès-verbaux des séances du Conseil de Perfectionnement*, November 21, 1825.
51. Here, Cauchy means the professor of applied analysis, Arago.
52. Archives E. P., *Registre des Procès-verbaux des séances du Conseil d'Instruction*, January 12, 1826.
53. A supplementary note, 'Sur la décomposition des fractions', was printed as a supplement to *Calcul Infinitésimal* of 1823 and distributed to the students during the academic year 1824–1825. It is paged from page 177 to page 182, following the

note 'Sur les formules de Taylor et de MacLaurin'. One copy of this unpublished note is kept in the library of the Ecole Polytechnique (Cote A3 a 57).

54. Cauchy stated in the foreword to the first volume that:

> This work, which is destined to follow the *Résumé des Leçons sur le Calcul Infinitésimal*, will present the applications of calculus to geometry. It will be divided into three volumes. The first two will examine the geometric applications of the differential and the integral calculus that are related to the first-year analysis course at the École Royale Polytechnique. The present volume covers the main applications of differential calculus.

Three articles of *Exercices de Mathématiques* completed the first volume of *Applications du Calcul Infinitésimal à la Géométrie*:

1. 'Sur les centres, les plans principaux et les axes principaux des surfaces du second degré', *Ex. Math.*, **3**, February 1828, pp. 1–22 (*O.C.*, 2, **8**, pp. 9–35).
2. 'Des surfaces que peuvent engendrer en se mouvant dans l'espace des lignes droites ou courbes de forme constante ou variable', *Ex. Math.*, **3**, April 1828, pp. 23–64, (*O.C.*, 2, **8**, pp. 36–82).
3. 'Discussion des lignes et des surfaces du second degré', *Ex. Math.*, **3**, June 1828, pp. 65–120 (*O.C.* 2, **8**, pp. 83–149).

55. These investigations connected some of the most general and abstract concerns that had commanded Cauchy's attention since his 1813 work on determinants: the transformation of homogeneous quadratic forms with arbitrarily many variables. Stimulated by Sturm's works, he presented to the Académie the study '*Sur l'équation à l'aide de laquelle on détermine les inégalités séculaires des mouvements célestes*' on July 27, 1829 [*Ex. Math.*, **4**, August 1829, pp. 140–160, (*O.C.*, 2, **9**, pp. 174–195)]. In this study, he proved that for any symmetrical linear mapping there exists a basis consisting of proper vectors, a theorem that he had already stated without proof in a note presented to the Académie on November 20, 1826 ['Sur l'équation qui a pour racines les moments d'inertie principaux d'un corps solide et sur diverses équations du même genre', *Mémoires Ac. Sci.*, **9**, (1826), 1830, pp. 111–113, (*O.C.*, 1, **2**, pp. 79–81)].

For more on this question, see Th, Hawkins, 'Cauchy, and the Spectral Theory of Matrices', *Historia Mathematica*, **2**, 1975, pp. 1–29 and A. Dahan, *Les Recherches Algébriques de Cauchy*, 3rd cycle thesis, typewritten, Paris, 1974, pp. 44–50.

56. During this period, 10 articles treating classical mechanics appeared in *Exercices de Mathématiques*, 1st and 2nd year; they were:

1. 'Sur la résultante et les projections de plusieurs forces appliquées en un seul point', *Ex. Math.*, **1**, May 1826, pp. 29–43 (*O.C.*, 2, **6**, pp. 23–37).
2. 'Sur les moments linéaires', *Ex. Math.*, **1**, May 1826, pp. 65–84 (*O.C.*, 2, **6**, 89–112).
3. 'Sur les moments linéaires de plusieurs forces appliquées à différents points', *Ex. Math.*, **1**, June 1826, pp. 117–124 (*O.C.*, 2, **6**, 149–158).
4. 'Usage des moments linéaires dans la recherche des équations d'équilibre d'un système invariable entièrement libre dans l'espace', *Ex. Math.*, **1**, June 1826, pp. 125–132 (*O.C.*, 2, **6**, pp. 59–168).
5. 'Sur les conditions d'équivalence de deux systèmes de forces appliquées à des points liés invariablement les uns aux autres', *Ex. Math.*, **1**, July 1826, pp. 151–154 (*O.C.*, 2, **6**, pp. 191–195).
6. 'Usage des moments linéaires dans la recherche des équations d'équilibre d'un système invariable assujetti à certaines conditions', *Ex. Math.*, **1**, July 1826, pp. 155–159 (*O.C.*, 2, **6**, pp. 196–201).

7. 'Recherche des équations générales d'équilibre pour un système de points matériels assujettis à des liaisons quelconques', *Ex. Math.*, **2**, March 1827, pp. 1–22 (*O.C.*, **7**, pp. 11–36).
8. 'Sur les mouvements que peut prendre un système invariable libre ou assujetti à certaines conditions', *Ex. Math.*, **2**, March 1827, pp. 70–90 (*O.C.*, 2, **7**, pp. 94–120).
9. 'Sur les moments d'inertie', *Ex. Math.*, **2**, April 1827, pp. 93–103 (*O.C.*, 2, **7**, pp. 124–136).
10. 'Sur la force vive d'un corps solide ou d'un système invariable en mouvement', *Ex. Math.*, **2**, April 1827, pp. 104–107 (*O.C.*, 2, **7**, pp. 137–140).

The library of the École Polytechnique has the first of *Notes sur quelques Parties de la Mécanique*, which deals with the composition of forces and linear moments (A3 a 57). This note, entitled 'Sur la résultante de plusiers forces appliquées à un seul points', was printed for distribution to the students, probably in the academic year 1826–1827. Its content corresponds almost exactly to the two articles, 1. and 2., of the *Exercices de Mathématiques.*

57. Archives E. P., *Registre des procès-verbaux des séances du Conseil de Perfectionnement*, December 1, 1826.
58. Archives E. P., *Registre des procès-verbaux des séances du Conseil de Perfectionnement*, November 17, 1826.
59. Archives E. P., *Registre des procès-verbaux des séances du Conseil de Perfectionnement*, December 14, 1827.
60. Archives E. P., *Registre des procès-verbaux des séances du Conseil de Perfectionnement*, May 28, 1828.
61. Archives E. P., *Registre des procès-verbaux des séances du Conseil de Perfectionnement*, December 26, 1828.
62. In the foreword to *Leçons sur le Calcul Différentiel*, Cauchy wrote:

> The edition of *Résumé des Leçons sur le Calcul Infinitésimal*, which appeared in 1823, being found to be limited, I decided to replace it by two separate works, one covering differential calculus and the other integral calculus. The present work, which treats differential calculus, is the first of the volumes.

63. Archives E. P., *Registre des procès-verbaux des séances du Conseil de Perfectionnement*, December 11, 1829.
64. See B. Belhoste, 'Le cours d'analyse de Cauchy à l'École Polytechnique en seconde année', *Sciences et Techniques en Perspective*, **9**, 1984–1985, pp. 101–178.
65. Cauchy had written the manuscript of the third volume of *Leçons sur les Applications du Calcul Infinitésimal à la Géométrie.* Moigno used this manuscript in preparing *Leçons de calcul différentiel*, published in 1840 (see the introduction to that work, pp. XVIII–XIX).

Notes to Chapter 6

1. On the history of wave theory, see H. Burkhardt, 'Entwicklungen nach oscillierenden Funktionen und Integration der Differentialgleichungen der mathematischen physik', *Jahresbericht der Deutschen Mathematiker-Vereinigung*, **10**, (1901–1908) paragraph 43ff, and H. Lamb, *Hydrodynamics*, Cambridge, 1932, esp. p. 373ff.
2. J. L. Lagrange, *Mécanique analytique*, part two, Dynamique, sect. XI, paragraph II, 'Applications au mouvement d'un fluide contenu dans un canal peu profond et presque horizontal, et en particulier au mouvement des ondes'.

3. *Procès-verbaux Ac. Sci.*, **5**, p. 530, and *Mém. Sav. Étr.*, **1**, 1827, p. 188 (*O.C.*, 1, **1**, p. 190).
4. These manuscript supplements are contained in Cauchy's *Cahier sur la Théorie des Ondes*, which belongs to Mrs. de Pomyers:

   pp. 90–97: 'Du cas où l'on a égard à la profondeur'.
   pp. 98–114: 'Observations sur le mémoire n^0 2 présenté au concours de 1815 relatif à la théorie des ondes'.

5. Poisson read his first paper on October 2, 1815, the closing day of the competition, and his second paper on December 18, 1815, i.e., a week before the awarding of the prize. In the second paper, he established the existence of a wave propagation with a constant velocity. Poisson published his papers in *Mém Ac. Sci.*, **1** (1816), 1818, pp. 69–186. Cauchy's paper came out in the *Mém. Sav. Étr.*, **1**, 1827, pp. 3–312 (*O.C.*, 1, **1**, pp. 5–318). The *Cahier sur la Théorie des Ondes*, belonging to Mrs. de Pomyers, contains the manuscript text of the 13 notes of the prize-winning paper and the text of notes XIV and XV, which Cauchy added in 1821. The manuscript text of undated note XVI is in another cahier, kept at the Sorbonne Library, Ms 2057. Note XVII is an improved version of a 1815's note (see *Cahier sur la Théorie des Ondes*, pp. 142–152).
6. *Procès-verbaux, Ac. Sci.*, **7**, p. 370.
7. On the history of the theory of elasticity, see A. Barré de Saint-Venant, 'Historique abrégé des recherches sur la résistance des matériaux et sur l'élasticité des corps solides', in his edition of Navier's *Résumé des Leçons sur l'Application de la Mécanique*, **1**, Paris, 1858; I. Todhunter and K. Pearson, *A History of Elasticity and of the Strength of Materials from Galilei to the Present Time*, **1**, Cambridge, 1886; A. E. H. Love. *A Treatise on the Mathematical Theory of Elasticity*, Oxford, 1927, Historical Introduction,' pp. 1–31; and S. P. Timoshenko, *History of Strength of Materials, with a Brief Account of the History of Theory of Elasticity and Theory of Structures*, New York, McGraw-Hill, 1953. On Cauchy's research works, see A. Dahan-Dalmedico, 'La mathématisation des théories de l'élasticité par A. L. Cauchy et les débats dans la physique mathématique française (1800–1840)', *Sciences et Techniques en Perspective*, **9**, 1984–1985, pp. 1–100.
8. See Ch. 1, p. 14.
9. *Procès-verbaux, Ac. Sci.*, **6**, pp. 474–477 (*O.C.*, 2, **15**, pp. 539–544).
10. *Mém. Institut*, **2**, (1812), 1816, pp. 167–226.
11. S. Germain, *Recherches sur la Théorie des Plaques Élastiques*, Paris, 1821. Short letter of acknowledgment from Cauchy to Sophie Germain under the date of July 24, 1821, BN, Ms ffr 9118.
12. The paper 'Sur la flexion des plans élastique', has never been printed, except as an abstract, published in the *Bull. Phil.*, 1823, pp. 95–102. The manuscript is not kept in the archives of the Académie, but the ENPC Library has a lithographic copy of the paper (Ms Navier 1820; see also Navier file Arch. Nat. F^{14} 2289^{1}).
13. L. Bucciarelli and N. Dworsky analyze this part of the paper in their book *Sophie Germain, an Essay on the History of Elasticity*, Dordrecht, Reidel Publishing Company, 1980, Ch. 9, n^0 3, pp. 139–140.
14. See C. Truesdell, 'The creation and unfolding of the concept of stress', in *Essays in the History of Mechanics*, New-York–Berlin–Heidelberg, Springer-Verlag, 1968, pp. 184–238.

15. See 'De la pression dans les fluides', *Ex. math.*, **2**, March 1827, pp. 23–24 (*O.C.*, **7**, pp. 37–39).
16. *Bull. Phil.*, January 1823, p. 10 (*O.C.*, 2, **2**, p. 301).
17. See 'Supplément au mémoire sur la double refraction' of January 13, 1822, presented to the Académie on January 22 (A. Fresnel, *Oeuvres*, **2**, no XLII, pp. 344–347) and 'Second supplément au memoire sur la double refraction' of March 31, 1822, presented to the Académie on April 1, 1822 (A. Fresnel, *Oeuvres*, **2**, no XLIII, pp. 369–442). Cauchy set forth Fresnel's argument in the addition to his article 'De la pression ou de la tension dans les corps solides', *Ex. Math.*, **2**, March 1827, pp. 56–59 (*O.C.*, 2, **7**, pp. 79–81).
18. *Bull. Phil.*, January 1823, pp. 11–12 (*O.C.*, 2, **2**, p. 303.
19. H. Freudenthal, 'Cauchy, Augustin-Louis', *Dictionary of Scientific Biography*, **3**, New York, 1971, pp. 131–148, especially p. 145.
20. Cauchy deduced the properties of the perfect fluids from the general laws of continuum mechanics in the paper 'Sur les équations qui expriment les conditions d'équilibre ou les lois du mouvement des fluides', *Ex. Math.*, **3**, March 1828, pp. 42–57 (*O.C.*, 2, **8**, pp. 128–146).
21. Abstract in *Bull. Phil.*, January 1823, pp. 9–13 (*O.C.*, 2, **2**, pp. 300–304). See also the *Registre des séances de la Société Philomatique*, 1822–1823, Ms 2086, fo 60 r. and fo 61 v, at the Sorbonne Library.
22. See the following articles:
    1. 'Sur la pression ou tension dans les corps solides', *Ex. Math.*, **2**, March 1827, pp. 42–57 (*O.C.*, 2, **7**, pp. 60–78). Cauchy deposited the manuscript with the bureau of the Académie on March 12, 1827. It is kept in the meeting's packet. The subtitle of the manuscript, 'Extrait de la première partie du mémoire sur l'équilibre et le mouvement intérieur des corps solides ou fluides élastiques ou non élastiques', proves that this article is an extract of the famous unpublished paper of September 30, 1822.
    2. 'Sur la condensation et la dilatation des corps solides', *Ex. Math.*, **2**, April 1827, pp. 60–69 (*O.C.*, 2, **7**, pp. 82–93).
    3. 'Sur les relations qui existent, dans l'état d'équilibre d'un corps solide ou fluide, entre les pressions ou tensions et les forces accélératrices', *Ex. Math.*, **2**, April 1827, pp. 108–111 (*O.C.*, 2, **7**, pp. 141–145).
    4. 'Sur les équations qui expriment les conditions d'équilibre ou les lois du mouvement intérieur d'un corps solide, élastique ou non élastique', *Ex. Math.*, **3**, September 1828, pp. 160–187 (*O.C.*, 2, **8**, pp. 195–226). In this article, Cauchy presented both of his continuum theories (for isotropic media), one with a single coefficient of elasticity (from 1822) and one with two coefficients (from 1828). The draft, deposited in a sealed envelope on August 18, 1828, is kept in the meeting's packet. The text is identical with the printed article.
23. See *Procès-verbaux, Ac. Sci.*, **7**, p. 420. The manuscript, which was initialed by Fourier on January 27, 1823, was written in 1815. It is inserted in the *Cahier sur la Théorie des Ondes* belonging to Mrs. de Pomyers, pp. 115–140.
24. The letter is kept in the meeting's packet of October 6, 1822.
25. Sorbonne Library, Ms 2086, *Registre des séances de la Société Philomatique*, 1822–1823, fo 62 v.
26. *Bull. Phil.*, March 1823, pp. 36–37.
27. All the authors who have written about the origins of the theory of elasticity and compared the works of Navier and Cauchy repeat this mistake. Only Prony, Poisson, and Fourier were actually members of the commission.

28. *Histoire de l'Académie*, partie mathématique, 1822, pp. 30–31.
29. *Bull. Phil.*, January 1823, p. 13 (*O.C.*, 2, **2**, p. 301), and *Histoire de l'Académie*, partie mathématique, 1822, p. 31.
30. *Mém. Ac., Sci.*, **7** (1824), 1827, pp. 375–394.
31. On November 17, 1823 (*Procès-verbaux, Ac. Sci.*, **7**, p. 585), while preparing the publication of his award-winning paper of 1815, 'Sur la théorie des ondes', he presented to the Académie the paper 'Sur les effets de l'attraction moléculaire dans le mouvement des ondes', in which the term molecule was used for the very first time to mean not an infinitely small material element but a center of infinitely small attractive action [the study, quoted in note XX of the paper 'Sur la théorie des ondes', is unpublished; the manuscript abstract is kept in the meeting's packet of November 17, 1823 (see the appendix I, p. 298)]. In October 1827, Cauchy specifically stated that he had been occupied 'for quite some time' with molecular theory [Analyse des Travaux de l'Académie... 1827, *Mém Ac. Sci.*, **10** (1827), 1831, iii–iv].
32. *Procès-verbaux, Ac. Sci.*, **8**, p. 603.
33. Sealed envelope no 126, kept in the meeting's packet of October 1, 1827. The sealed envelope was opened in October 5, 1977, and the paper published by C. Truesdell. See *C.R. Ac. Sci.*, **291**, Vie académique, pp. 33–46.
34. See *Procès-verbaux, Ac. Sci.*, **9**, p. 52 and p. 55.
35. Poisson offered an offprint of his paper to the Académie on November 3, 1828 (*Procès-verbaux, Ac. Sci.*, **9**, p. 137). The paper came out in the *Mém. Ac. Sci.*, **8**, 1829, pp. 751–780.
36. *Ex. Math.*, **3**, September 1828, pp. 188–212, and October 1828, pp. 213–236 (*O.C.*, 2, **8**, pp. 227–252 and 252–277).
37. Sealed envelope no 128, registered on August 18, 1828, and opened on October 5, 1977. Cauchy authorized unsealing the envelope on September 8, 1828, after having published the paper (*Procès-verbaux, Ac. Sci.*, **9**, p. 114).
38. Formally, Navier's equation for isotropic bodies can be deduced from Eq. (6.13) by assuming $G = 2R$.
39. See the following articles:

    1. 'Sur l'équilibre et le mouvement d'une lame solide' and 'Addition à l'article précédent', *Ex. Math.*, **3**, September 1828, pp. 245–326, and December 1828, pp. 326–327 (*O.C.*, 2, **8**, pp. 288–380). Cauchy presented the paper 'Sur la mouvement des lames élastiques ou non élastiques naturellement planes ou naturellement courbes, d'une épaisseur constante ou d'une épaisseur variable' to the Académie on October 6, 1828 (*Procès-verbaux, Ac. Sci.*, **9**, p. 130), and an addition to this paper on October 13, 1828 (*Procès-verbaux, Ac. Sci.*, **9**, p. 131). On November 17, 1828, he presented a new paper, 'Sur les vibrations des lames courbes'.
    2. 'Sur l'équilibre et le mouvement d'une plaque solide', *Ex. Math.*, **3**, December 1828, pp. 328–355 (*O.C.*, 2, **8**, pp. 381–411). Cauchy presented the paper 'Sur le mouvement des plaques et des verges élastiques ou non élastiques' on October 6, 1828 (*Procès-verbaux, Ac. Sci.*, **9**, p. 130). Two manuscripts are kept in the meeting's packet:

        'Première partie d'un mémoire sur les équations d'équilibre ou de mouvement d'une plaque solide naturellement droite et d'une épaisseur constante'.
        'Sur l'équilibre et le mouvement d'une plaque solide'.

    3. 'Sur l'équilibre et le mouvement d'une verge rectangulaire', *Ex. Math.*, **3**, December 1828, pp. 356–368 (*O.C.*, 2, **8**, pp. 412–423). Cauchy presented the paper 'Sur le

mouvement des plaques et des verges élastiques ou non élastiques' on October 6, 1828 (*Procès-verbaux, Ac. Sci.*, **9**, p. 130). A manuscript is kept in the meeting's packet:

'Sur l'équilibre et le mouvement d'une verge rectangulaire'.

On December 1, 1828, Cauchy presented a new paper, 'Sur l'équilibre et le mouvement des verges élastiques rectangulaires, droites ou courbes, d'épaisseur constante ou d'épaisseur variable', to the Académie.

40. See the following articles:

1. 'Sur l'équilibre et le mouvement d'une plaque élastique dont l'élasticité n'est pas la même dans tous les sens', *Ex. Math.*, **4**, February 1829, pp. 1–14 (*O.C.*, 2, **9**, pp. 9–22). Cauchy presented the paper to the Académie on February 26, 1829 (*Procès-verbaux, Ac. Sci.*, **9**, p. 187). A manuscript is kept in the meeting's packet:

   'Sur l'équilibre et le mouvement des corps élastiques des plaques élastiques, etc..., dont l'élasticité n'est pas la même dans tous les sens'.

2. 'Sur l'équilibre et le mouvement d'une verge rectangulaire extraite d'un corps solide dont l'élasticité n'est pas la même en tous sens', *Ex. Math.*, **4**, February 1829, pp. 15–29 (*O.C.*, 2, **9**, pp. 23–40). Cauchy presented the paper 'Sur le mouvement des lames de surface et des verges élastiques lorsque l'élasticité n'est pas la même dans tous les sens' to the Académie on December 22, 1828 (*Procès-verbaux, Ac. Sci.*, **9**, p. 167). Two manuscripts are kept in the meeting's packet of October 6, 1828:

   'Sur le mouvement des lames de surface et des verges élastiques lorsque l'élasticité n'est pas la même dans tous les sens'.
   'Sur l'équilibre et le mouvement d'une verge rectangulaire dont l'élasticité n'est pas la même en tous sens'.

   One manuscript is kept in the meeting's packet of December 22, 1828:

   'Lame élastique dans le cas ou l'élasticité varie dans toutes les directions'.

3. 'Sur les vibrations longitudinales d'une verge cylindrique ou prismatique à base quelconque', and 'Sur la torsion et les vibrations tournantes d'une verge rectangulaire', *Ex. Math.*, **4**, February 1829, pp. 43–46 and pp. 47–64 (*O.C.*, 2, **9**, pp. 56–60 and pp. 61–86). Cauchy presented the paper 'Sur la torsion et les vibrations tournantes d'une verge rectangulaire' on February 2 and 16, 1829 (*Procès-verbaux, Ac. Sci.*, **9**, p. 190 and p. 196). One manuscript is kept in the meeting's packet of October 6, 1828: 'Vibration tournante d'une verge rectangulaire'. Another manuscript is kept in the meeting's packet of February 16, 1829: 'Sur la torsion et les vibrations tournantes d'une verge rectangulaire'.

41. Cauchy explains his general molecular theory of elasticity in the article 'Sur les équations différentielles d'équilibre ou de mouvement pour un système de points matériels sollicités par des forces d'attraction ou de répulsion mutuelle', *Ex. Math.*, **4**, August 1829, pp. 129–139 (*O.C.*, 2, **9**, pp. 162–173).
42. See especially the unpublished notes of Cauchy and Poisson in the meeting's packets of April 6, 13, and 20, 1828.
43. On the theory of light, see E. T. Whittaker, *A History of the Theories of Aether and Electricity*, **1**, London, New-York, T. Nelson, 1951. On Cauchy's research works especially, see J. Z. Buchwald, 'Optics and the Theory of the Punctiform Ether', *Arch. Hist. Exact Sci.*, **21**, 1980, pp. 245–278.
44. See R. H. Silliman, 'Fresnel Augustin', *Dictionary of Scientific Biography*, **5**, New York, 1972, pp. 166–171.

45. 'Sur les mouvements d'un système de molécules qui agissent les unes sur les autres à de très petites distances et sur le mouvement de la lumière', *Bull. Férussac*, **11**, February 1829, pp. 111–112, and *Mém. Ac. Sci.*, **9** (1826), 1830, pp. 114–116.
46. 'Sur l'intégration d'une certaine classe des équations aux différences partielles et sur les phénomènes dont cette intégration sert à faire connaître les lois'. This unpublished paper is lost, but a note is kept in the meeting's packet of April 12, 1830. Cauchy published an abstract of the paper in *Bull. Férussac*, **13**, April 1830, pp. 273–279, and the first paragraph in *J.E.P.*, **13**, February 1831, pp. 175–221. It seems that Cauchy had already treated the subject in a paper he presented to the Académie on March 22, 1830 (*Procès-verbaux, Ac. Sci.*, **9**, p. 423).
47. 'Application des formules qui représentent le mouvement d'un système de molécules sollicitées par des forces d'attraction ou de répulsion mutuelle à la théorie de la lumière', *Ex. Math.*, **5**, September 1830, pp. 19–72 (*O.C.*, 2, **9**, pp. 390–450). See also *Mémoire sur la Théorie de la Lumière*. This detached paper is reproduced in the *Bull. Férussac*, **13**, 1830, pp. 414–427 (*O.C.*, 2, **2**, pp. 119–133), and in *Mém Ac. Sci.*, **10** (1827), 1830, pp. 293–316 (*O.C.*, 1, **2**, pp. 91–110), with some unimportant variants. Cauchy presented two parts of the paper to the Académie on June 7 and 14, 1830.
48. From his lectures given at the *Collège de France*, Cauchy published two lithographies:

    Extrait des leçons domnées au Collège de France. Résumé de la leçon de M. A. L. Cauchy du samedi 8 mai 1830. An exemplar of this lithography is kept in the meeting's packet of May 17, 1830. As for the manuscript, it is kept in the meeting's packet of May 10, 1830.

    Extrait des leçons données au Collège de France par M. A. L. Cauchy sur la théorie de la lumière à dater du 8 mai 1830. Réfraction et réflexion de la lumière. Dispersion de la lumière. An examplar of this lithography, with autographic corrections, is kept in the meeting's packet of June 21, 1830. A manuscript copy, not autographic and without the paragraph on light dispersion, and two notes to the compositor of mathematics for the *Mém Ac. Sci.* are kept in the same packet. In fact, Cauchy prepared the publication of his lectures in the *Mém Ac. Sci.* He intended to entitle his paper 'Second mémoire sur la lumière'.

49. Cauchy published the beginning of his paper 'Sur la dispersion de la lumière' in 1830. In 1835, he published the whole paper at Prague, in *Nouveaux Exercices de Mathématiques*, pp. 1–60 (*O.C.*, 2, **10**, pp. 196–260).
50. 'Sur l'équilibre et le mouvement intérieur des corps considérés comme des masses continues', *Ex. Math.*, **4**, May 1830, pp. 293–319 (*O.C.*, **9**, pp. 342–372).
51. *Sept Leçons de Physique Générale*..., Paris, 1868, published by F. Moigno (*O.C.*, 2, **15**, pp. 412–447).

Notes to Chapter 7

1. This study, the manuscript for which could not be located, was published only in 1827 in the *Mémoires des Savants Étrangers*, 2, **1**, 1827, pp. 599–799 (*O.C.*, 1, **1**, pp. 329–506) at the same time as the study 'Sur la théorie des ondes'. The long delay between the presentation and the publication of the study can be explained by the slowness that, since 1811, had characterized the Académie's publication of its

records and materials. The main results of Cauchy's study were known as far back as the end of 1824 because of a rather lukewarm report that was published by Poisson in the *Bull. Phil.*, December 1814, pp. 185–188 (*O.C.*, 2, **2**, pp. 194–198). The report of the Académie's evaluative commission was published in the *Analyse des Travaux de l'Académie des Sciences pour l'année 1814*. Finally, Cauchy added two supplements, which were written in 1814, to the manuscript, following a request from the commissioners. Moreover, a large number of important notes were appended to this work, probably in 1825. On September 14, 1825, Cauchy registered this paper with the secretariat of the Académie, so that it could be published along with the additions. An analysis of the study can be found in the general works on Cauchy; a more detailed study is given in the article by H. J. Ettlinger, 'Cauchy's 1814 paper on definite integrals', *Annals of Mathematics*, **23**, 1921–1922, pp. 255–270.

2. Shortly before Cauchy, Laplace and Poisson published works with the same titles: Laplace, 'Mémoire sur les intégrales définies et leur application aux probabilités', *Mém. Institut*, **10**, 1810–1811, pp. 279–347 (P. S. Laplace, *Oeuvres*, pp. 357–412); Poisson, 'Mémoire sur les intégrales définies', first part, *J.E.P.*, **9**, 16th cahier, 1813, pp. 215–246. As for Legendre, in 1811, he published the first volume of *Exercices de Calcul Intégral.*
3. Published in the *Mémoires Ac. Sci.*, 1782–1785, pp. 1–88 (*Oeuvres*, **10**, pp. 209–291). A sequel was published in *Mém. Ac. Sci.*, 1783–1786, pp. 423–467 (P. S. Laplace, *Oeuvres*, **10**, pp. 295–338).
4. Laplace, 'Mémoire sur les intégrales définies', *Mém. Inst.*, **11**, 1810–1811, p. 284 (P. S. Laplace, *Oeuvres*, **12**, p. 361).

   Cauchy cited this passage in his study, p. 612 (*O.C.*, 1, **1**, pp. 329–330). In his 1810 paper, Laplace used established, orthodox methods to recalculate certain formulas that he had derived from the passage from the real to the imaginary domain in 1782. He pursued this work in his additions to the second edition of *Théorie Analytique des Probabilités* (November 1814).
5. This study was published only in 1844 in the *J.E.P.*, **18**, 28th Cahier, pp. 147–248, (*O.C.*, 2, **1**, pp. 467–567). In a note, Cauchy pointed out that the study of January 2, 1815, had received certain additions 'about this same time', Cauchy, in fact, presented to the Académie an improved version of this study, which had the same title as the original work, on April 1, 1816 (*Procès-verbaux, Ac. Sci.*, **6**, p. 44). It was, no doubt, this later version that he published in 1844. The manuscript of the study could not be located. Meanwhile, Cauchy deposited two abstracts of it with the Académie on September 26, and November 7, 1825. The first reproduced the introduction to the paper, while the second reproduced the section 'Sur la conversion des différences finies des puissances en intégrales définies'. The manuscripts are kept in the meeting's packet.
6. See *Mémoires Sav. Étr.*, 2, **1**, 1827, note IV, pp. 140–145 (*O.C.*, 1, **1**, pp. 133–139).
7. On Cauchy's teaching at the Collège de France in 1817, see *Bull. Phil.*, Oct. 1822, p. 161 and p. 171 (*O.C.*, 2, **2**, p. 283 and p. 295); *J. E. P.*, **12**, 9th cahier, July 1823, p. 576 (*O.C.*, 2, **1**, p. 339); *Mémoires Sav. Étr.*, 2, **1**, 1827, p. 715, footnotes (*O.C.*, 1, **1**, p. 429, footnote); and *Ex. Math.*, **2**, 1827, p. 156 (*O.C.*, 2, **7**, p. 194).
8. 'Sur la réduction des intégrales finies et des sommes de séries en intégrales définies', cited by J. Mandelbaum, *La Société Philomatique de Paris*, **1**, p. 200. It remained

unpublished. As to its contents, see *Bull. Phil.*, August 1817, pp. 123–124 (*O.C.*, 2, **2**, pp. 226–227).

9. *Procès-verbaux, Ac. Sci.*, **6**, p. 201. The manuscript could not be located. Cauchy published this paper in August 1817: 'Sur une loi de réciprocité qui existe entre certaines fonctions', *Bull. Phil.*, August 1817, pp. 121–124 (*O.C.*, 2, **2**, pp. 223–227).
10. See Appendix I, *Registre d'instruction*, 1817–1818, p. 313.
11. See Fourier's letter to the Permanent Secretary of the Academy in Fourier's papers, BN Ms ffr 22529, p. 127.
12. 'Seconde note sur les fonctions réciproques', *Bull. Phil.*, December 1818, pp. 178–181 (*O.C.*, 2, **2**, pp. 228–237).
13. At first, Cauchy used an integral formula representing the cosine function, which had been obtained by use of the theory of singular integrals, in order to integrate the differential equations of the theory of waves. See 'Mémoire sur la théorie des ondes', note XVI, p. 187, and note XVIII, pp. 292–293 (*O.C.*, 1, **1**, pp. 189–190 and pp. 295–297; these notes were written for publication).
14. 'Sur la résolution analytique des équations de tous les degrés par le moyen des intégrales définies', unpublished (see *Procès-verbaux, Ac. Sci.*, **6**, p. 507). The manuscript could not be located. An announcement was published in *Analyse des Travaux de l'Académie des Sciences*, partie mathématiques, 1819, pp. 8–11, republished in *Mém. Ac. Sci.*, **4**, (1819–1820), 1824, pp. XXVI–XXIX (*O.C.*, 1, **2**, pp. 9–11). See also *Bull. Phil.*, October 1822, p. 168, (*O.C.*, 2, **2**, pp. 293) and *J.E.P.*, **12**, 9th cahier, July 1823, pp. 541–543 and pp. 580–581 (*O.C.*, 2, **1**, pp. 305–306 and pp. 343–345). Among the studies in the initial project on the publication of Cauchy's *Oeuvres Complètes* was a separate paper from 1819, entitled 'Observations sur les principes de la résolution des équations numériques', which was probably a copy of the paper of November 22, 1819. This text could not be located (see *O.C.*, 2, **15**, appendix, p. 584).
15. *Procès-verbaux, Ac. Sci.*, **7**, p. 380.
16. See *Bull. Phil.*, October 1822, pp. 161–174 (*O.C.*, 2, **2**, pp. 283–299).
17. 'Sur l'intégration des équations linéaires aux différences partielles et à coefficients constants', *J.E.P.*, **12**, 9th cahier, July 1823, pp. 571–592 (*O.C.*, 2, **1**, pp. 333–357).
18. *Résumé des Leçons données à l'École Royale Polytechnique sur le Calcul Infinitésimal, Calcul Intégral*, 24th, 25th, and 34th lectures.
19. *J.E.P.*, **12**, 19th cahier, July 1823, p. 574, footnote (*O.C.*, 2, **1**, p. 337, footnote).
20. *Mém. Sav. Étr.*, 2, **1**, 1827, note IX, p. 158 (*O.C.*, 1, **1**, p. 146).
21. See the 'Mémoire sur l'intégration de quelques équations linéaires aux différences partielles et particulièrement de l'équation générale du mouvement des fluides élastiques', which was presented by Poisson on July 19, 1819, to the Académie and published in *Mém. Ac. Sci.*, **3** (1818), 1820, pp. 121–176.
22. See *Procès-verbaux Ac. Sci.*, **7**, p. 231. The paper was published, in part, in *Analyse des Travaux de l'Académie des Sciences*, partie mathématique, 1821, pp. 25–32, and in *Bull. Phil.*, October 1821, pp. 101–112, and November 1821, pp. 145–152 (*O.C.*, 2, **2**, pp. 253–275).
23. This paper was presented to the Académie on January 22, 1822 (*Procès-verbaux, Ac. Sci.*, **7**, p. 271), and published in *Analyse des Travaux de l'Académie des Sciences*, partie mathématique, 1821, pp. 6–13 [see also *Mémoires Ac. Sci.*, **5** (1821), 1826, pp. 13–19], and in *Bull. Phil.*, April 1822, pp. 49–54 (*O.C.*, 2, **2**, pp. 276–282).
24. See *Bull. Phil.*, November 1821, p. 152 (*O.C.*, 2, **2**, p. 275). This issue of the *Bulletin* appeared at the beginning of 1822.

25. *Procès-verbaux, Ac. Sci.*, **7**, p. 366. This study constitutes the substance of the article of the *J.E.P.* cited above in note 17; see also the announcement in Appendix I, p. 296.
26. 'Moyen d'intégrer les équations linéaires aux différences totales ou partielles, finies ou infiniment petites, avec un dernier terme variable et d'un ordre quelconque, dans tous les cas possibles, lorsque les coefficients du premier membre sont constants et dans certains cas, lorsque les coefficients varient, sans être obligé de résoudre aucune équation algébrique' (*Procès-verbaux, Ac. Sci.*, **7**, p. 503), for which the first three sections were given in the article of the *J.E.P.* cited above in note 17. The fourth section appeared in *Mémoires Ac. Sci.*, **9** (1826), 1830, pp. 97–103 (*O.C.*, 1, **2**, pp. 67–72), and in *Bull. Fer.*, **4**, August 1825, pp. 71–75 (*O.C.*, 2, **2**, pp. 66–71). On the same day, May 26, 1823, Cauchy presented another study, 'Sur la détermination des intégrales définies et sur la résolution des équations algébriques ou transcendantes par le moyen de ces mêmes intégrales', as 'a complement to the papers that the author presented in 1814, 1819 and 1822' (*Procès-verbaux, Ac. Sci.*, **7**, p. 503). This study remained unpublished and the manuscript could not be located. One month later, on July 21, 1823, Cauchy presented to the Académie a paper entitled 'Divers théorèmes servant à intégrer les équations propres à la théorie analytique de la chaleur', which remained unpublished.
27. 'Mémoire sur la théorie des ondes', *Memoires Sav. Étr.*, 2, **1**, 1827, note XVIII, pp. 281–293 (*O.C.*, 1, **1**, pp. 288–299).
28. S. D. Poisson, 'Sur les intégrales des fonctions qui passent à l'infini entre les limites de l'intégration, et sur l'usage des imaginaires dans la détermination des intégrales définies', *J.E.P.*, **11**, 18th cahier, January 1820, pp. 295–341.
29. See *Bull. Phil.*, November 1821, pp. 171–174 (*O.C.*, 2, **2**, pp. 296–299), and *J.E.P.*, **12**, 19th cahier, July 1823, P.S., pp. 590–591 (*O.C.*, 2, **1**, pp. 354–355).
30. See R. Taton, Brisson, Barnabé, *Dictionary of Scientific Biography*, **2**, New York, 1970, pp. 473–475.
31. As to the dispute between Cauchy and Poisson, see Ch. 4, pp. 51–52. Cauchy's report is published in the *Procès-verbaux, Ac. Sci.*, **8**, pp. 223–226 (*O.C.*, 2, **15**, pp. 560–565).
32. 'Sur l'intégration des équations linéaires et leur application à divers problèmes de physique', was published without the applications in *Mem. Ac. Sci.*, **22**, 1850, pp. 39–130 (*O.C.*, 1, **2**, pp. 195–281) under the title 'Mémoire sur le calcul intégral'. A part of this study was published in lithograph on May 2, 1825, under the title 'Mémoire sur l'analogie des puissances et des différences et sur l'intégration des équations linéaires' (*O.C.*, 2, **15**, pp. 23–40). As to the applications, see *Mém. Ac. Sci.*, **7** (1824), 1827, *Histoire de l'Académie*, partie mathématique, pp. XLV–XLVI (prepared by Cauchy from the handwritten copy kept in meeting's packet of May 30, 1825), and the unpublished announcement of the study 'Sur l'intégration des équations linéaires et sur le mouvement des plaques élastiques rectangulaires' of January 17, 1825 (See Appendix I, pp. 300–301).
33. *Procès-verbaux, Ac. Sci.*, **8**, p. 225 (*O.C.*, 2, **15**, p. 564).
34. Undoubtedly, the 'Recherches sur la détermination des séries qui doivent representer des fonctions données dans une partie seulement de leur étendue' was presented to the Académie on August 27, 1827, and never published. See *Mémoire*

Sur les intégrales définies prises entre des limites imaginaires, p. 2 (*O.C.*, 2, **15**, p. 42), and Appendix I, p. 302.

35. *Mémoire sur les intégrales définies prises entre des limites imaginaires*, p. 2 (*O.C.*, 2, **15**, p. 42), and Appendix I, p. 302.

    See A. Iushkevitch, Ostrogradski, *Dictionary of Scientific Biography*, **10**, New York, Chales Scribner's Sons, 1974, pp. 247–251, and *Michel Ostrogradski et le progrès de la science au XIXème siècle Conférence du Palais de la Découverte.* Paris 1966.
36. Unpublished. The handwritten manuscript is kept in the meeting's packet for February 13, 1826.
37. Unpublished. The handwritten manuscript is kept in the Ostrogradski file at the Académie des Sciences.
38. See Appendix I, p. 302.
39. *Ex. Math.*, **1**, March 1826, pp. 11–24 (*O.C.*, 2, **6**, pp. 23–37). As to Cauchy's teaching at the Collège de France in 1824–1825, see Chapter 4, p. 49.
40. Unpublished. However, there is a handwritten announcement in the meeting's packet of February 25, 1825 (see Appendix I, p. 301).
41. In the meeting's packet of February 28, there is a handwritten manuscript of the paper, with the corrections. It is identical to the printed edition, but does not contain the addition. Undoubtedly, this manuscript was deposited with the bureau of the Académie at the meeting of February 28. The minutes of the meeting only indicate that Cauchy read a paper on analysis that day. Moreover, the manuscript was neither dated nor initialled by the secretary of the Académie, as was customary. This manuscript is quite likely the study registered with the Académie on August 8, 1825, the printed edition having been sent the following week (see *Procès-verbaux, Ac. Sci.*, **8**, p. 189 and p. 192). Cauchy published a résumé of the paper in *Bull. Fér.*, **3**, April 1825, pp. 214–221 (*O.C.*, 2, **2**, pp. 57–65 and *O.C.*, 2, **15**, pp. 41–89).
42. *Mémoire sur les intégrales définies prises entre des limites imaginaires*, p. 26 (*O.C.*, 2, **15**, p. 59).
43. In these notes, Cauchy discusses the progress that had been made since 1814 on the theory of singular integrals, and he used his theory of imaginary integrals from 1822–1823 to rewrite the double equations connecting the real integrals.
44. 'Mémoire sur les intégrales définies où l'on donne une formule générale de laquelle se déduisent les valeurs de la plupart des intégrales définies déjà connues et celles d'un grand nombre d'autres', *Annales de mathématiques*, **16**, no 4, October 1825, pp. 97–108 (*O.C.*, 2, **2**, pp. 343–352).
45. 'Sur diverses relations qui existent entre les résidus des fonctions et les intégrales définies', *Ex. Math.*, **1**, June 1826, pp. 95–124 (*O.C.*, 2, **6**, pp. 124–145).
46. The method of decomposing rational fractions by means of the calculus of residues is given in the article 'Sur un nouveau genre de calcul analogue au calcul infinitesimal', *Ex. Math.*, **6**, March 1826, pp. 11–24 (*O.C.*, 2, **6**, pp. 23–37). Cauchy resumed the theory of extraordinary integrals by using the formalism of the calculus of residues in the article 'Sur un nouveau genre d'intégrales', *Ex. Math.*, **1**, May 1826, pp. 57–65 (*O.C.*, 2, **6**, pp. 78–88). He extended the method of decomposing to cover the case of functions that have an infinite number of poles in the article 'Sur diverses relations qui existent entre les résidus des fonctions et les

intégrales définies', *Ex. Math.*, **1**, June 1826, pp. 95–124 (*O.C.*, 2, **6**, pp. 124–145). See also the two following articles in *Ex. Math.*:

1. 'Usage du calcul des résidus pour la sommation de plusieurs suites composées d'un nombre fini de termes', which was presented to the Académie on December 26, 1825. The manuscript is contained in the meeting's packet for that day. A corrected and completed version of this study was published in *Ex. Math.*, **1**, May 1826, pp. 44–53 (*O.C.*, 2, **6**, pp. 62–73).
2. 'Sur le développement des fonctions d'une seule variable en fractions rationnelles', was presented to the Académie on December 10, 1827. The handwritten manuscript is contained in the meeting's packet for that day. The study was published in *Ex. Math.*, **2**, November 1827, pp. 277–296 (*O.C.*, 2, **7**, pp. 324–344). In 1843, Cauchy designated the function $\omega(x)$ as a complementary function (*C.R.*, **19**, p. 138; *O.C.*, 1, **8**, p. 361).

47. 'Sur les limites placées à droite et à gauche du signe ε dans le calcul des résidus', *Ex. Math.*, **1**, September 1826, pp. 205–232 (*O.C.*, 2, **6**, pp. 256–286).
48. 'Sur quelques propositions fondamentales du calcul des résidus', *Ex. Math*, **2**, November 1827, pp. 245–276 (*O.C.*, 2, **7**, pp. 291–323), was presented to the Académie on November 5, 1827. The handwritten announcement is in the meeting's packet for that day. See also the articles, 'Usage du calcul des résidus pour la sommation ou la transformation des séries dont le terme général est une fonction paire du nombre qui représente le rang de ce terme', *Ex. Math.*, **2**, December 1827, pp. 298–314 (*O.C.*, 2, **7**, pp. 345–362), which was presented to the Académie on December 17, 1827 (the manuscript is contained in the meeting's packet for that day), and 'Méthode pour développer des fonctions d'une ou de plusieurs variables en séries composées de fonctions de même espèce', *Ex. Math.*, **2**, December 1827, pp. 317–340, (*O.C.*, 2, **7**, pp. 366–392).
49. 'Usage du calcul des résidus pour déterminer la somme des fonctions semblables des racines d'une équation algébrique ou transcendante', *Ex. Math.*, **1**, January 1827, pp. 339–357 (*O.C.*, 2, **6**, pp. 401–420).
50. See the following articles in *Ex. Math.*:

    1. 'Application du calcul des résidus à l'intégration des équations différentielles linéaires à coefficients constants', *Ex. Math.*, **1**, July 1826, pp. 202–204 (*O.C.*, 2, **6**, pp. 252–255).
    2. 'Application du calcul des résidus à l'intégration de quelques équations différentielles linéaires à coefficients variables', *Ex. Math.*, **1**, October 1826, pp. 262–264 (*O.C.*, 2, **6**, pp. 316–319).
    3. 'Sur la détermination des constantes arbitraires renfermées dans les intégrales des équations différentielles linéaires', *Ex. Math.*, **2**, March 1827, pp. 25–27 (*O.C.*, 2, **7**, pp. 40–54).
    4. 'Sur la transformation des fonctions qui représentent les intégrales générales des équations différentielles linéaires', *Ex. Math.*, **2**, October 1827, pp. 211–220 (*O.C.*, 2, **7**, pp. 255–266).

51. See the following two papers:

    1. The paper 'Usage du calcul des résidus pour la solution des problèmes de physique mathématique' was presented to the Académie on December 26, 1826 (a handwritten abstract dated December 26, 1826, is contained in the meeting's packet for that day), and the full version of this study was presented on February 5, 1827. The incomplete manuscript is in the meeting's packet for the February 5, 1827. The first part of this study was published separately in January 1827, the second part appeared in February 1827 (*O.C.*, 2, **15**, pp. 90–137).
    2. The paper 'Deuxième mémoire sur l'application des résidus aux questions de

physique mathématique' was presented to the Académie on September 17, 1827. The manuscript is contained in the meeting's packet for the day. The study was published in *Mémoires Ac. Sci.*,**7** (1824), 1827, pp. 463–472 (*O.C.*, 1, **2**, pp. 20–28).

52. See Chapter 6, p. 105.
53. 'Mémoire sur les développements des fonctions en séries périodiques', presented to the Académie on February 27, 1826 (the manuscript is in the meeting's packet for that day). This study was published in *Mémoires Ac. Sci.*, **6** (1823), 1827, pp. 603–612 (*O.C.*, 1, **2**, pp. 12–19).
54. G. P. Dirichlet 'Sur la convergence des séries trigonométriques qui servent à représenter une fonction arbitraire entre des limites données', *Journal de Crelle*, **4**, 1829, pp. 157–169.

Notes to Chapter 8

1. See Ch. Magnin, 'Note biographique sur M. J. de Bure', *Journal général de l'Imprimerie et de la Librairie, Feuilleton*, July 17, 1847 p. 240, and A. Delavenne, *Recueil Généalogique de la Bourgeoisie Ancienne*, Paris, 1954.
2. Marriage contract between A. L. Cauchy and A. de Bure, Arch. Nat., Minutier central des notaires, Etude LXXIII, 1260, April 4, 1818.
3. *Vers à l'occasion du mariage de M. A. L. Cauchy avec Melle Aloïse de Bure*, Paris, undated.
4. C. A. Valson, *La Vie et les Travaux du Baran Cauchy*, **1**, p. 69.
5. F. R. de Lamennais, *Correspondance Générale*, **3** (1825–June 1828), 1971, document 28, letter from Abbé Gerbet to M. de Senfft, dated August 19, 1826, Paris.
6. See the letter from Cauchy to Libri, March 28, 1828, in Appendix III, p. 324.
7. See Arch. Nat. F_1dIVC^4, request for the Légion d'Honneur, and Arch. Nat. F^{14}11872. Letters from L. F. Cauchy requesting the Légion d'Honneur and promotion to the rank of engineer ordinary first class on behalf of A. L. Cauchy are found here. Since 1816, Cauchy was no longer in practice as engineer.
8. See J. B. Duroselle, 'Les "filiales" de la Congrégation', *Revue d'Histoire Ecclésiastique*, 1950, p. 867, and Arch. Nat. F^7 6699, file 1, Société des Bonnes Études.
9. F. R. de Lamennais, *Correspondance Générale*, **1**, (1805–1819), Paris, 1971, p. 423, letter 276 to Brute, dated July 22, 1818, where he wrote 'Teysseyre, Binet, Cauchy, etc., are busy from dawn to dusk with charitable works'. According to C. A. Valson, *La Vie et les Travaux du Baron Cauchy*, **1**, pp. 195–196, Cauchy was especially involved in the religious instruction of young chimney sweeps.
10. J. B. Duroselle, 'Les "filiales" de la Congrégation', pp. 878–880.
11. For more on this school of counterrevolutionary thought, see D. Bagge, *Les Idées Politiques en France sous la Restauration*, Paris, 1952.
12. The following handwritten note is kept at the Académie in the folder of the meeting's packet for July 19, 1824:

> I make no pretensions to having a knowledge of anatomy. But, I do swear that I was singularly shocked to have here heard praise given to the system propounded by Doctor Gall as concerns the brain. We all know how often his blunders have made him truly ridiculous in the eyes of the whole world. I do not think they will be accepted at the Académie; and what convinced me of this is precisely that the Académie has never put him down on its list of candidates for the sections (departments) of medicine and anatomy. Now, precisely what is the principle that M. Gall claims to have discovered? It has long been known that the material brain is made of a great number of parts, and

this fact was never in question—even before Doctor Gall. But, what Doctor Gall did not prove—and what he will never prove—is that the diversity of the parts of the material brain destroys the unity of the 'self', that thought can be dissected and various geometrical forms assigned to it. Such a principle is as repulsive to true philosophy as it is to the very foundations on which the peace and happiness of society rests.

13. See *Procès-verbaux, Ac. Sci.*, **7**, p. 5.
14. See *Procès-verbaux, Ac. Sci.*, **8**, pp. 136–138 (*O.C.*., 2, **15**, pp. 557–560).
15. *Mémorial Catholique*, 1st year, **2**, 1824, p. 192.
16. *New Monthly Magazine*, June 1825 (see also Stendhal, *Courrier anglais*, **3**, Paris, 1935, pp. 228–229).
17. *New Monthly Magazine*, November 1825 (see also Stendhal, *Courrier anglais*, **3**, Paris, 1935, pp. 228–229).
18. The following note is in Cauchy's handwriting: it is contained in the meeting's packet for January 22, 1827, of the Académie:

    M. Cauchy made some observations that have to do with this study. It should be noted that the results obtained with respect to the departments that constitute Old Brittany confirm the statements he made at one of the preceding meetings by showing the beneficial influence that religious instruction has exerted on the habits and manners of people. In fact, as to the province in question, the study indicates that the department of Ille-et-Vilaine presents the most natural and unaffected manners and habits. But, it is precisely in this department that schools directed by a religious order known as the Petits Frères were were founded and operated most successfully; the Petits Frères go forth into the countryside to be received by the poorest communities. The study that was just read, M. Cauchy added, goes to show to what degree the religious establishment of the Petits Frères is really worthy of the government's protection.

    Cauchy showed less mistrust with regard to statistics than he did in the introduction of *Analyse Algébrique*, stating that the statistical results seem to agree with his own opinions.
19. See Geoffroy de Grandmaison, *La Congrégation (1801–1830)*, Paris, 1889.
20. For more on the association, J.B. Duroselle, art. cit., pp. 880–882, and G. Bertier de Savigny, 'Le rôle des laïcs dans l'Église de France sous la Restauration', *L'Anneau d'Or*, **32**, March–April 1950, pp. 98–104.
21. See the prospectus of the Collège de Juilly, from the summer of 1828, in F. R. de Lammenais, *Correspondance Générale*, **6**, 1834–1835, Paris, 1977, document 33, p. 32:

    Instruction in mathematics would come next. There would be established special classes for those students who had the intent of taking the examinations for the naval school, the military school, or the École Polytechnique. Two scholars to whom we owe our recognition, M. Ampère and M. Cauchy, both of whom are professors at the École Polytechnique, are willing to take charge of this important part of the instructional program: they will enlighten the professors with their advice and they will keep abreast of the students' progress by regular examinations.
22. F. R. de Lammenais *Correspondance Générale*, **6**, (1834–1835), Paris, 1977, annexe, p. 488 and p. 520.

Notes to Chapter 9

1. Liste des présences, *Procès-verbaux, Ac. Sci.*, **9**, p. 748. Nevertheless, this list indicates that Cauchy did not attend the meeting of August 23, 1830, even though,

according to the minutes of the meeting, he read a paper 'Sur la dispersion de la lumière', on that day. This obvious contradiction leaves us uncertain about the length of his absence.

2. According to the *Procès-verbaux, Ac. Sci.*, **9**, pp. 748–753, from September 6 on, Cauchy regularly missed the meetings of the Académie. Moreover, we know from the Registry of Permits of Fribourg that on August 11, 1830, he obtained a passport that was delivered in Paris and was valid until July 15, 1831. (See G. Castella, 'Documents inédits sur un projet de fonder une "Académie Helvétique" à Fribourg en 1830', *Revue d'Histoire Ecclesiastique Suisse*, **21**, 1927, p. 309, note 2.)
3. See, for example, C. A. Valson, *La Vie et les Travaux du Baron Cauchy*, **1**, pp. 73–77.
4. The circular from the Ponts et Chaussées requiring the loyalty oath is dated September 4, 1830. At the Faculté des Sciences, the professors took the oath on September 18, 1830.
5. The expression is from R. Taton, 'Sur les relations scientifiques d'Augustin Cauchy et d'Évariste Galois', *Rev. Hist. Sci.*, **24**, 1971, pp. 123–148, esp, p. 125. This article gives a portrayal of A. L. Cauchy in 1830.
6. Undated letter written shortly after the death of A. L. Cauchy's youngest brother, Amédée, who was born in 1802 and died in 1831. Cited by C. A. Valson, *La Vie et les Travaux du Baron Cauchy* Valson, **1**, p. 76.
7. Arch. Nat. F^{14} 2187^2, Cauchy file.
8. On September 18, in a letter that Alexandre Cauchy wrote to the Dean of the Faculté des Sciences, the 'short trip' became an 'extended trip' (Arch. Nat. AJ 16 5120, *Procès-verbaux des délibérations de la Faculté des Sciences*, September 18, 1830). On November 28, 1830, Aloïse wrote that the 'extended trip' was now 'an extended stay in Italy' (Arch. Nat. F^{14} 2187^2, Cauchy file).
9. For example, we know that François Moigno left Paris for Switzerland on September 6, 1830.
10. From Valson, *La Vie et les Travaux du Baron Cauchy*, **1**, p. 75.
11. From his letters to the Emperor of Austria and to the Czar, published by A. Terracini, 'Cauchy a Torino', note 53 bis, p. 195, and by G. Castella, 'Documents inedits pour un Projet de fonder une Académie Helvétique...', pp. 312–313.
12. *Nouveau Dictionnaire des Girouettes...*, by an immovable weathervane, 1832, pp. 200–201; cited by Dr. Delaunay, 'Le Parnasse du temps de Napoléon...', p. 139.
13. See also Cauchy's letter of October 7, 1830, to the Emperor of Austria: 'Some members of the Institut and some professors, because of their decision to remain true to their oaths, are obliged to give up their professorial chairs.'
14. A letter from Lamennais to the Abbé Gerbet, dated October 8, 1830, gives us an understanding and appreciation of the state of mind of the promoters of the project:

    > Herewith are two letters for you from Fribourg. There is nothing in them to say in reply. They have completely lost their minds; they are every persuaded that within two years events will lead them to success. They are now busy trying to set up some kind of academy or university and expect the most beautiful things to result from it.

    F. R. de Lamennais *Correspondance Générale*, **4**, (July 1828–June 1831), Paris, 1973, p. 362, letter 1693).
15. See A. Terracini, 'Cauchy a Torino', note 53 bis, pp. 193–199.

16. On March 8, 1831, Cauchy's name was written in the *Registre des permis de séjour du canton de Fribourg* (see G. Castella, 'Documents inédits pour un projet de fonder une Académie Helvétique', p. 309).
17. Confirmed by the *Procès-verbaux, Ac. Sci.*, **9**, p. 750, Cauchy was present on March 14, 1831.
18. Cauchy's letter to the president of the Académie is kept in the meeting's packet of July 4, 1831. See Appendix III, p. 325.
19. 'Sui metodi analitici', *Biblioteca italiana*, **60**, pp. 202–219; **61**, pp. 321–324; and **62**, pp. 373–386 (*O.C.*, 2, **15**, pp. 149–181).

    The French manuscript of the last two articles sent by Cauchy in Geneva to the Académie des Sciences in Paris has not been published. This manuscript, entitled 'Sur le calcul différentiel et le calcul des variations présentant le résumé de ces deux calculs', has been preserved in the meeting's packet of July 4, 1831.
20. See A. Terracini, 'Cauchy a Torino', p. 182.
21. Bibliographical studies of the first Turin memoir have been published by A. Terraccini, 'Cauchy a Torino', pp. 183–185; R. Taton, in *O.C.* 2, **15**, pp. 262–263; I. Grattan-Guinness, 'On the publication of the last volume of the works of Augustin Cauchy', *Janus*, **62**, 1975. pp. 179–191, and J. Peiffer, *Les Premiers Exposés Globaux de la Théorie des Fonctions de Cauchy*, 1840–1860, 3rd cycle Thesis, typewritten, Paris, 1978.

    Here is a bibliographical analysis of the first Turin memoir:

    1. An introductory résumé was read before the Academy of Turin on October 11, 1831 (the handwritten manuscript is in the archives of that Academy, and it was lithographed several days later and presented to the Academy of Turin on November 27, 1831). This résumé was published in France in the *Bull. Fér.*, **15**, May 1831, pp. 260–269 (*O.C.*, 2, **2**, pp. 158–168).
    2. *Extrait du mémoire présenté à l'Académie de Turin le 11 Octobre 1839*, 153 pages, lithographed in Turin in 1832, was presented to the Academy of Turin on January 27, 1833. An *Addition* of 51 pages, dated March 6, 1833, was presented by Cauchy on April 14, 1833, as the third and final part of the study. An Italian translation of the *Extrait* was printed in Milan in the *Opuscoli Matematici e Fisici*, **2**, 1834. The third paragraph of the first part, the second part, and the addition of the *Extrait* are printed in *O.C.*, 2, **15**, pp. 262–411. For the first and second paragraphs of the first part, see below, 3 and 7.
    3. In 1837, section 1 of the first part and the first lines of section 1 of the second part of the 1832 *Extrait* appeared in *J.L*, **2**, 1837, pp. 406–412 (*O.C.*, 2, **2**, pp. 1823).
    4. In a letter to Coriolis, dated January 29, 1837, an extract of which appeared in *C.R. Ac. Sci.*, **4**, February 13, 1837, pp. 216–18 (*O.C.*, 1, **4**, pp. 38–41), Cauchy gave the statement of the Turin theorem and indicated possible applications. He also referred to 'a new study in analysis in which I will give a greater study of the methods developed in the preceding.' It appears that this study was never published, however.
    5. In *C.R. Ac. Sci.*, **9**, August 5, 1839, pp. 184–190 (*O.C.*, 1, **4**, pp. 483–490), Cauchy gave a proof for the Turin theorem. For the first time, he required that the derivative of the function should be continuous. This proof was published in *Ex. An. Phy. Math.*, **1**, 1840, pp. 27–32 (*O.C.*, 2, **11**, pp. 43–50).
    6. In *C.R. Acad. Sci.*, **11**, April 20, 1840, pp. 640–650 (*O.C.*, 1, **5**, pp. 180–191), Cauchy presented a new proof of the Turin theorem using the mean value of a function of a complex variable on a circle with center 0. This proof was

published in the *Ex. An. Phy. Math.*, **1**, 1840, pp. 269–287 (*O.C.*, 2, **11**, pp. 331–353).

7. In 1841, the introductory résumé of 1831, along with section 2 of the first part of the *Extrait* of 1832, with some changes, which were given in a note, appeared in the *Ex. An. Phys. Math.*, **2**, 1841, pp. 41–98 (*O.C.*, 2, **12**, pp. 48–112).

22. H. Burckhardt (see 'Entwicklungen nach oscillierenden Funktionen und Integration der Differentialgleichungen der mathematischen Physik', *Jahresbericht der Deutschen Mathematiker Vereinigung*, **10** (1904–1908, p. 24) and H. Freudenthal (see *Dictionary of Scientific Biography*, **3**, p. 140) found in Cauchy's papers 'Sur la détermination du reste de la série de Lagrange par une intégrale définie' and 'Règles de convergence de la série de Lagrange et d'autres séries du même genre', *Mémoires Ac. Sci.* **8** (1825), 1829, pp. 97–129 (*O.C.*, 1, **2**, pp. 29–66), a first outline of the method presented at Turin in 1831. Nevertheless, the calculus of residues was not used in these studies and the methods of proof were very different.
23. Here is a bibliographical analysis of the second Turin memoir:

    1. An introductory résumé, read before the Academy of Turin on November 27, 1831, was lithographed in Turin and dated December 17, 1831. The text of the lithograph was reproduced in France in the *Bull. Fér.*, **16**, Sept. 1831, pp. 116–130 (*O.C.*, 2, **2**, pp. 169–183).
    2. The complete memoir was lithographed at Turin and dated August 8, 1832. It was presented to the Académie des Sciences in Paris on October 8, 1832 (*O.C.*, 2, **15**, pp. 182–261). An Italian translation of this memoir was published in the *Memorie della Societa Italiana delle Scienze in Modena*, 1, **22**, 1838, pp. 91–183.

24. This study, 'Sur un certain type d'équations', which was presented to the Académie des Sciences in Paris on October 8, 1832, has not been published. A résumé of the work in Italian (no doubt the translation of the presentation made by Cauchy on September 10, 1832) was published in the *Gazzetta Piemontese*, no. 113, for September 22, 1832. This text has been reproduced by A. Terracini, 'Cauchy a Torino', pp. 178–179. In 1837, in several letters and notes that appeared in the *C.R. Ac. Sci.* between May and September (*O.C.*. 1, **4**, pp. 42–99), Cauchy came back to the method of solving equations that he had discussed in the study of September 10, 1832. See *C.R. Ac. Sci.*, **12**, June 21, 1841, pp. 1133–1145 (*O.C.*, 1, **6**, pp. 175–186) and *Ex. An. Phys. Math.*, **2**, November 1841, pp. 109–136 (*O.C.*, 2, **12**, pp. 125–156).
25. The text of this lithograph constitutes section 1 of the article 'Calcul des indices des fonctions', *J.E.P.*, **15**, 25th cahier, 1837, pp. 176–226 (*O.C.*, 2, **1**, pp. 416–466).
26. A. Terracini, 'Cauchy a Torino', p. 160, and note 3, pp. 170–171.
27. C. A. Valson, *La Vie et les Travaux du Baron Cauchy*, **1**, p. 75.
28. Cited by A. Terracini, 'Cauchy a Torino', p. 160.
29. B. Boncompagni, 'La vie et les travaux du Baron Cauchy par C. A. Valson...', *Bulletino di bibliografia e di storia delle scienze matematiche e fisiche*, **2**, 1869, p. 22, note 10.
30. C. A. Valson, *La Vie et les Travaux du Baron Cauchy*, **1**, p. 77. Valson probably paraphrased certain letters that he had examined but are now no longer available.
31. The chronology of events that we propose is, in fact, a hypothesis that we infer from the following passage of C. A. Valson, *La Vie et les Travaux du Baron Cauchy*, **1**, p. 77:

    His family, in fact, had not been able to even think of his decision to go into exile; but, all their efforts to dissuade him from doing so bore no fruit. Two short trips that Cauchy made to Paris and some 'reciprocal' visits by various members of his family were not

> enough to compensate for the emptiness caused by his absence. His parents and his friends were always waiting for his next visit... [Valson here tells of the trip made by Eugène Cauchy]. He seems to have been deeply impressed by all of this and appears to have been on the point of agreeing to come back when new trouble broke out in Paris and this caused him to put away any thoughts of returning.

What the troubles were that Valson refers to is not clear, unless the reference is to the dramatic events of June 1832. As for the two trips that Cauchy made to Paris, which were undertaken earlier, they undoubtedly were those of March 1831 and of March 1832, as is confirmed by the *Procès-verbaux, Ac. Sci.*

32. See Archives Ac. Sci., meeting packet of October 19, 1832.
33. Published by F. Moigno in 1868 (*O.C.*., 2, **15**, pp. 412–447).
34. A. Terracini, 'Cauchy a Torino', note 33, pp. 180–181.
35. Ibid., p. 165, and note 43, p. 190.
36. L. Menabrea, *Memorie*, published by the Centro per la storia della tecnica in Italia del Consiglio nazionale delle ricerche, 1971, pp. 14–15.
37. *Procès-verbaux de la classe des sciences physiques et mathématiques de l'Académie des Sciences*, **3**, October 11, 1831, cited by A. Terracini, 'Cauchy a Torino', note 35, p. 186.
38. From L. Menabrea, *Memorie*, p. 15:

> In the ultramontain sense, Cauchy was a saint; this, however, does not mean that he did not have certain 'little faults' or that he was not lacking in charitable spirit. Thus, Plana enjoyed a position in Turin that, though modest, sufficed for him to devote himself to science. Aside from the chair in analysis at the university, he was responsible for the direction of the observatory as well as for the direction of studies at the military academy where he also lectured. While the total payment that he received was not much, it is nevertheless true that he was content with it. Cauchy thought this was excessive and asked that Plana be stripped of one of his positions so that it could be awarded to him [i.e., to Cauchy]. This behavior raised quite a stir, because Plana was very popular, and Cauchy, despite the support of the Jesuits, was never able to get one of Plana's positions awarded to him.

39. From L. Menabrea, *Memorie*, p. 11:

> When any geometer came up with some new transcendental theory, strewn through with some more or less elegant formulas, he [Plana] would say, by way of judging, 'I expect numbers to be used'. Frequently, there would be no numbers in the formulas.... His opposition to what is called the new transcendental analysis, which Cauchy had promoted, grew in intensity as he himself grew older.

40. 'I will always recall with true delight the scientific conversations that we had while you were in Turin' wrote Bidone in a letter of thanks to Cauchy on November 4, 1835. Cauchy had sent him his study, 'Sur l'interpolation', along with a letter cited by A. Terracini, art. cit., note 69, p. 202.
41. See the letter of September 24, 1833, published in Appendix III, pp. 327–328. In 1833, after Cauchy left for Prague, Father La Chèze took charge of the printing of the editions of the *Résumés Analytiques.*
42. This letter from Sophie Germain, dated April 18, 1831, was published by C. Henry. See C. Henry, 'Les manuscrits de Sophie Germain et leur récent éditeur. Documents nouveaux', *Revue Philosophique de la France et de l'Étranger*, **8**, 1879 pp. 619–641, esp. p. 632.

Notes to Chapter 10

1. From a copy written by Cauchy, undated, kept in the *Musée National de l'Éducation* in Rouen, Ms A 10822–1.
2. See the letter by Cauchy in Appendix III, p. 326.
3. Barande remained in Bohemia after his dismissal, where he studied geology, gaining a reputation and name for himself as a geologist.
4. For more on this affair, see Comte de Damas d'Anlézy, 'L'Éducation du duc de Bordeaux', *Revue des Deux Mondes*, October 1902, pp. 602–640, in particular, pp. 609–611. In this article, the author who had access to private papers, stated that the Marquis de Foresta, 'always tireless in serving his prince', called on Cauchy in Turin to ask him to serve as Barande's replacement (pp. 619–620). This assertion does not agree with the letters from Baron de Damas that Cauchy received in Turin.
5. See Appendix III, pp. 327–328.
6. Brochure published in Prague in September 1833. This declaration appeared in Fribourg, in *L'Invariable*, **4**, pp. 64–78, with the following foreword:

   > We are all the more happy to publicize such noble thoughts because the so-called royalist newspapers have raged with insolent violence and perfidious hypocracy against the new teachers of the Duke of Bordeaux and have carefully omitted M. Cauchy. That European name was an obstacle to their declamation; and, in order to lament over the departure of M. Barande and present it as irreparable, it was necessary not to tell how it has, in fact, been repaired.

7. A. d'Hautpoul, *Quatre Mois à la Cour de Prague*, published by Count de Fleury, Paris, 1902, p. 129, and p. 147.
8. Ibid., p. 160.
9. Ibid., p. 242:

   > Of all the people around me, the one who most obstructed my principle of education was Cauchy.

10. Ibid., p. 327.
11. C. A. Valson, *La Vie et les Travaux du Baron Cauchy*, **1**, p. 90.
12. Ibid., 1, p. 90. It is difficult to identify the paper to which Aloïse referred. Perhaps it is 'Sur l'interpolation', which was lithographed in September 1835, or 'Mémoire sur l'intégration des équations différentielles', which was lithographed at the end of the same year.
13. A. d'Hautpoul, *Quatre Mois à la Cour de Prague*, p. 160.
14. F. Moigno, *Leçons de Calcul Différentiel et de Calcul Intégral*, **1**, 1840, Introduction, p. XIV:

   > M. Cauchy has developed, on a totally new basis, an elementary treatment of arithmetic and geometry. The reason for his having done so is, of course, known; and, it is a pleasure to see a great thinker, inspired with self-sacrifice, suspend the pursuit of his own brilliant discoveries and research in order to make the most important secrets of science available to a young royal person who is in exile.

   Unfortunately, the archives of the Duke of Bordeaux have been entirely destroyed. Two of Cauchy's notebooks relating to this elementary level teaching are kept in the Sorbonne Library: a notebook of arithmetic exercices (Ms 1760) and a notebook containing Section 1 of Chapter 7, 'Exponentielle et logarithme', from an elementary treatise referred to by Moigno. That section is devoted to powers using arbitrary exponents or exponentials (11 pages, unfinished, Ms. 1761).

15. Method published in *C.R. Ac. Sci.*, **11**, November, 16 and 23, 1840, pp. 789–798 and (*O.C.*, 1, **5**, pp. 431–454). Cauchy described its origins in the following terms:

    Because of the honorable mission that was conferred upon me at the time that I was publishing these tables [that is, in 1836], having presented me with an opportunity to investigate whether it is possible to make the various rules of computing more easy and more certain, I came to realize that some very simple procedures might be of advantage in this regard, even in arithmetic.

16. Those assertions by General d'Hautpoul that we have been able to verify all appear to be correct; and this compels us, regardless of d'Hautpoul's resentment of Cauchy, to take account of his testimony.
17. A. d'Hautpoul, *Quatre Mois à la Cour de Prague*, pp. 246–292.
18. Ibid., p. 291.
19. Ibid., p. 292.
20. Ibid., pp. 245–246.
21. Ibid., p. 244.
22. Ibid., p. 243.
23. P. L. F. de Villeneuve, *Charles X et Louis XIX en Exil. Mémoires Inédites*, Paris, 1889, pp. 242–243.

    We also note the imprecise testimony dated February 24, 1826, in the *Invariable* (Fribourg), **8**, 1836, p. 196:

    We then moved on to the examination in the physical sciences. The prince responded correctly to the questions that were addressed to him, on applications to astronomy and navigation. His very learned teacher was M. Cauchy, to whom the sciences were indebted for his analytical treatment of the theory of light, a work that had the unexpected result of leading this famous geometer to create some new interpolation formulas, which have at once made an immense step in analysis as a whole and in the theory of light.

24. A. Terraccini, 'Cauchy a Torino', note 33, pp. 180–181.
25. W. R. Hamilton, 'On a general method in dynamics', *Mathematical Papers*, **2**, Cambridge (England), 1940, p. 103.
26. Cauchy maintained that he had been thinking about this reduction for a long time. In 1835, he wrote:

    The fine study by M. Hamilton, in which he made the integration of the differential equations of dynamics depend on the determination of a single function, represented by a definite integral that satisfies two second-order partial differential equations, has brought my ideas to a point that occurred to me a long time ago and is worthy of being examined with particular care. I thought that there perhaps would be some advantage in reducing the integration of a system of differential equations to the problem of integrating a single partial difference equation of the first order.

    (*Ex. An. Phys. Math.*, **1**, December 1840, p. 331, *O.C.*, 2, **11**, p. 404). Five years latter he wrote:

    The method of reduction that I have just applied to a system of differential equations does not differ from the one that I gave in the 1835 study; and it is one that I have been considering for a long time. In fact, I have just found it again in a note dated August 31, 1824, placed behind various studies that were presented to the Académie during 1823.

(*C.R. Ac. Sci.*, **10**, June 29, 1840, p. 957, *O.C.*, 1, **5**, p. 236). The note could not be found. Cauchy probably gave this method of reduction in his analysis teaching at the École Polytechnique.

27. The first installment of *Nouveaux Exercices*, pp. 1–24 (*O.C.*, 2, **10**, pp. 195–220), goes back to the 'Mémoire sur la dispersion de la lumière' of 1830, As to financing the publication, see the advice to the reader, dated June 10, 1836.
28. Lithograph dated September 26, 1835, published in 1837 in *J. L.*, **2**, pp. 193–205 (*O.C.*, 2, **2**, pp. 5–17). On this study, see C. C. Heyde and E. Seneta, *I. J. Bienaymé, Statistical theory anticipated*, Springer-Verlag, New York, Heidelberg, Berlin, 1977, pp. 71–76.
29. This study, which is not widely known, is not contained in the *Oeuvres Complètes* (*O.C.*). A copy of the lithograph is preserved in the Houghton Library at Harvard (FCB. 02103. B843d).
30. From the 'observation' published at the end of his lithographed study *Sur la Théorie de la Lumière* (August 1836). Cauchy brought several notebooks back to France with him; these notebooks contained manuscripts devoted to his research on light; some of these are dated 1836. An entire notebook, entitled 'Researches nouvelles sur la lumière' and dated January 1838, is kept in the Sorbonne Library (Ms. 1762).

    According to *C.R. Ac. Sci.*, **15**, September 26, 1842, p. 606 (*O.C.*, 1, **7**, p. 157), another notebook, which is now lost, contained two unpublished investigations:

    1. a first study, 'Sur la théorie de la lumière', in 4 sections: the first three dealt with the theory of spherical and cylindrical waves (see *C.R. Ac. Sci.*, **9**, November 18, 1839, pp. 637–649 (*O.C.*, 1, **5**, pp. 5–20); the 4th section presented a discussion of the two phenomena of shadows and diffraction (*C.R. Ac. Sci.*, **7**, September 26, 1842, pp. 157–170).
    2. the manuscript of the lithographed study *Sur la théorie de la lumière.*

31. These letters have been published almost completely in the *Compte Rendus Hebdomadaires des Séances de l'Académie des Sciences.* Handwritten copies containing some modifications compared to the printed versions are contained in the manuscript book *Mélanges*, which is kept in Ivoy-Le-Pré at the residence of Madame de Pomyers. The dates on which these letters were written are given in that manuscript:

    1. February 12, 1836: letter to Ampère, published in the *C.R. Ac. Sci.*, **2**, February 22, 1836, pp. 182–185 (*O.C.*, 1, **4**, pp. 5–8);
    2. February 19, 1836; letter to Ampère, published in the *C.R. Ac. Sci.*, **2**, February 29, 1836, pp. 207–209 (*O.C.*, 1, **4**, pp. 9–11);
    3. March 19, 1836; letter to Libri, published in the *C.R. Ac. Sci.*, **2**, April 4, 1836, pp. 341–343 (*O.C.*, 1, **4**, pp. 11–13);
    4. March 28, 1836; letter to Libri, published in the *C.R. Ac. Sci.*, **2**, April 4, 1836, pp. 343–349 (*O.C.*, 1, **4**, pp. 13–21);
    5. April 1, 1836: letter to Ampère, published in the *C.R. Ac. Sci.*, **2**, April 11, 1836, pp. 364–371 (*O.C.*, 1, **4**, pp. 21–27);
    6. April 16, 1836: letter to Ampère, published as addressed to Libri, in the *C.R. Ac. Sci.*, **2**, May 2, 1836, pp. 424–428 (*O.C.*, 1, **4**, pp. 30–32);
    7. April 22, 1836: letter to Libri, published in he *C.R. Ac. Sci.*, **2**, May 9, 1836, pp. 455–456 (*O.C.*, 1, **4**, pp. 32–34):

8. April 26, 1836: letter to Libri, published in the *C.R. Ac. Sci.*, **2**, May 9, 1836, pp. 456–461 (*O.C.*, 1, **4**, pp. 34–36).

32. The handwritten manuscript book *Mélanges*, which is kept in Ivoy-le-pré at the residence of Madame de Pomyers, contains 'Recherches sur la théorie de la lumière' (unpublished), with 63 pages, signed on November 4, 1837 by Augustin Cauchy. These 'Recherches' are divided into 5 sections:

    1. polarisation rectiligne (sent to Moigno on December 15, 1837);
    2. axes optiques dans la polarisation rectiligne;
    3. surface des ondes;
    4. ombre et diffraction; and
    5. suite du 4ème §.

    Aside from this, one can also find section 9 of 'Théorie de la lumière', entitled 'Lois de propagation de la lumière dans le vide et dans les milieux qui ne dispersent pas les couleurs' (4 p.), and 'Notes sur la théorie de la lumière envoyées à M. L'abbé Moigno' (27 p.):

    Note 1 'Equations du mouvement de l'éther dans un milieu où la propagation de la lumière s'effectue de la même manière en tous sens autour de l'axe' (sent to Moigno on October 6, 1837).
    Note 2 'Sur les vibrations de l'éther dans un milieu transparent' (sent to Moigno on October 6, 1837).
    Note 3 'Ellipsoïde'.

33. See p. 285, note 20.
34. The manuscript of the 1835 study, 'Sur l'intégration des équations différentielles's, is kept at the Academy of Sciences of Czechoslavakia. See K. Rychlik, 'Un manuscrit de Cauchy aux Archives de l'Académie tchécoslovaque des Sciences', *Revue d'Histoire des Sciences*, **10**, 1958, pp. 259–261.
35. On relations between Bolzano and Cauchy in Prague, see R. and D. J. Struik, 'Cauchy and Bolzano in Prague', *Isis*, **11**, 1928, pp. 364–366, and K. Rychlik, 'Sur les contacts personnels de Cauchy et Bolzano', *Revue d'Histoire des Sciences*, **15**, 1962, pp. 163–164 and H. Sinaceur, 'Cauchy et Bolzano', *Revue d'Histoire de Sciences*, **20**, 1973, pp. 97–112.
36. See C. A. Valson, *La Vie et les Travaux du Baron Cauchy*, **1**, p. 91, who gives as the places where Cauchy resided Hradshin, Toeplitz in 1835 (probably during the summer), Budweis, Kirchberg, Goeritz in 1836, and Kirchberg in 1838. See also L. Bader, *Les Bourbons de France en Exil à Gorizia*, Paris, Perrin 1977, p. 100, and appendix 10, *Etat du personnel des maisons et de la suite de la famille royale à Goeritz en Novembre 1837* pp. 372–374.
37. Two notes, one from September 18 and the other from October 23, 1837, have not been published in the *C.R. Ac. Sci.* The unpublished manuscript is kept in the meeting's packet of September 18, 1837. In it, Cauchy suggested that he was working on a study of Ampère's interpolation function. As to these investigations, see also the three manuscripts by Cauchy that are kept at the Sorbonne Library, Mss. 1759, 1760, and 1786.
38. From C. A. Valson, *La Vie et les Travaux du Baron Cauchy*, **1**, p. 94.

Notes to Chapter 11

1. See G. Bigourdan, 'Le Bureau des Longitudes, son histoire et ses travaux de l'origine à ce jour', 3rd part, *Annuaire du Bureau des Longitudes*, 1930, A 18–26.
2. Published in the *C.R. Ac. Sci.*, **9**, August 5, 1839, pp. 184–190 (*O.C.*, 1, **4**, pp. 483–490). Repeated in the *Ex. An. Phys. Math.*, **1**, September 1839, p. 27 (*O.C.*, 2, **2**, p. 43).
3. *C.R. Ac. Sci.*, **9**, August 5, 1839, p. 190 (*O.C.*, 1, **4**, p. 490).
4. *Procès-verbaux des séances du Bureau des Longitudes*, November 6, 1839, cited by G. Bigourdan, 'Le Bureau des Longitudes,...', A 29.
5. C. A. Valson, *La Vie et les Travaux du Baron Cauchy*, **1**, pp. 97–98. The problem at the Bureau des Longitudes is treated extensively in this work (pp. 97–104).
6. See *C.R. Ac. Sci.*, **8**, April 15, 1839, pp. 553–661 [Cauchy] (*O.C.*, 1, **4**, pp. 312–321); April 22, 1839, pp. 581–582 [Poisson]; and pp. 582–589 [Cauchy], (*O.C.*, 1, **4**, pp. 322–330).
7. *Procès-verbaux des séances du Bureau des Longitudes*, November 13, 1839, cited by G. Bigourdan, 'Le Bureau des Longitudes,...', A 29–30.
8. *Procès-verbaux des séances du Bureau des Longitudes*, November 27, 1839, cited by G. Bigourdan, 'Le Bureau des Longitudes,...', A 30.
9. From the *Procès-verbaux des séances du Bureau des Longitudes*. The discussion of this problem is handled very discretely. Arago seems to have played a very important role in these developments. G. Bigourdan,'Le Bureau des Longitudes,...', A 30.
10. J. B. Biot, 'Lettre à M. de Falloux', abstract from *Le Correspondant*, 1857, p. 9. See also J. Bertrand, *Éloges Académiques*, **2**, Paris, 1902, pp. 116–117:

    It is in jest told that, being pressed to accept an unimportant formality, he replied: 'May they chop off my head!' That was his way of saying 'No!' most emphatically.

11. A remark by J. B. Biot reported in the *Procès-verbaux des séances du Bureau des Longitudes*; cited by G. Bigourdan, 'Le Bureau des Longitudes,...', A 30.
12. In a study on celestial mechanics that was presented to the Académie on August 3, 1840 (*C.R. Ac. Sci.*, **11**, August 3, 1840, pp. 179–185 (*O.C.*, 1, **5**, pp. 260–276), he expressed regret at not being able to present his works 'to the gathering of scholars who had done so much to advance celestial mechanics' and 'to offer to have them [his works] placed in the *Connaissance des Temps*'. He concluded his study with these words:

    The only thing I can do is exert every possible effort to respond to the kindness that the friends of science have shown toward my works. Moreover, if it is possible, I hope to be able to prove that the title of geometer is not altogether at odds with the ususal concerns of the old professor, who, in times past, was kindly consulted by the masters of science (celestial mechanics).

    (See also the foreword of the first volume of the *Exercices d'Analyse et de Physique Mathématique*, probably written in the spring of 1841.)

13. F. Hoefer, *Nouvelle Biographie Universelle*, Paris, 1852–1866, **9**, article on Cauchy.
14. 'Discussion des lignes et des surfaces du second degré', *Ex. math.*, **3**, June 1828, pp. 65–127 (*O.C.*, 2, **8**, pp. 83–149):

    This manner of arriving at the equation of the tangent plane [Cauchy wrote in a note] has been pointed out to us by a young ecclesiastic, who is as learned in the divine sciences as he is in the human sciences. He is a member of that illustrious society that has rendered great services to civilization in both hemispheres.

15. In the preface to his *Leçons de Mécanique Analytique*, which was published in 1867, Moigno pointed out that he had used the lectures that Cauchy had given prior to 1830 in his courses at the Faculté des Sciences as the basic text for his own teaching at the *École Normale Écclésiastique* from 1838 to 1843.
16. F. N. M. Moigno, *Leçons de Calcul Différentiel et de Calcul Intégral*, **1**, 1840, introduction, p. XIII.
17. Ibid., **1**, introduction, pp. XVIII–XIX and Lessons 42 and 45.
18. *C.R. Ac. Sci.*, **11**, November 2, 1840, pp. 687–693 (*O.C.*, 1, **5**, pp. 424–430).
19. See J. B. Duroselle, *Les Débuts du Catholicisme Social en France, 1822–1870*, Paris, PUF, 1951, pp. 248–249, and Archives of the Compagnie de Jésus, Bibliothèque du Centre des Fontaines, Chantilly, Moigno file.
20. F. N. M. Moigno, *Leçons de Calcul Différentiel et de Calcul Intégral*, **2**, 1844, introduction.
21. *C. R. Ac. Sci.*, **23**, November 16, 1846, pp. 911–914 (*O.C.*, 1, **10**, pp. 202–205).
22. See C. A. Valson, *La Vie et les Travaux du Baron Cauchy*, **1**, pp. 201–213.
23. Many authors have wrongly confused the two conference centers (L. Baunard, *Frédéric Ozanam d'après sa Correspondance*, Paris, 1912, p. 345, J. B. Duroselle, op. cit., pp. 247–248). Valson also makes this error. Jeanne Caron, however, correctly distinguishes between the Institut Catholique and the Cercle Catholique in the critical edition of the *Lettres de Frédéric Ozanam*, **2** (1841–1844), Paris, CELSE, 1971, p. 295, note (2). The Cercle Catholique met in the rue de Grenelle and published its own bulletin.
24. C. A. Valson, *La Vie et les Travaux du Baron Cauchy*, **1**, p. 205.
25. *Notice sur l'Institut Catholique*, February 1842, published with the *Bulletin de l'Institut Catholique* containing the 'Séance d'ouverture de l'année 1842'.
26. C. A. Valson, *La Vie et les Travaux du Baron Cauchy*, **1**, p. 205 and p. 206. See also *Bulletin de l'Institut Catholique*, p. 42, meeting of March 3, 1842.
27. In this *Bulletin*, the lectures given at the opening meeting and the general meetings were published. Collection of the *Bulletin*, that is kept at the *Bibliothèque Nationale*, concludes with the meeting of March 13, 1844.
28. See *Notice sur l'Institut catholique*, February 1842.
29. C. A. Valson, *La Vie et les Travaux du Baron Cauchy*, **1**, pp. 205–206.
30. Ibid., **1**, p. 199.
31. See G. Bertier de Sauvigny, *Le Comte Ferdinand de Bertier, 1782–1864, et l'Énigme de la Congrégation*, Paris, 1949, pp. 536–599, and *Lettres de F. Ozanam*, **2**, p. 74, note 6.
32. Letter from Dominique Meynis, March 8, 1841, in *Lettres de F. Ozanam*, **2**, p. 96.
33. Letter to Dominique Meynis, March 17, 1841, ibid., p. 96.
34. Letter to Domique Meynis, March 23, 1841, ibid., p. 119.
35. Letter from Cauchy to Father Roothan, A. R. S. J., *Franc.* 1007–XXXIV, 8... Rome. Reactions that it provoked in Rome, A. R. S. J., Reg. Prov. Franciae, II, 239.
36. G. Monod, 'Les troubles du Collège de France en 1843', *Séances et Travaux de l'Académie des Sciences Morales et Politiques*, 1909, **2**, pp. 407–423.
37. Cauchy had known Libri since the 1820s. In his letter of June 19 (See Appendix IV, p. 343), he noted the importance that was attached to his candidacy by the other candidates—Liouville and Libri—who had attended his lectures in the past.

38. Letter from Cauchy to Letronne, dated June 11, 1843, Archives of the *Collège de France*, B-II Mathématiques, b-1. See Appendix IV, p. 339.
39. Michelet wrote in his diary for May 6, 1843; 'A visit from Libri: rubbiano. Shocking disclosures!'
40. G. Libri, 'Lettres sur le clergé': 'I. De la liberté de conscience', *Revue des Deux Mondes*, 1843, **3**, pp. 329–356; II. 'Y-a-t-il encore des jésuites?', *Revue des Deux Mondes*, 1843, **3**, pp. 968–981.
41. Archives of the Collège de France, Libri file.
42. Archives of the Collège de France, B-II, Mathématiques, b-1, letter from Cauchy to Letronne, undated. See Appendix IV, p. 340.
43. Archives of the Collège de France, CXII, Liouville, 1B, letter from Liouville to Letronne, dated June 19, 1843. See Appendix IV, p. 341.
44. Letter from Cauchy to the members of the Académie des Sciences, June 19, 1843, Archives Ac. Sci., Libri file, rough draft in the Archives of the Compagnie des Jésus, Bibliothèque du Centre des Fontaines, Chantilly, Moigno file. See Appendix IV, p. 343.
45. Archives Ac. Sci., Libri file. See Appendix IV, p. 345. Relations between Libri on the one hand and Sturm and Liouville on the other hand had been very bad since about 1838–1839.
46. This declaration was published in the *C.R. Ac. Sci.*, **16**, p. 1365.
47. In the end, Liouville only published the paper by Galois in his journal in 1846. He declined to publish the commentaries that were proposed.
48. *Procès-verbaux des séances du Bureau des Longitudes*, November 15, 1843, cited by G. Bigourdan, 'Le Bureau des Longitudes', A 31.
49. *Procès-verbaux des séances du Bureau des Longitudes*, November 29, 1843, cited by G. Bigourdan, 'Le Bureau des Longitudes', A 31.
50. C. A. Valson, *La Vie et les Travaux du Baron Cauchy*, **1**, pp. 101–104.
51. For more on the relationship between Father François de Ravignan and Cauchy, see A. de Ponlevoy, *Vie du R. P. Xavier de Ravignan* **2**, Paris, 1876, pp. 386–387, and C. A. Valson, op. cit., **2**, pp. 111–112.
52. Ibid. **2**, pp. 189–190.
53. E. Gossin, *Vie de M. Jules Gossin, 1789–1855*, Paris, 1907, pp. 325–328, and *C.R. Ac. Sci.*, **12**, May 11, 1846, p. 751.
54. E. Gossin, *Vie de M. Jules Gossin*, p. 326, and C. A. Valson, *La Vie et les Travaux du Baron Cauchy*, **2**, pp. 188–189.
55. H. de Riancey, article in the *Union*, January 16, 1857, cited by C. A. Valson, *La Vie et les Travaux du Baron Cauchy*, **2**, pp. 190–191.
56. Ibid., **2**, pp. 191–193.

Notes to Chapter 12

1. *Le National*, October 19, 1842.
2. See the letter from Cauchy to Moigno dated October 26, 1842, Appendix III, p. 331.
3. J. B. Biot, 'Comptes rendus hebdomadaires des séances de l'Académie des sciences, publiés par MM. les secrétaires perpétuels, commençant au 3 août 1836', *Journal des Savants*, November 1842, pp. 641–661. See especially pp. 659–660.
4. *C.R. Ac. Sci.*, **15**, November 14, 1842, pp. 910–916 (*O.C.*, 1, **7**, pp. 200–207).

5. *C.R. Ac. Sci.*, **15**, December 19, 1842, p. 1075 (*O.C.*, 1, **7**, p. 212). The scientific writer for the *National*, always ridiculing Cauchy, had the following to say in the December 21, 1842 issue:

   On reading in the *Journal des Savants*, his criticism on publications abuses, which was addressed to an illustrious member of the Académie, M. Cauchy exclaimed: 'Well, truly, if the article was not signed, then one would swear that the author intended to speak of M. Biot'. Is the remark authentic? That could not be ascertained. However, it might well be.

6. From December 1842 to December 1843, Cauchy published 43 articles in the *Comptes Rendus Hebdomadaires des Séances de l'Académie des Sciences* (which in the *O.C.* amounts to 370 pages), that is, about the same number of articles as from December 1841 to December 1842 (or about 310 pages in the *O.C.*).
7. J. Bertrand, 'La Vie et les travaux du Baron Cauchy, par C. A. Valson', *Journal des Savants*, 1869, p. 210.
8. The following sarcastic remark about Cauchy is due to Poinsot and is cited by Barré de Saint-Venant in a letter to Boussinesq on July 9, 1876 (Bibliothèque de l'Institut, Ms. 4227):

   Poinsot said that the majority of Cauchy's works were like a notebook with blank pages.

9. From June 1839 to February 1848, Cauchy and Liouville worked together on 51 evaluation committees; similarly, from October 1838 to February 1848, he worked on 31 committees with Sturm; from October 1838 to February 1848 he served on 22 committees with Poncelet; from July 1843 until February 1848, he worked with Binet on 30 committees; and from March 1843 to February 1848, he worked on 21 committees with Lamé.
10. See R. Taton, 'Liouville, Joseph', *Dictionary of Scientific Biography*, **8**, New York, 1973, pp. 381–387, and P. Speziali, 'Sturm, Charles François', ibid., **13**, 1976, pp. 126–132.
11. In a letter written in 1840, Liouville wrote to Dirichlet concerning two notes by Cauchy (*C.R. Ac. Sci.*, **10**, March 16, 1840, pp. 437–452, and May 11, 1840, pp. 719–731 (*O.C.*, 1, **5**, pp. 135–152 and pp. 199–212)) on a question that had been studied by Dirichlet in his paper 'Sur l'usage des séries infinies dans la théorie des nombres' and that Liouville presented in a course at the Collège de France.

    You can imagine my astonishment at seeing that I had been cited by M. Cauchy in the two notes that he ventured to publish with reference to yours. This especially concerns the one of May 11. I see scarcely anything in these notes that is not like an amplification of the teacher's paper by a good student. In the second note, the plagiarism is particularly evident. I clearly explained to him that he had forced me to play a dirty role by citing me thusly without my knowledge and [I also spoke of] the injustice that he had committed with regard to you. I must tell you that during the conversation he appeared to be much more fair-minded that he had been on paper. He promised me that he would not mention me and would be more mindful of your rights in the second edition, which I think will appear in the memoirs of the Académie. In any event, we will see what he will do. If he should fail to acknowledge you, then I strongly advise you to take up the matter yourself when you publish a new work.

12. In January 1836, Liouville founded the *Journal de Mathématiques Pures et Appliquées*. Practically all of the leading mathematicians of the day—especially young authors—published in it. Liouville was closely connected with many

mathematicians and corresponded regularly with them, thus providing a link between the French and German scientific communities.

13. The handwritten report by Cauchy accompanying a letter to the president of the Académie is kept in the Bertrand collection at the Académie des Sciences.
14. *Bulletin de l'Institut Catholique*, 1st issue, 'Séance d'ouverture', January 13, 1842, p. 13.
15. Letter from Cauchy to J. B. Dumas, June 20, 1843, Archives Ac. Sci., Libri file.
16. Following are some figures: from 1 December 1843 to 1 December 1844, Cauchy presented to the Académie 19 communications and 4 reports (or 186 pages in the *O.C.*) from December 1, 1842 to December 1, 1843, and 44 communications and 4 reports (or 536 pages in the *O.C.*), as opposed to 39 communications and 5 reports (or 362 pages in the *O.C.*) from December 1, 1844 to December 1, 1845. On the other hand, from December 1, 1843 to December 1, 1844, he served on 11 academic commissions, 5 of which took place during November 1844, as opposed to 20 academic commissions from December 1, 1842 to December 1, 1843, and on 19 academic commissions from December 1, 1844, to December 1, 1845.
17. A. L. Cauchy, *Recueil de Mémoires sur la Physique Mathématique*, Paris, Bachelier, 1839.
18. In the first volume of the new series of *Exercices*, 11 of 15 papers were devoted to questions relative to mathematical physics and, more particularly, to the theory of light.
19. See *C.R. Ac. Sci.*, **7**, October 29, November 19 and 26, and December 10, 1838, pp. 751–759, pp. 865–867, pp. 907–912, and pp. 985–992; *C.R. Ac. Sci.*, **8**, January 7, 14, and 28, and February 4, 11, 18, and 25, 1839, pp. 7–13, pp. 39–46, pp. 114–119, pp. 146–155, pp. 189–196, pp. 229–231 and pp. 985–1000 (*O.C.*, 1, **4**, pp. 99–186 and pp. 427–443); *C.R. Ac. Sci.* April 8, 22, and 29, May 20, and June 24, 1839, pp. 505–522, pp. 589–597, pp. 659–673, and pp. 767–778 (*O.C.*, 1, **4**, pp. 237–312); *C. R. Ac. Sci.*, **9**, July 1, 8, and 15, 1839, pp. 1–9, pp. 59–68, and pp. 91–103 (*O.C.*, 1, **4**, pp. 427–483); and the three papers, 'Mémoire sur les mouvements infiniment petits d'un système de molécules sollicitées par les forces d'attraction et de répulsion mutuelle', *Ex. An. Phys. Math.*, **1**, September 1839, pp. 1–15 (*O.C.*, 2, **11**, pp. 11–27); 'Mémoire sur les mouvements infiniment petits dont les équations présentent une forme indépendante de la direction des trois axes coordonnées, supposés rectangulaires, ou seulement de deux de ces axes', ibid., September 1839, pp. 101–132 (*O.C.*, 2, **11**, pp. 134–174); and 'Mémoire sur la polarisation rectiligne et la double réfraction', May 20, 1839, *Mém. Ac. Sci.*, **18** (1840), 1842, pp. 153–216 (*O.C.*, 1, **2**, pp. 111–166).
20. See *C.R. Ac. Sci.*, **8**, April 22, and May 20, and 27, 1839, pp. 597–598, pp. 719–731, and pp. 779–783 (*O.C.*, 1, **4**, pp. 343–369); *C.R. Ac. Sci.*, **9**, November 11, 1839, pp. 589–590 (*O.C.*, 1, **4**, pp. 520–523); and 'Mémoire sur les mouvement infiniment petits de deux systèmes de molécules qui se pénètrent mutuellement', *Ex. An. Phys. Math.*, **1**, September 1839, pp. 33–52 (*O.C.*, 2, **11**, pp. 51–74).
21. See, for example, *C.R. Ac. Sci.*, **9**, August 26, 1839, pp. 283–288 (*O.C.*, 1, **4**, pp. 491–496).
22. *C.R. Ac. Sci.*, **9**, November 25, and December 2, 1839, pp. 676–691 and pp. 726–730 (*O.C.*, 1, **5**, pp. 20–42), and 'Mémoire sur la réflexion et la réfraction d'un mouvement simple transmis d'un système de molécules à un autre, chacun des deux systèmes étant supposé homogène et tellement constitué que la propagation des mouvements infiniment petits s'y effectue en tous sens suivant les mêmes lois', *Ex. An. Phys. Math.*, **1**, December 1839, pp. 133–177 (*O.C.*, 2, **11**, pp. 173–226).

23. An exception, the 'Mémoire sur la théorie de la polarisation chromatique', *C.R. Ac. Sci.*, **18**, May 27, 1844, pp. 961–972 (*O.C.*, 1, **8**, pp. 213–224).
24. 'Mémoire sur l'intégration des équations linéaires', *C.R. Ac. Sci.*, **8**, May 27, and June 3, 10, and 17, 1839, pp. 827–830, pp. 845–865, pp. 889–907, and pp. 931–937 (*O.C.*, 1, **4**, pp. 369–427), and *Ex. An. Phys. Math.*, **1**, September 1839, pp. 53–100 (*O.C.*, 2, **11**, pp. 75–133).
25. 'Mémoire sur la réduction des intégrales générales d'un système d'équations linéaires aux différences partielles', *C.R. Ac. Sci.*, **9**, August 26, and November 18, 1839, p. 288 and pp. 637–649 (*O.C.*, 1, **4**, p. 497, and **5**, pp. 5–19), and *Ex. An. Phys. Math.*, **1**, December 1839 and April 1840, pp. 178–211 (*O.C.*, 2, **11**, pp. 227–264).
26. P. H. Blanchet, 'Mémoires sur la propagation et la polarisation du mouvement dans un milieu homogène indéfini, cristallisé d'une manière quelconque', *J.L.*, **5**, 1840, pp. 1–30.
27. Communications each week from July 5 until September 13, 1841 (*O.C.*, 1, **6**, pp. 202–340); numerous other communications until March 1842 (*O.C.*, 1, **6**, pp. 367–421).
28. *C.R. Ac. Sci.*, **23**, August 9, 1841, pp. 339–340.
29. *C.R. Ac. Sci.*, **23**, November 15, 1841, pp. 958–960.
30. *C.R. Ac. Sci.*, **23**, December 20, 1841, p. 1152.
31. *C.R. Ac. Sci.*, **10**, April 20, 1840, pp. 640–650 (*O.C.*, 1, **5**, pp. 180–191), and *Ex. An. Phys. Math.*, **1**, August 1840, pp. 269–287 (*O.C.*, 2, **11**, pp. 331–353).
32. *C.R. Ac. Sci.*, **12**, December 16, 1844, p. 339 (*O.C.*, 1, **8**, pp. 336).
33. *C.R. Ac. Sci.*, **9**, August 5, and November 11, 1839, pp. 184–190 and 587–588 (*O.C.*, 1, **4**, pp. 483–491 and pp. 518–519); *C.R. Ac. Sci.*, **10**, April 20, 1840, pp. 650–656 (*O.C.*, 1, **5**, pp. 191–198); and *Ex. An. Phys. Math.*, **1**, August 17, 1840, pp. 279–187 (*O.C.*, 1, **11**, pp. 343–353).
34. *C.R. Ac. Sci.*, **15**, July 11 and 25, 1842, pp. 44–59 and pp. 85–102 (*O.C.*, 1, **7**, pp. 17–49 and pp. 62–67).
35. *C.R. Ac. Sci.*, **17**, October 30, 1843, pp. 938–942 (*O.C.*, 1, **8**, pp. 115–119).
36. *J.L.*, **1**, 1836, pp. 197–210, and *C.R. Ac. Sci.*, **11**, September 14 and 21, 1840, pp. 453–475 and pp. 501–511 (*O.C.*, 1, **5**, pp. 288–311 and pp. 311–331).
37. *C.R. Ac. Sci.*, **13**, August 9, 1841, p. 317 (*O.C.*, 1, **6**, p. 282). See also *C.R. Ac. Sci.*, **13**, October 4 and 24, 1841, pp. 682–687 and pp. 850–854 (*O.C.*, 1, **6**, pp. 341–346 and pp. 354–358). A letter from Leverrier to Cauchy about Lexell's comet (*C.R. Ac. Sci.*, **18**, April 29, 1844, pp. 826–827) shows that relations between the two men remained close during the following years.
38. *C.R. Ac. Sci.*, **15**, August 8, August 16, August 22, August 29, and September 5, 1842, pp. 255–269, pp. 303–305, pp. 357–366, pp. 411–418, and pp. 487–483 (*O.C.*, 1, **7**, pp. 86–126).
39. *C.R. Ac. Sci.*, **20**, March 17, March 24, March 31, April 7, and April 21, 1845, pp. 767–796, pp. 825–847, pp. 907–927, pp. 996–999, pp. 1166–1180 and pp. 1612–1626 (*O.C.*, 1, **9**, pp. 121–129). See J. B. Biot, *Mélanges scientifiques et littéraires*, **3**, Paris, 1858, p. 153, and C. A. Valson, *La Vie et les Travaux du Baron Cauchy*, **2**, pp. 167–169. Cauchy was criticized for having published studies on planetary perturbations, because this was the subject of a competition of the Académie. See *C.R. Ac. Sci.*, **20**, April 7, 1845, pp. 996–999 (*O.C.*, 1, **9**, pp. 186–190).
40. *C.R. Ac. Sci.*, **21**, September 15, 22, and 29, October 6, 13, 20, and 27, November 3, 10, 17, and 24, and December 1, 8, 15, 22, and 29, 1845, pp. 593–607, pp. 668–679, pp. 727–742, pp. 779–797, pp. 835–852, pp. 895–902, pp. 931–933, pp. 1025–1041,

pp. 1042–1044, pp. 1093–1101, pp. 1123–1134, pp. 1188–1201, pp. 1238–1255, pp. 1287–1300, pp. 1356–1369, and pp. 1401–1409; and *C.R. Ac. Sci.*, **22**, January 5, 12, 19, and 26, February 2 and 9, and April 11, 1846, pp. 2–31, pp. 53–63, pp. 99–107, pp. 159–160, pp. 193–196, pp. 235–238, and pp. 630–632 (*O.C.*, 1, **9**, pp. 277–505, and **10**, pp. 5–68).

41. Bertrand's postulates states that for any integer $n > 1$, there exists at least one prime between n and $2n$.
42. *Ex. An. Phys. Math.*, **3**, p. 250 (*O.C.*, 2, **13**, p. 280).
43. These papers were finally published by Liouville in the November 1846 edition of the *J.L.*, **11**, pp. 381–444.
44. *C.R. Ac. Sci.*, **21**, December 8, 1845, and January 5, 1846, **21**, pp. 1247–1248, and **22**, pp. 30–31 (*O.C.*, 1, **9**, p. 459, and **10**, p. 34).
45. See A. Dahan, 'Les travaux de Cauchy sur les substitutions. Etude de son approche du concept de groupe', *Arch. Hist. Ex. Sci.*, **23**, 1980, pp. 296–310.
46. A. Lamarle, 'Note sur le théorème de M. Cauchy relatif au développement des fonctions en séries', *J.L.*, **11**, April 1846, pp. 129–141.
47. *J.L.*, **11**, August 11, 1846, pp. 313–330 (*O.C.*, 2, **2**, pp. 35–54). Lamarle responded to Cauchy in 1847 in *J.L.*, **12**, August 1847, pp. 305–342.
48. *J.L.*, **11**, August 1846, p. 323 (*O.C.*, 2, **2**, p. 46).
49. *C.R. Ac Sci.*, **23**, August 8, 1846, pp. 251–255 (*O.C.*, 1, **10**, pp. 70–75). Following Saint-Venant's paper, 'Sur les sommes et les différences géométriques et sur leur usage pour simplifier la mécanique', for which he had been appointed commissioner, Cauchy had defined the senses, direct or backward, of a rotation. See *C.R. Ac. Sci.*, **23**, August 3, 1846, p. 251 (*O.C.*, 1, **8**, pp. 69–71), and 'Mémoire sur les résultantes que l'on peut former, soit avec les cosinus des angles, soit avec les coordonnées de deux ou trois points', *Ex. An. Phys. Math.*, **4**, pp. 5–86 (*O.C.*, 1, **14**, pp. 1–92).
50. *C.R. Ac. Sci.*, **23**, September 21, 1846, pp. 557–563 (*O.C.*, 1, **10**, pp. 135–143).
51. *C.R. Ac. Sci.*, **17**, October 2, 9, 16, 23, and 30, 1843, pp. 572–581, pp. 640–651, pp. 693–702, pp. 921–925, and pp. 1159–1164 (*O.C.*, 1, **8**, pp. 65–115).
52. *C.R. Ac. Sci.*, **19**, December 16, 1844, pp. 1262–1263 (*O.C.*, 1, **8**, pp. 366–375). Another proof is in the *C.R. Ac. Sci.*, **19**, December 23, 1844, pp. 1377–1384 (*O.C.*, 1, **8**, pp. 143–153). It is known that this theorem can easily be deduced from the Cauchy inequalities.
53. *C.R. Ac. Sci.*, **23**, October 12, 1846, pp. 689–702 (*O.C.*, 1, **10**, pp. 152–168).
54. Without explanation, Cauchy ceased publishing on his theory on October 26, 1846. Puiseux, nevertheless, indicates the existence of a study in which Cauchy recovered the known double periodicity of elliptic functions.
55. *C.R. Ac. Sci.*, **23**, September 21, 1846, p. 566 (*O.C.*, 1, **10**, p. 146).
56. 'Mémoire sur les fonctions de variables imaginaires', *C.R. Ac. Sci.*, **23**, August 10, 1846, pp. 271–275 (*O.C.*, 1, **10**, pp. 75–80), and *Ex. An. Phys. Math.*, **3**, October 1846, pp. 361–387 (*O.C.*, 2, **13**, pp. 405–436).
57. 'Mémoire sur la théorie des équivalences algébriques, substituée à la théorie des imaginaires', *C.R. Ac. Sci.*, **24**, June 28, 1847, pp. 1120–1130 (*O.C.*, 1, **10**, pp. 312–324), and *Ex. An. Phys. Math.*, **4**, pp. 87–110 (*O.C.*, 2, **14**, pp. 93–120).
58. The two polynomials $\chi(x)$ and $\rho(x)$ are equivalent moduli of the polynomial $\omega(x)$ [notation: $\chi(x) \equiv \rho(x)$ mod. $\omega(x)$] if upon division by the polynomial $\omega(x)$, the remainders are equal.
59. See *Ex. An. Phys. Math.*, **4**, pp. 97–100 (*O.C.*, 2, **14**, pp. 104–108).

60. *C.R. Ac. Sci.*, **24**, March 1, 1847, p. 310. Lamé had long been interested in Fermat's last theorem and had competed in 1818 in the Grand Prix de Mathématiques of the Académie des Sciences. In 1839, he showed the impossibility of solving, in integers, the equation $x^7 + y^7 = z^7$. Cauchy wrote the report, which was highly laudatory. *C. R. Ac. Sci.*, **9**, September 16, 1839, pp. 359–363 (*O.C.*, 1, **4**, pp. 499–504).
61. *C.R. Ac. Sci.*, **24**, March 15, 1847, pp. 430–434.
62. Sealed messages no. 726 and no. 727.
63. *C.R. Ac. Sci.*, **24**, March 22 and 29, and April 5, 12, and 19, 1847, pp. 469–481, pp. 516–528, pp. 578–584, pp. 633–636, and pp. 661–666 (*O.C.*, 1, **10**, pp. 240–285).
64. *C.R. Ac. Sci.*, **24**, May 17, 1847, pp. 899–900.
65. *C.R. Ac. Sci.*, **24**, May 24 and 31, and June 7, 14, and 28, 1847, pp. 885–887, p. 943, pp. 996–999, pp. 1022–1030, and pp. 1117–1130, and *C.R. Ac. Sci.*, **25**, July 5, 12, 19, and 26, and August 2, 9, and 23, 1847, p. 6, pp. 37–54, pp. 93–99, pp. 129–136, pp. 177–182, pp. 242–243, and pp. 285–288 (*O.C.*, 1, **10**, pp. 292–371).
66. *C.R. Ac. Sci.*, **24**, May 31, 1847, p. 88 (*O.C.*, 1, **10**, p. 295).
67. For example, he showed that a polynomial formed from a 23rd root of unity could not be factored into prime factors. *C.R. Ac. Sci.*, **24**, June 14, 1847, pp. 1022–1030 (*O.C.*, 1, **10**, pp. 299–308).
68. *C.R. Ac. Sci.*, **24**, June 28, 1847, p. 1121 (*O.C.*, 1, **10**, p. 312), and 'Mémoire sur la théorie des équivalences algébriques, substituée à la théorie des imaginaires', *Ex. An. Phys. Math.*, **4**, pp. 87–110 (*O.C.*, 1, **14**, pp. 93–120).
69. J. B. Biot, 'M. le Baron Cauchy', *Mélanges Scientifiques et Littéraires*, **3**, Paris, 1858, p. 152.

Notes to Chapter 13

1. F. N. M. Moigno, preface to A. L. Cauchy, *Sept Leçons de Physique Générale*..., Paris, 1868.
2. A. L. Cauchy, 'La Chandeleur', *Bulletin de l'Institut Catholique*, February 1, 1843.
3. A. Fourcy, *Histoire de l'École Polytechnique*, Paris, 1828, Summary Table, pp. 255–257.
4. See Chapter 5, pp. 61–63.
5. Cited by G. Bertier de Sauvigny, *La Restauration*, Paris, Flammarion, 1955, p. 346.
6. A. L. Cauchy, 'Sur la recherche de la vérité', *Bulletin de l'Institut Catholique*, April 14, 1842, p. 21.
7. A. L. Cauchy, *Sept Leçons de Physique Générale*..., *O.C.*, 2, **15**, p. 413.
8. Ibid., p. 413.
9. A. L. Cauchy, 'Sur les limites des connaissances humaines', *O.C.*, 2, **15**, p. 7.
10. M. Jullien, 'Quelques souvenirs d'un étudiant jésuite à la Sorbonne et au Collège de France, 1852–1856', *Les Études*, **127**, 1911, pp. 329–348, especially, p. 336.
11. A. L. Cauchy, *Quelques Réflexions sur la Liberté de l'Enseignement*, Paris, 1844, especially Chapter 3, 'De l'enseignement scientifique et de l'enseignement religieux', pp. 14–16.
12. A. L. Cauchy, *Considérations sur les Moyens de Prévenir les Crimes et de Réformer les Criminels*, Paris, 1844.
13. A. L. Cauchy, 'Sur la recherche de la vérité', *Bulletin de l'Institut Catholique*, April 14, 1842, p. 21.

14. A. L. Cauchy, *Considérations sur les Moyens de Prévenir les Crimes et de Réformer les Criminels*, Paris, 1844.
15. *Sept Leçons de Physique Générale...*, *O.C.*, 2, **15**, p. 413.
16. Practically in the same terms in *Sur les limites...* of 1811, *O.C.*, 2, **15**, pp. 5–6; *Sept Leçons de Physique Générale...* of 1833, ibid., pp. 412–413; and 'Sur la recherche de la vérité', *Bulletin de l'Institut Catholique*, April 14, 1842, p. 20.
17. *Sept Leçons de Physique Générale...*, *O.C.*, 2, **15**, p. 412.
18. A. L. Cauchy, 'Sur quelques préjugés contre les physiciens et les géomètres', *Bulletin de l'Institut Catholique*, March 3, 1842, p. 43.
19. *Analyse Algébrique*, introduction, pp. V–VI (*O.C.*, 2, **3**, pp. V–VI).
20. 'Sur quelques préjugés...', *Bulletin de l'Institut Catholique*, March 3, 1842, p. 43.
21. *Sept Leçons de Physique Générale...*, *O.C.*, 2, **15**, p. 418.
22. Ibid., p. 415.
23. Ibid., p. 419.
24. 'Sur quelques préjugés...', *Bulletin de l'Institut Catholique*, March 3, 1842, p. 45.
25. A. L. Cauchy, 'Epître d'un mathématicien à un poète, ou la leçon d'astronomie', *Bulletin de l'Institut Catholique*, January 13, 1842, p. 20.
26. 'Sur la recherche de la vérité', *Bulletin de l'Institut Catholique*, April 14, 1842, p. 21.
27. *Analyse Algébrique*, introduction, p. VII (*O.C.*, 2, **3**, p. VII).
28. The manuscript for this unpublished note from 22, January 22, 1837, is kept in the meeting's packet for that day.
29. *C.R. Ac. Sci.*, **21**, July 14, 1845, pp. 134–143 (*O.C.*, 1, **9**, pp. 240–253).
30. Ibid., p. 134 (*O.C.*, 1, **9**, pp. 240).
31. 'Sur les limites...', *O.C.*, 2, **15**, p. 6.
32. *Analyse Algébrique*, introduction, pp. III–V (*O.C.*, 2, 3, pp. III–V).

Notes to Chapter 14

1. *Procès-verbaux du Bureau des Longitudes*, cited by Bigourdan, art. cit., 1930, A 33.
2. *C. R. Ac. Sci.*, **26**, April 3 and 17, and May 1, 1848, pp. 404–408, p. 429, pp. 441–443, pp. 448–449, and pp. 469–472 (*O.C.*, 2, **11**, pp. 30–49).
3. See C. A. Valson, *La Vie et les Travaux du Baron Cauchy*, **1**, p. 105.
4. Eugène Briffault wrote in *Dictionnaire de la Conversation et de la Lecture*, Paris, 2nd ed., 1852–1876, article on Cauchy:

   > During the reign of Louis-Philippe, the Cauchys were not just a family, but a triple—almost a dynasty. This race of bureaucratic climbers was attached to whichever party happened to be on top at the Palais du Luxembourg; there, they formed a colony; they were identified with the place, part and parcel, to such an extent that they themselves constituted a party. It was said that in order to get rid of the Cauchys you would have to tear the place (i.e., the Palais du Luxembourg) down. Thus, for a long time, revolutionary events would come and go without apparently touching the Cauchy's pleasant position.
   >
   > The heredity of a peerage failed to materialize, but aside from this, the Cauchy heredity remained untouched. This succession seemed destined to continue indefinitely into the future, with no end in sight. Vanitas vanitatum! Then! All at once, the fatal hour sounded on February 24, 1848. The Republic was proclaimed, and the first blows of the clock's hammer scattered the Cauchys into flight. Nothing about this revolution was so unbelievable as that they had been forced to clear out of the Palais du Luxembourg.

5. See Arch. Nat. F^{17} 20356.

6. These letters were written by Cauchy after certain accusations had been made against the Jesuits at the meetings of February 23 and 25, 1850. In the first letter, Cauchy took up the arguments that had been used in the opuscules of 1844; the second letter contains a historical account of the suppression of the order during the 18th century.
7. *O.C.*, 2, **15**, pp. 511–513.
8. See Chapter 11, pp. 185–186.
9. See A. L. Cauchy, *Note* read before the Académie on January 6, 1851, Appendix IV, p. 357.
10. See Appendix IV, p. 346.
11. Archives of Collège de France, G II 5, *Registre des délibérations de l'assemblée des professeurs du collège*, December 1, 1850. See E. Neuschwander, 'Joseph Liouville (1809–1882): correspondance inédite et documents biographiques provenant de différentes archives parisiennes', *Bolletino di Storia delle Scienze Matematiche*, **4**, 1984, pp. 55–132, especially p. 125.
12. See Appendix IV, p. 347.
13. Archives of Collège de France, G II 5, *Registre des délibérations de l' assemblée des professeurs du collège*, December 8, 1850. See E. Neuschwander, 'Joseph Liouville...', p. 125.
14. See Appendix IV, p. 354.
15. Letter from Cauchy to the administration of the Collège de France, Appendix IV, p. 355.
16. *C.R. Ac. Sci.*, **32**, January 6, 1851, p. 3.
17. Report by Cauchy, *C.R. Ac. Sci.*, **32**, March 31, 1851, pp. 442–450 (*O.C.*, 1, **11**, pp. 363–373); remarks by Liouville, ibid., pp. 450–452; note by Cauchy, ibid., pp. 452–454 (*O.C.*, 1, **11**, pp. 373–376). The manuscript of Liouville's course on doubly periodic functions is kept in the meeting's packet for that day.
18. See B. Belhoste et J. Lützen, 'Joseph Liouville et le Collège de France', *Rev. Hist. Sci.*, 1984, **37**, pp. 255–304, especially pp. 274–278.
19. Drafts of letters from Liouville to Cauchy, Bibliothèque de l'Institut, Ms 3623, book note 1, dated July 9, 1856. See E. Neuenschwander, 'Joseph Liouville...', p. 112.
20. See Appendix III, p. 336.
21. The course of mathematical astronomy was held during the second semester of the school year (the first semester was devoted to physical astronomy). Faye was appointed on March 13, 1853, as a substitute for Cauchy and Leverrier on December 24, 1853 (*Bulletin administratif du Ministère de l'Instruction Publique*, 1, **4**, 1853, p. 89 and p. 649). Moigno in his preface to *Sept Leçons de Physique Générale*... of 1885 wrongly stated that Cauchy gave up his chair in 1851 and resumed it in 1853.
22. See *Arch. Nat.* F[17] 20356 and Moigno, preface to *Sept Leçons de Physique Générale*..., p. VII.
23. M. Jullien, 'Quelques souvenirs d'un étudiant jésuite...', p. 127, p. 191, and p. 336.
24. C. A. Valson, *La Vie et les Travaux du Baron Cauchy*, **2**, introduction, p. VIII.
25. Reports presented in July 1848 by Laurent; in January 1849 and July 1850 by La Provostaye and Desains. During this time, Cauchy also examined, but did not make evaluation reports on, several important studies by Saint-Venant on the theory of elasticity.

26. *C.R. Ac. Sci.*, **27**, July 31, August 14, 21, and 28, October 9, and 16, November 13, 20, and 27, December 4, 11, and 18, 1848, p. 105, p. 133, p. 162, p. 198, p. 225, p. 356, p. 373, p. 433, p. 499, p. 525, p. 537, p. 572, and p. 596 (*O.C.*, 1, **11**, p. 73 and pp. 76–91).
27. *C.R. Ac. Sci.*, **27**, December 18, 1848, pp. 621–622; *C.R. Ac. Sci.*, **28**, January 2 and 8, 1849, pp. 2–6 and pp. 25–28 (*O.C.*, 1, **11**, pp. 92–104); and *Mém. Ac. Sci.*, **22**, pp. 29–37 (*O.C.*, 1, **2**, pp. 187–194).
28. *C.R. Ac. Sci.*, **28**, January 15, 1849, pp. 57–65 (*O.C.*, 1, **11**, pp. 104–113), and *C.R. Ac. Sci.*, **31**, July 29, August 5, 19, and 26, September 2, 9, and 16, October 7, and 14, November 11, and December 2 and 23 1850, pp. 112–114, pp. 160–166, pp. 225–232, pp. 257–262, pp. 297–306, pp. 331–342, pp. 532–533, pp. 666–667, and p. 766 (*O.C.*, 1, **11**, pp. 245–289). Cauchy based the theory of reflection and refraction of light on two principles: that of corresponding motions and the continuity of motion in the ether.
29. *C.R. Ac. Sci.*, **27**, January 2, 1849, p. 2 (*O.C.*, 1, **11**, p. 95).
30. C. A. Valson, *La Vie et les Travaux du Baron Cauchy*, **1**, p. 252.
31. *C.R. Ac. Sci.*, **25**, September 20, October 4, 18, 25, November 8, 15, and 29, and December 13 and 27, 1847, pp. 401–413, pp. 475–478, pp. 531–538, pp. 572–579, pp. 650–656, pp. 700–705, pp. 775–781, pp. 879–883, and pp. 953–959, and **26**, January 10, 17, 24, and 31, and February 21, 1848, pp. 29–33, pp. 57–61, pp. 133–136, pp. 157–162, and pp. 236–240 (*O.C.*, 1, **10**, pp. 403–499).
32. *C.R. Ac. Sci.*, **29**, July 23 and 30, 1849, pp. 65–67 and pp. 103–106 (*O.C.*, 1, **11**, pp. 141–147).
33. *C.R. Ac. Sci.*, **33**, December 29, 1851, pp. 649–709; *C.R. Ac. Sci.*, **34**, January 5, 19, and 26, and February 2, 1852, pp. 8–9, pp. 70–77, pp. 121–124, and pp. 156–159 (*O.C.*, 1, **11**, pp. 385–403).
34. *C.R. Ac. Sci.*, **38**, May 22 and 29, and June 5, 12, and 26, 1854, pp. 910–913, p. 945, p. 952, pp. 990–993, and p. 1033 (*O.C.*, 1, **12**, pp. 148–167).
35. On August 11, 1856, Cauchy deposited a sealed envelope (no 1591) containing the study 'Note sur l'intégration des équations différentielles qui renferment les mouvements des astres dont se compose le système solaire'. This unpublished paper is in the meeting's packet for that day.
36. *C.R. Ac. Sci.*, **44**, April 27, and May 4, 1857, p. 528, p. 595, and pp. 805–807 (*O.C.*, 1, **12**, pp. 445–446). These were the last papers that Cauchy wrote; a few days before his death, he was working on a paper in astronomy (C. A. Valson, *La Vie et les Travaux du Baron Cauchy*, **1**, p. 259).
37. J. Bertrand, 'La Vie et les Travaux du Baron Cauchy par C. A. Valson', p. 214.
38. 'Astronomie mathématique, cours de M. Cauchy', contained in the handwritten notebook *Mathématiques. Cours Inédits* by G. Lespiault, Bibliothèque de la faculté des sciences de Bordeaux, Ms 52 (old mark). These notes, filling eight leaves recto verso, concern the theory of geometric quantities and the theory of the functions of geometric quantities.
39. This fragment, which Professor Roos of the University of Stockholm has communicated to me, is kept in the papers of C. A. B. Bjerknes in the library of the University of Oslo.
40. A copy of *Report to the ministry for the church and education about a journey abroad to study pure mathematics*, in Norwegian, is kept in the papers of Bjerknes, at the University of Oslo, Professor Roos has translated for me in English the passage of the report concerning Cauchy:

I arrived in Paris on October 15 [1856]. The winter semester had not yet started there....

On November 15, the courses started at the Sorbonne.... Cauchy was supposed to lecture about the movements of the celestial bodies, when they are attracted by the sun. In this course, which in fact was the last one that the eminent mathematician taught, he developed a whole series of mathematical theories. He developed in a particularly instructive and original way the foundations for almost all of those beautiful and important theories, that current science owes to him, e.g., his theory of indices, his calculus of residues, the theory about the criteria for convergence of the Taylor series, and the theory of the so-called geometric quantities. In view of the completeness of his treatment of the preliminaries, he only had a few opportunities to treat the real subject of his lectures, that by the way for me only would have had a minor interest, as being something that was outside my real subject of interest....

Incidentally, the lectures that I attended had a very small audience. While the elementary lectures about, e.g., integral calculus and analytic geometry, were attended by rather many auditors, it turned out that the higher, purely mathematical subjects, even in Paris, only attracted a small audience. Thus, Liouville's and Cauchy's lectures were only attended by 3, 4, or 5 persons, and the lectures of Bertrand had an audience of 6–8.

41. See J. Peiffer, *Les Premiers Exposés Globaux de la Théorie des Fonctions de Cauchy*, 3rd cycle Thesis, typewritten, 1978, pp. 94–123.
42. C. A. Valson, *La Vie et les Travaux du Baron Cauchy*, **1**, p. 253.
43. See J. Peiffer, *Les Premiers Exposés Globaux...*, pp. 82–93.
44. *Ex. An. Phys. Math.*, **4**, 1849, pp. 157–180 (*O.C.*, 2, **14**, pp. 175–202), and *Mém. Ac. Sci.*, **22**, 1850, pp. 131–180 (*O.C.*, 1, **2**, pp. 282–328).
45. Two letters from Saint-Venant to Cauchy from the end of 1845, which are kept in Archives E. P., respond to certain objections that Cauchy had raised and further reveal that he (Cauchy) was not yet convinced of the importance of this new geometrical calculus. In a letter dated December 27, 1845, Saint-Venant defended his calculus in the following terms:

I realize that all that is proved by way of the kind of geometrical calculus that I can develop here, perhaps, is established without recourse to it. But, by use of it, proofs can be made more simple when, once and for all, the terms have been defined, and it has been shown that geometrical equations are to be treated in exactly the same way as ordinary equations. It is rare that a new algorithm solves questions that cannot be solved in some other way; in general, it will only provide a simplification. If it is to be granted that I may compare my work to something that is clearly superior, then I should compare it to M. Poinsot's theory of couples, or to other analytical theories that focus on the same matters.

46. *Ex. An. Phys. Math.*, **4**, 1849, p. 157 (*O.C.*, 2, **14**, p. 175).
47. *C.R. Ac. Sci.*, **29**, November 19, 1849, p. 594, Charles Hermite, *Oeuvres*, **1**, Paris, 1905, p. 74.
48. See C. Houzel, 'Histoire de la théorie des fonctions elliptiques' in *Abrégé d'Histoire des Mathématiques*, **2**, Ch. VII, 1–113, especially p. 22. The study by Hermite was not published. It is not in his *Oeuvres*, 'having disappeared from the archives of the Académie' (Emile Picard). However, today it is kept in the meeting's packet of November 19, 1849, except for the first few pages, which are missing.
49. See G. Darboux, *Notice Historique sur Charles Hermite*, Paris, 1905, particularly p. 29.
50. C. A. Valson *La Vie et les Travaux du Baron Cauchy*, **1**, p. 253.

51. M. Jullien, 'Quelques Souvenirs d'un étudiant jésuite...', p. 340.
52. Memoir published in the *J.L.*, **15**, 1850, pp. 365–480.
53. *C.R. Ac. Sci.*, **32**, January 20, 1851, pp. 68–75 (*O.C.*, 1, **11**, pp. 292–300).
54. *C.R. Ac. Sci.*, **32**, February 10, 1851, pp. 160–162 (*O.C.*, 1, **11**, pp. 301–304). In 1829, Cauchy had already given the definition of the derivative of a function of a complex variable, but he had not given the conditions for the differentiability of such a function.
55. *C.R. Ac. Sci.*, **32**, April 7, 1851, pp. 484–487 (*O.C.*, 1, **11**, pp. 376–380).
56. *C.R. Ac. Sci.*, **32**, May 12, 1851, pp. 704–705 (*O.C.*, 1, **11**, pp. 384–385).
57. *C.R. Ac. Sci.*, **34**, February 23, 1852, pp. 265–273 (*O.C.*, 1, **11**, pp. 406–415).
58. *Ex. An. Phys. Math.*, **4**, November 1853, pp. 308–313 (*O.C.*, 2, **14**, pp. 359–366). Cauchy first defined a real function of a real variable:

    Two real variables are said to be functions of each other when they vary simultaneously in such a way that one of them determines the value of the other.

    For functions of geometrical quantities, he adopted the following definition:

    ... Z is to be deemed a function of z when the value of z determines the value of Z.... Then, the position of the moving point A always determines the position of the moving point B.

59. *Ex. An. Phys. Math.*, **4**, November 1853, pp. 314–335 (*O.C.*, 2, **14**, pp. 367–392).
60. *Ex. An. Phys. Math.*, **4**, November 1853, pp. 336–347 (*O.C.*, 2, **14**, pp. 393–406).
61. See the papers of January 15, February 12, and December 31, 1855; the reports by Cauchy of March 12, 1855, *C.R. Ac. Sci.*, **40**, March 12, 1855, pp. 557–567 (*O.C.*, 1, **12**, pp. 243–246) and of July 7, 1856, *C.R. Ac. Sci.*, **43**, July 7, 1856, pp. 26–29 (*O.C.*, 1, **12**, pp. 330–333); and *C.R. Ac. Sci.*, **43**, pp. 13–20 and pp. 69–75 (*O.C.*, 1, **12**, pp. 323–330 and pp. 333–342).
62. *C.R. Ac. Sci.*, **40**, April 2, 1855, pp. 787–789.
63. From a letter that Méray wrote to the Minister of Public Instruction in 1863. It was brought to our attention by Professor Pierre Dugac:

    Among my scientific works, I can mention to your Excellency, aside from my thesis, ... the publication of the *Cours d'Analyse*, by M. Cauchy, which was undertaken at the request of his family and of M. Saint-Venant; this work is presently being published.

    The book obviously never appeared.
64. *Ex. An. Phys. Math.*, **4**, November 1853, p. 356 (*O.C.*, 1, **14**, p. 417). The first mention of the calculus of algebraic keys appears on January 10, 1853, *C.R. Ac. Sci.*, **36**, January 10, 1853, pp. 70–75 (*O.C.*, 1, **11**, pp. 439 ff). Cauchy also used his calculus of keys in the course he taught at the Collège de France.
65. Not knowing Saint-Venant's address, Grassmann wrote to Cauchy in April 1847, asking him to forward his letter to Saint-Venant, along with one of the two copies of the *Ausdehnungslehre* of 1844. On June 17, 1847, Saint-Venant replied to Grassmann who sent him another letter, along with several papers, in January 1848 (H. G. Grassmann, *Gesammelte mathematische und physikalische Werke*, **3**, Leipzig, 1911, pp. 120–122).
66. See the meeting's packet for that day. Grassmann claimed priority for the theory of algebraic keys that Cauchy presented in 1853, as well as on the geometrical calculus that Saint-Venant presented in 1845 (H. G. Grassmann, *Gesammelte mathematiche und physikalische Werke*, **3**, pp. 174–198).

67. The draft of the note from Saint-Venant to Cauchy is kept in the Fonds Saint-Venant, Archives E. P., and the letter from Saint-Venant to Grassmann is published along with his reply in Grassmann, *Gesammelte Werke*, **3**, pp. 199–200.
68. See Archives Ac. Sci., Cauchy file. The list gives the investigations that the Académie obtained from the Cauchy family.
69. See the critical analysis of this dispute in H. Freudenthal, 'Cauchy, Augustin-Louis', *Dictionary of scientific biography*, **3**, pp. 141–142.
70. See M. Jullien, 'Quelques souvenirs d'un étudiant jésuite...', p. 338.
71. See the account by Guyon, Mayor of Sceaux, at the funeral services for Cauchy:

    Almost every day he comes to visit me; often sometimes several times a day. They are short visits and are free of all foolish chatter. Time was too valuable for someone who was making such a valuable use of it.

    Also see that of M. Jullien 'Quelques souvenirs d'un étudiant jésuite...', pp. 334–335:

    From time to time, he would come to the little chateau (a tiny building from wood at the bottom of the entrance to the house across from the door, the pressed and busy famous Parisien) and find us in our little study. He came like a puff of wind, without knocking, without uttering any compliments, and started to talk about his business... On another day, as I was quietly meditating at five o'clock in the morning, I suddenly heard someone coming hurriedly up the wooden stairs. I thought something untoward was happening. M. Cauchy came in like someone who was about to miss his train: 'Father, read this quickly; it is a hymn that I was asked to give for the fête de Sainte-Geneviève at Saint-Étienne du Mont. It is to be sung at the seven o'clock mass. I forgot to show it to any theologian. Maybe it contains some heresay or some incorrect, improper expression. Read it over, and I will correct it right away'.

    Then, see that of Armand de Melun, cited by Alexis Chevalier in *Le Vicomte Armand de Melun, Fondateur de l'Oeuvre des Jeunes Apprentis et des Jeunes Ouvriers*, Versailles, 1893, p. 25:

    How considerable were the efforts—the canvassing and the discussions—that he went through to assure this 'conquest' [the works of the Écoles d'Orient]. Cauchy often woke me up at daybreak to advise me about ways of getting around this inconvenience or of overcoming that obstacle.

72. See C. A. Valson, *La Vie et les Travaux du Baron Cauchy*, **1**, pp. 223–242.
73. See the letter from Armand de Melun cited by A. Chevalier, *Le Vicomte Armand de Melun...*, p. 25:

    In the Middle Ages, Baron Cauchy would have been the first person to take up the cross to march to the conquest of the holy places. But, in the present age, he has taken up a new crusade: to establish the Écoles d'Orient.

    See also C. A. Valson, *La Vie et les Travaux du Baron Cauchy*, **1**, p. 226.
74. Ibid., **1**, p. 236.
75. H. de Lacombe, *Note sur l'Oeuvre d'Orient à l'Occasion du Cinquantenaire de sa Fondation (1856–1906)*, Paris, 1906.
76. C. A. Valson, *La Vie et les Travaux du Baron Cauchy*, **1**, p. 237, and the letter from Cardinal Lavigerie to E. Beluze, which serves as a preface to E. Beluze, *La Vie de Mgr Dauphin, Prélat de la Maison de leurs Saintetés Pie IX et Léon XIII, 1806–1882*, Paris, 1886.

77. From the speech by Guyon, Mayor of Sceaux, at the funeral services of Cauchy, cited by C. A. Valson, *La Vie et les Travaux du Baron Cauchy*, **1**, pp. 272–274.
78. For more about this controversy, see *C.R. Ac. Sci.*, **43**, December 8, 1856, p. 1065 (remarks by Bertrand), p. 1067 (remarks by Cauchy); December 22, 1856, pp. 1137–1139 (*O.C.*, 1, **12**, pp. 395–398, notes by Cauchy); December 29, 1856, pp. 1165–1166 (remarks by Duhamel), and pp. 1166–1167 (*O.C.*, 1, **12**, pp. 398–401, reply by Cauchy). *C.R. Ac. Sci.*, **44**, January 5, 1854, pp. 3–5 (new remarks by Duhamel), January 19, 1857, pp. 80–81 (*O.C.*, 1, **12**, p. 405, reply by Cauchy), pp. 81–82 (reply by Duhamel), pp. 82–89 (remarks by Poncelet), pp. 89–91 (remarks by Morin), pp. 101–104, 'Sur quelques propositions de mécanique rationelle', by Cauchy (not published in the *O.C.*), p. 104 (reply by Duhamel), and pp. 104–107 (new intervention by Poncelet).
79. C. A. Valson, *La Vie et les Travaux du Baron Cauchy*, **1**, p. 255.
80. Ibid. **1**, p. 259. Cauchy's mother died in 1839, and his father died in 1848. His two sisters were married, the eldest to Felix de l'Escalopier, an official at the Court of Accounts, and the younger to Alfred de Saint-Pol. His younger brother Eugène had been widowed since 1839. Aloïse Cauchy died on June 13, 1863 at Sceaux.
81. Augustin Cauchy, *Discours aux Funérailles de Jacques Binet*, Paris, 1856.
82. Letter written by Alicia to Père Coué, S. J., on May 23, 1857, cited by C. A. Valson, that reports, in vivid terms, the pain and exemplary death of the great scientist. See C. A. Valson, *La Vie et les Travaux du Baron Cauchy*, **1**, pp. 259–267.

Appendix I

Presentation Notes of Various Works Presented by Cauchy to the Académie des Sciences (1816–1830)

Study on the Integration of Linear Partial Differential Equations with Constant Coefficients (Mémoire sur l'intégration des équations linéaires aux différences partielles et à coefficients constants)[1]

This study, in which I discuss the main results of my lectures at the Faculté des Sciences, is an expansion of the material that I presented, under the same title, to the Académie des Sciences on October 8, 1821.[2] There is, however, a difference in the way that the main question is examined, and it includes some important additions. The work is divided into two parts. In the first part, I establish certain formulas which will be used later on, either to solve the partial differential equations or to reduce the integrals obtained. Among these formulas should be noted those that, in the third section, are used to transform the values of the multiple integral

$$\iiint \cdots f(\alpha^2 + \beta^2 + \gamma^2 + \cdots) \cos a\alpha \cos b\beta \cos c\gamma \cdots d\alpha \, d\beta \, d\gamma^* \tag{1}$$

taken between the limits $-\infty$ and $+\infty$ over all the variables, for a simple or double integral, and those that are used to determine one or several real or

[1] Presentation of the study to the Académie; the reading was on September 16, 1822. The manuscript is in the meeting's packet for that day. The study is published in *J.E.P.*, **12**, 19th Cahier, July 1823, pp. 510–570 (*O.C.*, 2, **1**, pp. 275–333).

[2] The paper mentioned here is 'Sur l'intégration générale des équations linéaires à coefficients constants'; published in *Bull. Phil.*, October 1821, pp. 101–112 (*O.C.*, 2, **2**, pp. 253–266) and November 1821, pp. 145–152 (*O.C.*, 2, **2**, pp. 267–275).

*If the number of variables... is odd, by using n to denote this number and writing $a^2 + b^2 + c^2 + \cdots = s$, I find that the integral in Eq. (1) is equivalent to the product:

$$(-1)^{n-1/2} 2^{n-1} \pi^{n-1/2} \frac{\partial^{(n-1)/2}}{\partial s^{(n-1)/2}} \int \cos(s^{\frac{1}{2}}\alpha) f(\alpha^2) d\alpha \qquad (\alpha = -\infty, \alpha = +\infty).$$

[Note by Cauchy.]

imaginary roots of an arbitrary algebraic or transcendental equation—or more generally, the sum of functions similar to several of the systems of values that may have various variables subjected to certain given equations.

In the second part of the study, I apply certain formulas that are found in the first part to the solution of linear equations with constant coefficients whose last term is a variable. I determine the general value of the principal variable expressed by means of the functions that represent the initial values of this variable and its derivatives; and I show that in this determination one can dispense entirely with the solution of the algebraic equations. Finally, I point out the reductions that hold for a special class of partial differential equations that are related to a large number of problems in physics and mechanics.

Research on Definite Integrals That Contain Imaginary Exponentials (Recherches sur les intégrales définies qui renferment des exponentielles imaginaires)[3]

These investigations were undertaken in 1819 following the study on the solution of algebraic equations by means of definite integrals.[4] These questions lead me to some important formulas, which are related to the transformation of the functions and serve to replace an arbitrary function of the variable x by a definite integral in which there enters no more than one rational fraction of this variable whose denominator is of the first degree.

Research on the Motion of Two Superimposed Fluids, One of Which Is Compressible While the Other Is Incompressible (Recherche sur le mouvement de deux fluides superposés, l'un compressible, l'autre incompressible)[5]

After having finished the study on the theory of waves, which won a prize from the Académie in 1815, I undertook some investigations on the motion of two superimposed fluids when one of them is compressible and the other is incompressible; and I was able to establish the formulas that serve to determine the waves produced by the contact between the two fluids. These formulas are found in the notebook that includes the notes appended to my first study. I thought about removing them and to develop a separate; but, time would not allow me to do so, and this subject, not being suitable for extensive, detailed treatment in a lecture, I am bound to offer them to the Académie in

[3] Presentation of the study to the Académie. The reading was on September 16, 1822; the manuscript is in the meeting's packet for that day. The study is lost.

[4] The study is 'Sur la résolution analytique des équations de tous les degrés par le moyen des intégrales définies' of November 22, 1819 (*Mémoires Ac. Sci.*, **4** (1819–1820), 1824, pp. XXVI–XXIX; *O.C.* 1, **2**, pp. 9–11).

[5] Presentation of the study was on January 27, 1823. The manuscript is in the meeting's packet for that day. The study, which is unpublished, is inserted in the *Cahier sur la Théorie de Ondes*, belonging to Mrs. de Pomyers, pp. 115–140.

their original form. I do not know if the analysis that I make use of and that, it seems to me, should merit the consideration of geometers is related in some way to the one by M. Poisson in which he established the laws of reflection and refraction of light in systems of undulations, as he stated in the secret committee at the last meeting.[6]

On the Effects of Molecular Attraction in the Motion of Waves (Sur les effets de l'attraction moléculaire dans le mouvement des ondes)[7]

The more or less considerable viscosity of a liquid gives rise to the attractions that its molecules exercise on each other at very small distances and necessarily exerts an influence on the motion of waves at the surface of the liquid. I propose to determine the nature of this influence. In order to accomplish this, it is necessary to integrate the equations of motions for fluids, after having introduced new terms that depend on the molecular attraction and to then suitably determine the arbitrary functions, based on the initial data of the problem. These are, in effect, the questions that I have resolved by means of an analysis that offers the following noteworthy fact: it provides a way of integrating the equations of motion of a fluid even in the situation when the accelerations along axes multiplied, respectively, by the differentials of the coordinates do not yield exact differentials. This analysis, when applied to the case in which the viscosity is zero, yields the formulas that M. Poisson and I obtained by other methods in earlier studies. Moreover, when one takes the viscosity of the liquid into consideration, the variable that represents the mutual attractions of the two molecules and is regarded as a function of the distance (which decreases very rapidly) vanishes as soon as the integrations have been performed and the arbitrary functions determined—and, in their places, in the definitive formulas, only two constants are left. One of these constants is proportional to the attractions exerted in a point on the surface of the liquid by the neighboring molecules, while the other one is proportional the sum of the actions exerted on this point and on those that are below it along the same vertical. In particular, in order to obtain the equation that shall determine the vertical ordinate of the surface of the liquid, at the end of a given

[6] The study is *Mémoire sur la propagation du mouvement dans les fluides élastiques*, read by Poisson to the Académie on March 24, 1823. Some extracts were published in *Annales de Chimie*, 2, **22**, 1823, and *Mémoires. Ac. Sci.* (1827) **1**, 1831.

[7] Presentation of the study to the Académie. The reading was on November 17, 1823; the manuscript is in the meeting's packet for that day. At the end of note XX of the study 'Sur la théorie de la propagation des ondes', which was added again for publication in 1824, Cauchy declared: In another study, we will derive by the preceding methods the formulas that we presented to the Académie Royale des Sciences on November 17 last, and by means of which we have determined the motion of waves on the surface of a heavy liquid, keeping in mind the adhesion that exists between its molecules.

This study was never published.

time t, when the initial ordinate is known, it suffices to take the equation that furnishes this ordinate in the case in which the viscosity is zero, and then (1) the weight of a quantity equal to the first constant and (2) the initial ordinate and the ordinate at the end of time t by a quantity equal to the second.

A Study on Various Points in Analysis (Mémoire sur divers points d'analyse)[8]

I have the honor of presenting to the Académie a study on various points in analysis. In this study, I begin by deriving certain formulas that I have previously given for definite integrals, not only the Taylor and Lagrange series, but also a considerable number of other series of the same type and on the remainder terms needed to complete these various series. The same principles furnish the means for expressing the sum of similar functions of the roots of an algebraic equation by means of the coefficients of the (given) equation. I then show how it is possible to directly establish (1) the formulas that serve to determine the sum of the functions in question and (2) those formulas that are used to determine their product. In making use of these formulas, it is possible to directly compose, without recourse to the theory of symmetric functions, the final equation that should result from the elimination of several unknowns between the algebraic equations.

A. L. Cauchy

A Study on the Integration of Linear Equations and Their Application to Various Problems in Physics (Mémoire sur l'intégration des équations linéaires et leur application à divers problèmes de physique)[9]

I am honored to present to the Académie two new papers on the integration of linear equations and the determination of the arbitrary functions that they involve. In these studies, I establish several notable properties of M. Fourier's formula, and I show how, by applying to this formula a system of notation proposed by M. Brisson, the results that he obtained can be rigorously proved. I then propose a direct method by means of which one can determine, in a large number of cases, the arbitrary functions by assuming only that the initial functions are known between certain limits, and I apply this method to the solution of several problems in mathematical physics. Among these problems,

---

[8] Presentation of the study to the Académie. The reading was on August 9, 1824. The manuscript is in the meeting's packet of August 9, 1824.

[9] Note of presentation read to the Académie on December 27, 1824. The manuscript is in the Staatsbibliothek Preussischer Kulturbesitz, Dokumentensammlung Darmstaedter F 1c 1836 (1). This note was used almost completely by Fourier in the mathematics part of the *Histoire de l'Académie* for 1824 (*Mémoire Ac. Sci.*, **7**, (1824), 1827, pp. XLV–XLVI).

I mention the propagation of waves in a canal of finite length or in a rectangular basin, regardless of the depth of the liquid. One result following from my formulas is that the waves produced in a rectangular basin are absolutely the same as if the basin were extended indefinitely in all directions; the initial surface of the liquid in the basin is continually reflected off its sides in such a way that two continuous rectangles with sides equal to those of the basin are always recovered by two symmetrical and symmetrically placed surfaces on all sides of the vertical plane formed by the joint side of these two rectangles. If, for the sake of greater convenience, one should assume that the surface of the basin is contained by four vertical plane mirrors, the surfaces we will construct outside the basin will not be different from the images of the original surface of the liquid reflected an infinity of times by the four mirrors.

This conclusion holds even when one or several of the mirrors recede to infinity, that is to say, when the basin is extended indefinitely in one or several directions. If three of the mirrors recede to infinity, there will be no more than one image; and, at the same time, one would obtain the motion of the waves on the surface of a liquid bounded by a vertical plane. If, in this last case, the ordinate of the initial surface of the liquid is always zero except in the neighborhood of a certain point, then this surface, when extended, has an ordinate that is zero except for the neighborhood of a second point, which will coincide precisely with the image of the first, and the surface of the liquid will be as if it were extended indefinitely beyond the plane and if the two points in question would be two centers of the motion of the waves. From this, it can be immediately concluded that the circumferences of the circles formed by different incident and reflected waves are cut along the given plane in such a way that the radius of the incident wave and the radius of the reflected wave always intersect at the same angle with the normal to the plane. This is what takes also place, as is known, in the theory of light where it is regarded as a product of the motions of the waves of an etheral fluid.

December 27, 1823 (sic)
A. L. Cauchy

A New Study on the Integration of Linear Equations and on the Motion of Elastic Plates (Nouveau mémoire sur l'intégration des équations linéaires et sur le mouvement des plaques élastiques)[10]

I am honored to offer to the Académie a new study on the integration of linear equations in which I present the general solution for several equations of this type; particular attention has been given to those that are related to the motion

[10] Note of presentation read before the Académie on January 17, 1825. The manuscript is in the meeting's packet for that day. This paper has never been published.

of an elastic plate that is held at its borders or to the motion of a rectangular elastic plate held fixed along its four sides.* Those equations that I have obtained have been presented in a form such that it will be easy to recognize those cases in which the motion becomes periodic in the nature of the sounds that correspond to motions of this type. The method by which I obtained this integral can be applied to a large number of questions of this type, and for this reason merits some attention.

A. L. Cauchy

A New Study on the Calculus of Residues and Definite Integrals (Nouveau mémoire sur le calcul des résidus et les intégrales définies)[11]

I am honored to present to the Académie certain new applications of the calculus of residues along with some new formulas having to do with the determination and the transformation of certain definite integrals. These formulas complete the method that I proposed in a study in 1814 and in which I presented the evaluation of certain definite integrals as depending on the theory of singular integrals, that is, integrals taken between limits but in which the function under the $\int$ sign becomes either indeterminate or infinite. In the 1814 study, I was restricted to the consideration of the case in which the terms divided by the function under the $\int$ sign gives, as a quotient, an expression that becomes infinitely small in the first order when an infinitely small increase of order one is attributed to the variable. In the other studies that I presented during this period, I considered the case in which the quotient in question is an infinitely small quantity of order greater than unity, but is represented by an integer. In my new investigation, I examined the case in which this order is replaced by a fraction, and I exhibited the formulas that are to replace those that I obtained earlier. The new formulas, of course, include as special cases certain results that are already known, as well as a multitude of other formulas that would appear to be noteworthy. Aside from this, in order to recapture the previously known formulas, it suffices to substitute integers diminished by very small quantities (which are soon to be reduced to zero) for the fractions I referred to earlier.

[Paris, February 7, 1825
A. L. Cauchy][12]

---

*These two motions have already been considered in the last study by M. Brisson. But his solutions are incomplete, even with regard to elastic plates, considering that the integrals that he gave or indicated also contain arbitrary functions [note by Cauchy].

[11] Note of presentation read before the Académie on February 14, 1825. The manuscript is in the meeting's packet of that day. The study has never been published.

[12] The words in brackets are cancelled by Cauchy.

The last paragraph of the new study is devoted to the investigation of the number and the nature of the values that can be assumed by a definite integral taken between imaginary limits. It is known that M. Laplace has drawn some rather curious results from the consideration of such integrals. M. Ostrogradsky, a young Russian, who is very talented and wise, as well as very learned in infinitesimal analysis, had deduced from there one of the general formulas that we published in *Journal de l'École Polytechnique.* Finally, M. Brisson has declared to us that in a work that he is presently engaged in he has used these integrals for the summation of periodic series and for those whose different terms include exponentials obeying a known law.

Nevertheless, I do not know that if at the present anyone has determined in a sufficiently precise way the sense that ought to be attached to integrals taken between imaginary limits and the different values that they may assume. Such, then, is the question that I now propose to resolve in the fourth section of my new study. The solution is derived quite easily from the calculus of variations combined with the theory of singular integrals. Aside from this, these questions led me to certain new formulas, which include those that I gave in my earlier investigations and can be quite fruitfully applied, either to the determination or the transformation of definite integrals taken between real limits.

Paris, February 15, 1825
A. L. Cauchy

A Study on the Equilibrium and the Motion of Fluids (Mémoire sur l'équilibre et le mouvement des fluides)[13]

On finishing the note that I read at the last meeting of the Académie,[14] I observed that the theory of fluids can be deduced from the principles established in *Exercices de Mathématiques.* I am now going to give a few expansions on this subject.

When one considers a system of molecules that attract or repel each other and if one supposes that the interior pressures reduce to zero in a natural state, the equations of equilibrium or of motion of this system include fifteen coefficients that depend on the way in which the molecules are distributed about each point. The same equations include six coefficients more, if in a natural state the interior pressures are assumed to be nonzero. These various coefficients will be represented by certain triple sums with finite differences, and if the total action exerted on a given molecule by those around it can be regarded as essentially independent of the more neighboring molecules, then the sums in question can be transformed, with no trouble, into simple integrals of infinitesimally small differences.

---

[13] Study presented to the Académie on May 4, 1829, and read on May 25, 1829. The manuscript is in the meeting's packet of May 4, 1829.

[14] Addition to the study 'Sur la dilatation et la condensation linéaire des corps solides ou fluides, et sur l'équilibre ou le mouvement des fluides', read on April 27, 1829.

Appendix II

Documents on Cauchy's Analysis Course at the École Polytechnique (1816–1819)

First-Year Program of the Analysis Course by Augustin Cauchy (November 1816)[1]

Algebraic Analysis

General topics. Review of the theory of exponential, logarithmic, and circular quantities and the most useful of the trigonometric formulas.

Imaginary expressions. Proof of the formula

$$(\cos\varphi \pm \sqrt{-1}\sin\varphi)^n = \cos n\varphi \pm \sqrt{-1}\sin n\varphi).$$

Some notions on functions in general; the distinction between continuous and discontinuous functions; the distinction between entire functions and fractional functions, rational or irrational, simple or composite, explicit or implicit.

Decomposition of the first member of an equation into real factors of the second degree. Solutions of equations of degree three and four by algebraic methods or by means of tables of sines; solutions of binomial equations of the form $x^m - 1 = 0$.

Theorem of Côtes

Decomposition of rational fractions.

Degree of the final equation that results from an elimination performed on several algebraic equations.

[To determine the functions that verify the equations $f(x+y) = fx\,fy$, $f(x+y) = fx + fy$, etc.] (cancelled by Cauchy.)

[1] Archives Ac. Sci., Fônds Ampère, box 4, chapter 4, file 75, The manuscript, which was written in Cauchy's hand, is undated, but examination of the records of the deliberations of the Conseils and of an abridged version of this project, which is kept in the Archives suggest a date of November 1816 with great certitude.

Expression of integral powers of sines and cosines by series of sines and cosines of multiple arcs.
Rules on the convergence of series.
Expansion of certain functions in a series by means of the method of undetermined coefficients.
Theory of recurrent series.
Proof based on a series for the formula

$$e^{\phi\sqrt{-1}} = \cos\phi + \sqrt{-1}\sin\phi.$$

Interpolation formula obtained by the method of undetermined coefficients.

Calculus of (Finite) Differences

General topics. Finite differences and integrals of the first order. Finite differences and integrals of different orders.
Analogies of powers and differences. The most simple notions on solving certain finite difference equations.
On the values that correspond to certain integral indices and on the number of arbitrary constants that appear in these same values.
Interpolation formula.

Differential and Integral Calculus

Basic principles.
Differentials of the first order of functions and integrals.
Differentials of the first order of functions of functions and of composite functions. Formulas for the integration of these same functions and particularly of rational fractionals; differentials affected by a second-order radical; differentials of binominals and of radicals that contain exponential, logarithmic, or circular functions of a single variable. Changes of the independent variable.
Differentials of the first and higher orders for functions of several variables. The conditions for integrability and the integration of the same differentials when they satisfy these conditions.
Taylor's theorem. Theory of minima and maxima for functions of a single variable.
Values of fractions that appear in an indeterminate form.
Taylor's theorem extended to functions of several variables. Maxima and minima of these classes of functions. Theorem on homogeneous functions.
Differentials of the first and higher orders for implicit functions.
Elimination of constants between an equation and its differentials of different orders. The number of arbitrary constants that appear in the integral [solution] of a differential equation.
Determining the factor that renders a first-order differential equation integrable.
Particular solution of first-order differential equations.
Complete integration of the equation $y - px = f(p)$, where $p = \dfrac{dy}{dx}$.

Integration of simultaneous differential equations, in particular, of those equations that are linear with constant coefficients.
The most simple analytical notions on the integral calculus of partial differences. Integration of the first-order equation or of linear equations with constant coefficients of arbitrary order.
Integration by series of differential formulas, certain differential equations of different orders, and partial differential equations.

Applications of Integral Calculus to Geometry

Analytical formulas for the determination of tangents, normals, asymptotes, etc.
Various expressions for the radius of curvature. Properties of values. Applications to conic sections, cycloids, etc.
Geometrical considerations on the rectification of curves, the quadrature of curves and of curved surfaces; the curvature of solids.
Analytical formulas for the solution of the preceding questions, including formulas related to solids of revolution and those that have as their boundaries arbitrary surfaces by application of interpolation formulas to these same questions.

Official First-Year Program of the Analysis Course (1816)

In order to provide a deeper appreciation of the implications of Cauchy's project for the first-year course, we will now present the official first-year program, which was finally adopted in 1816.

Algebraic Analysis

Solution of algebraic equations of the 3rd and 4th degree. Demonstration of how any equation of the 3rd or 4th degree can be solved by use of tables of sines.
Series expansions of certain functions by means of the method of undetermined coefficients.
General law of recurrent sequences as observed in the development of rational fractions.
Special examination of recursive sequences depending on two terms, their decomposition into two geometric progressions, and their general terms.
Review of the most useful trigonometric formulas and of the exponential equation that arises between a number and its logarithm. To derive the general properties of logarithms from this equation. To compare the different systems and how to 'pass' from one system to another. To establish the formulas

$$(\cos\varphi + \sqrt{-1}\sin\varphi)^m = \cos m\varphi + \sqrt{-1}\sin m\varphi$$

and

$$e^{\varphi\sqrt{-1}} = \cos\varphi + \sqrt{-1}\sin\varphi$$

Use of the first formula in obtaining the roots of the equations $x^m - 1 = 0$, $x^m + 1 = 0$, which leads to the theorem of Côtes.
Use of the same formulas to express the powers of sines and cosines by series of sines and cosines of multiple arc.
Some notions on functions in general and on their classification as entire, rational, etc.

Differential Calculus

To exhibit the principles underlying differential calculus by consideration of infinitesimals. To show, in the most simple cases, the agreement between this method and methods based on limits and series expansions.
To determine the differentials of x^m, xy, and $\frac{x}{y}$ after which it is easy to determine the derivative of any algebraic function, whether of one or several variables, implicit or explicit.
Differentials of circular, logarithmic, and exponential functions, both simple and compound.
Second and third derivatives.
Proof of Taylor's theorem.
Proof of the binominal formula for the case of negative or fractional exponents.
Theory of equal roots using differential calculus.
Application of Taylor's theorem to series expansions, which determine logarithms, exponentials, sines, and cosines, as arc functions, and conversely.
The same theorem extended to two variables.
Notions on partial derivatives.
Theory of the maxima and minima of functions of one or two variables.
Ways of distinguishing maxima and minima.
Applications to some selected examples.
Formulas for subtangents, subnormals, tangents, etc. The determination of asymptotes.
Expression for the radius of curvature.
General properties of the involute and the method of finding its equation.
Application to conic sections, the cycloid, etc., to determine the involute of the parabola and its rectification.
To change a function of a differential equation of the second order, in which a first derivative has been assumed to be constant, into one in which no differential is assumed constant.
To show succinctly that, if in a fraction $\frac{P}{Q}$, the two terms vanish when $x = \alpha$, then it is equal to $\frac{dP}{dQ}$. This is almost always sufficient to determine the value [of the fraction].

Integral Calculus

General notions on integration.
Integration of monomial differentials and entire functions.
Methods of making rational differentials that are of the form

$$\sqrt{a + bx + cx^2}.$$

Integrability of binominals.
To exhibit the basic formulas to which other formulas are related and to show how these integrals may be expressed.
Formulas on the quadrature of curves, rectification, quadrature of surfaces, and the curvature of solids of revolution. Emphasis on applications and the determination of constants.

Cauchy's analysis course, first-year, 1816–1817.[2]

Date	Lecture number	Subject of the lessons
Thursday, January 16		Point de leçon. L'installation de l'Ecole n'ayant eu lieu que le 17
Saturday, January 18	1	Considérations générales sur les nombres et les quantités, les variables et les fonctions.
Monday, January 20	2	Suite des considérations générales sur les fonctions, somme et différence. La somme de plusieurs variables a pour limite la somme de leurs limites.
Tuesday, January 21		Point de leçon à cause de l'anniversaire de la mort de Louis XVI (cette leçon a eu lieu Lundi de $1h$ à $2h\,\frac{1}{2}$).
Thursday, January 23	3	Suite des considérations générales. Produits et quotiens. Lorsque $B, B' \ldots$ sont des quantités de même signe, la fraction $\frac{A + A' \ldots}{B + B' \ldots}$ est comprise entre la plus petite et la plus grande des fractions $\frac{A}{A'}, \frac{B}{B'} \ldots$ Le produit de plusieurs variables a pour limite le produit des limites de ces mêmes variables.
Saturday, January 25	4	Revue de la théorie des quantités exponentielles.
Tuesday, January 28	5	Suite des matières traitées dans la leçon précédente. $a, a', a'' \ldots b, b', b'' \ldots$ désignant des nombres quelconques, $\sqrt[b+b'+b''+\cdots]{a + a' + a'' + \cdots}$ est comprise entre la plus petite et la plus grande des racines

(*Continued*)

[2]Archives E. P., X II C7, Registre d'instruction 1816–1817.

(*Continued*)

Date	Lecture number	Subject of the lessons
		$\sqrt[b]{a}, \sqrt[b']{a'}, \sqrt[b'']{a''} \ldots$ Lorsque les variables x et y ont respectivement pour limites X et Y, x^y a pour limite X^Y. Distinction entre les puissances arithmétiques des nombres et les puissances algébriques des quantités.
Thursday, January 30	6	Application des principes établis dans les leçons précédentes. Limite du développement du binôme.
Saturday, February 1	7	Développment du produit des différences $a-b, a-c, a-d \ldots, b-c, b-d \ldots, c-d$, etc. Comment on peut reconnaître les signes des différents termes. Transformation des exposants en indices.
Tuesday, February 4	8	Application des formules trouvées dans la leçon précédente aux problèmes de l'interpolation.
Thursday, February 6	9	Revue de la théorie des logarithmes.
Saturday, February 8	10	Revue des fonctions circulaires.
Tuesday, February 11	11	Continuation de même sujet. Dans un triangle le rapport d'un côté au sinus de l'angle opposé est le diamètre du cercle circonscrit.
Thursday, February 13	12	Relations diverses entre les lignes trigonométriques des trois angles d'un triangle. Lorsque les côtés sont représentés par les sinus, les distances des sommets au point d'intersection des trois perpendiculaires sont représentées par les cosinus.
Saturday, February 15	13	Exprimer dans un triangle les angles, la surface et le rayon du cercle circonscrit au moyen des trois côtés. Résolution des triangles rectilignes en général.
Tuesday, February 18	14	On a établi la formule $(\cos a + \sqrt{-1}\sin a)(\cos b + \sqrt{-1}\sin b) = \cos(a+b) + \sqrt{-1}\sin(a+b)$. Application de cette formule à la détermination de sin *na* et cos *na* exprimés en fonction de sin *a* et cos *a*.
Thursday, February 20	15	Exprimer les puissances du sinus et du cosinus au moyen des sinus et cosinus des arcs multiples.
Saturday, February 22	16	Des diverses valeurs de l'expression $(f+g\sqrt{-1})^{\frac{m}{n}}$ (*m* étant premier à *n*) Si l'on fait $f = r\cos Q, g = r\sin Q, r$ étant positif, on aura $= (f+g\sqrt{-1})^{\frac{m}{n}} =$

(*Continued*)

Date	Lecture number	Subject of the lessons
		$f\left(\cos\frac{mQ}{n}+\sqrt{-1}\sin\frac{mQ}{n}\right)(1)^{\frac{1}{n}}$ $(1)^{\frac{1}{n}}=\cos\frac{2K\pi}{n}\pm\sqrt{-1}\sin\frac{2K\pi}{n}$ K étant $= ou < \frac{n}{2}$.
Tuesday, February 25	17	Quotient de deux expressions imaginaires. Puissances négatives de ces mêmes expressions.
Thursday, february 27	18	Puissances irrationnelles des expressions imaginaires. On les déduit des puissances irrationnelles de l'unité. Ces dernières sont en nombre indéfini. Si b étant irrationnel, on suppose $(1)^b=\cos x+\sqrt{-1}\sin x$, x pourra recevoir à peu près dans cette équation toutes les valeurs possibles.
Saturday, March 1	19	Classement des diverses fonctions d'une seule variable. Distinction des fonctions continues et discontinues. Des valeurs des fonctions d'une seule variable dans quelques cas particuliers.
Tuesday, March 4	20	Résolution génerale de l'équation $f(u)-b=0$, $f(x)$ étant une fonction continue de x, lorsque l'on connaît deux valeurs de x qui, substituées dans le premier membre, donnent des résultats de signes contraires. Classement des fonctions de plusieurs variables. La limite d'une fonction continue de plusieurs variables est la même fonction de leurs limites. Conséquence de ce théorème relativement à la continuité des fonctions composées qui ne dépendent que d'une seule variable.
Thursday, March 6	21	Toute équation qui n'a pas de racines réelles a nécessairement des racines imaginaires de la forme $r(\cos Q+\sqrt{-1}\sin Q)$. Démonstration des différents lemmes qui conduisent à cette proposition.
Saturday, March 8	22	Continuation du même sujet Décomposition des polynômes x^n+1, $x^{2n}-2a^nx^n\cos Q+a^{2n}$ en facteurs réels.

(*Continued*)

(*Continued*)

Date	Lecture number	Subject of the lessons
Tuesday, March 11	23	Théorème de Moivre et de Côtes. Résolution générale de l'équation du 3e degré.
Thursday, March 13	24	Résolution générale de l'équation du 4e degré. Discussion des racines.
Saturday March 15	25	Dans une équation quelconque, le produit des quarrés des différences entre les racines est positif ou négatif suivant que les couples des racines imaginaires sont en nombre pair ou en nombre impair. Déterminer les fonctions qui satisfont à des conditions données.
Tuesday, March 18	26	Suite de la leçon précédente.
Thursday, March 20	27	Règles sur la convergence des séries.
Saturday, March 22	28	Suite de la leçon précédente.
Tuesday, March 25	29	Développemens en séries des exponentielles, des sinus et des cosinus. Les séries obtenues sont convergentes pour toutes les valeurs possibles de la variable. Des séries imaginaires convergentes.
Thursday, March 27	30	On a établi au moyen des séries la formule $e^{r\sqrt{-1}} = \cos r + \sqrt{-1}\sin r$. Définition des expressions $\cos(x\sqrt{-1})$, $\sin(x\sqrt{-1})$, $(f + g\sqrt{-1})^{x+y\sqrt{-1}}$, $\log(u + v\sqrt{-1})$.
Saturday, March 29	31	Développemens en séries des fonctions $(1+x)^\mu$, $(1 + x\cos\varphi + x\sin\varphi)^\mu$, $\log(1+x)$, $\log(1 + 2x\cos\varphi + x^2)$, $\mathrm{arc}(\mathrm{tang}\, x)$, $\mathrm{arc}\left(\mathrm{tang}\,\frac{x\sin\varphi}{1 + x\cos\varphi}\right)$ Décomposition des fractions rationnelles.
Tuesday, April 1	32	Théorie des séries récurrentes. Définition du coëfficient différentiel.
Thursday, April 3	33	Principes fondamentaux du calcul différentiel et intégral.
Saturday, April 5		Point de leçon (Samedi Saint).
Tuesday, April 8	34	Des intégrales définies, considérées comme des sommes d'éléments.
Thursday, April 10	35	Différentielles des fonctions simples, des fonctions de fonctions et des fonctions composées.
Saturday, April 12	36	Des intégrales définies. Diverses méthodes d'intégration.

(*Continued*)

Date	Lecture number	Subject of the lessons
Tuesday, April 15	37	Différentielles de divers ordres. Changement de variable indépendante.
Thursday, April 17	38	Différentiation et intégration sous le signe ∫ Intégrales définies. Des intégrales prises entre des limites infinies, et de celles où la fonction sous le signe ∫ devient infinie entre les limites données. Décider si une intégrale prise entre des limites données est finie ou infinie.
Saturday, April 19	39	Suite des matières précédentes.
Tuesday, April 22	40	Idem.
Thursday, April 24	41	Intégration par séries. Théorème de Taylor. Maxima et minima pour les fonctions d'une seule variable. Valeurs des fractions qui se présentent sous une forme indéterminée.
Saturday, April 26	42	Différentielles des fonctions de plusieurs variables. Conditions d'intégrabilité et intégration de ces mêmes différentielles.
Tuesday, April 29	43	Théorème de Taylor étendu aux fonctions de plusieurs variables.
Thursday, May 1	44	Maxima et minima des fonctions de plusieurs variables. Théorème des fonctions homogènes.
Saturday, May 3	45	Revue générale du calcul infinitésimal
Tuesday, May 6	46	Idem.
Thursday, May 8	47	Idem.
Saturday, May 10	48	Différentielles des fonctions implicites. Théorie des tangentes.
Tuesday, May 13	49	Théorie des contacts de divers ordres. Application au cercle.
Thursday, May 15		Point de leçon. Fête de l'Ascension.
Saturday, May 17	50	Sur le rayon de courbure générale. Application aux courbes du second degré. Théorie des développées. Application à l'ellipse.
Tuesday, May 20	51	Développée de la parabole. Discussion des paraboles des divers ordres. Points de rebroussement de ler et de 2e espèces. Cycloïde.
Thursday, May 22	52	Développée de la cycloïde. Logarithmique. Asymptote. Points multiples et points isolés. Coordonnées polaires des spirales en général et en particullier de la spirale logarithmique.

(*Continued*)

(*Continued*)

Date	Lecture number	Subject of the lessons
Saturday, May 24	53	Contact des divers ordres entre les surfaces courbes. Equation du plan tangent.
Tuesday, May 27	54	Sur les rayons de plus ou moins grande courbure d'une surface courbe. Considérations géométriques sur la quadrature des surfaces courbes et la cubature des solides.
Thursday, May 29	55	Formules analytiques pour les quadratures et les cubatures.
Saturday, May 31	56	Rectification des courbes et quadratures des surfaces courbes.

Cauchy's analysis course, second-year 1817–1818.[3]

Date	Lecture number	Subject of the lessons
Tuesday, December 2	1	Considérations générales sur les intégrales définies.
Thursday, December 4	2	Valeur de l'intégrale $\int \frac{P}{Q} dx \left\{ \begin{matrix} -\infty \\ +\infty \end{matrix} \right\}$, $\frac{P}{Q}$ étant une fraction rationnelle. Application au cas ou l'on suppose $\frac{P}{Q} = \frac{x^{2m}}{1 + x^{2m}}$.
Saturday, December 6	3	Valeur de l'intégrale $\int z^{a-1} \frac{dz}{1+z} \left\{ \begin{matrix} 0 \\ \infty \end{matrix} \right\}$, des intégrales $\int x^{a-1} e^{-x} dz \left\{ \begin{matrix} 0 \\ \infty \end{matrix} \right\} = \Gamma(a)$ et $\int z^{a-1} \frac{dz}{(+z)^b}$. Valeurs de $\Gamma(n)$ et de $\Gamma(n+\frac{1}{2})$, n étant un nombre entier quelconque.

[3] Archives E. P., XII C7, Registre d'instruction, 1817–1818.

(*Continued*)

Date	Lecture number	Subject of the lessons
Tuesday, December 9	4	Passage du réel à l'imaginaire. Valeurs des intégrales $\int x^{n-1} e^{-ax} \cos bx\, dx$, $\int x^{n-1} e^{-ax} \sin bx\, dx$, entre les limites $x=0$, $x=\infty$.
Thursday, December 11	5	Suite de la leçon précédente. Valeur de l'intégrale $\int e^{-x^2} \cos 2bx\, dx \begin{Bmatrix} -\infty \\ +\infty \end{Bmatrix}$.
Saturday, December 13	6	Usage de l'équation $dp/dy = dq/dx$ pour déterminer les relations qui existent entre certaines intégrales définies. Pour trouver des fonctions p et q qui satisfassent à l'équation précédente, il suffit de poser $f(x+y\sqrt{-1}) = p + q\sqrt{-1}$. Sommation des séries $1 + \frac{1}{4} + \frac{1}{9} + \frac{1}{16}$, etc. Etablir les formules $\int \frac{\sin bx}{x} dx \begin{Bmatrix} 0 \\ \infty \end{Bmatrix} = \frac{\pi}{2}$, $e^{\frac{-x^2-b^2}{4x^2}} dx \begin{Bmatrix} 0 \\ \infty \end{Bmatrix} = \frac{1}{2}\pi^{\frac{1}{2}} e^{-b}$.
Tuesday, December 16	7	Usage des intégrales définies pour la sommation des séries convergentes.
Thursday, December 18	8	Sur les fonctions réciproques de la première et de seconde espèces.
Saturday, December 20	9	Sommation des séries convergentes par le moyen des fonctions réciproques. Limite du produit $\{1+(x_1-x_0)F(x_0)\}$ $\{1+(x_2-x_1)F(x_1)\}\ldots$ $\{1+(x_n-x_{n-1})F(x_{n-1})\}$, etc., tandis que les élémens de la différence $X - x_0$ décroissent indéfiniment.
Tuesday, December 23	10	Considérations générales sur les équations différentielles. Intégration des équations différentielles du premier ordre $dy = F(x)\,dx$, $dy = yF(x)\,dx$.
Thursday, December 25		Point de leçon. Fête de Noël.
Saturday, December 27	11	L'intégrale d'une équation différentielle du premier ordre renferme une constante. Méthode pour obtenir par approximation une intégrale particulière correspondante à une valeur déterminée de la constante.

(*Continued*)

(*Continued*)

Date	Lecture number	Subject of the lessons
Tuesday, December 30	12	Suite de la leçon précédente.
Thurday, January 1		Point de leçon.
Saturday, January 3	13	Sur la distinction des intégrales générales, des intégrales particulières et des solutions particulières.
Tuesday, January 6	14	Du facteur propre à rendre une équation intégrable. On peut trouver autant de facteurs que l'on veut propres à remplir cette condition. Relation entre ces facteurs.
Thursday, January 8	15	Séparation des variables. Intégration de l'équation linéaire du premier ordre et de l'équation homogène.
Saturday, January 10	16	Suite de l'intégration des équations homogènes. Solution particulière de ces équations. Application à l'équation du premier degré.
Tuesday, January 13	17	Intégrale générale et solution particulière de l'équation $y = f(y') + xf(y')$ dans laquelle $y' = \frac{dy}{dx}$. Intégration de l'équation $$\frac{dx}{\sqrt{(1+\alpha x^2+\beta x^4)}} + \frac{dy}{\sqrt{(1+\alpha y^2+\beta y^4)}} = 0.$$
Thursday, January 15	18	Toutes les solutions particulières de l'équation différentielle $dy - pdx = 0$, vérifient la formule $\frac{1}{(dp/dy)} = 0$. Mais la réciproque n'est pas vraie.
Saturday, January 17	19	Méthode générale pour distinguer les solutions particulières des intégrales particulières.
Tuesday, January 20	20	Considérations générales sur l'intégration des équations différentielles simultanées du premier ordre.
Thursday, January 22	21	Suite de la leçon précédente.
Saturday, January 24	22	Les intégrales générales de plusieurs équations différentielles simultanées du premier ordre renferment autant de constantes arbitraires qu'il y a d'équations. Méthode pour obtenir par approximation les intégrales particulières correspondantes

(*Continued*)

Date	Lecture number	Subject of the lessons
		à des valeurs déterminées de ces mêmes constantes.
Tuesday, January 27	23	Sur la distinction des intégrales générales des intégrales particulières et des solutions particulières dans les équations différentielles simultanées du premier ordre. Méthode pour la détermination de toutes les solutions particulières de ces mêmes équations.
Thursday, January 29	24	Détermination des intégrales générales et des solutions particulières de quelques équations différentielles simultanées du premier ordre. Intégration des équations linéaires simultanées.
Saturday, January 31	25	Suite de la leçon précédente. Cas des racines égales et des racines imaginaires.
Tuesday, February 3	26	Considérations générales sur l'intégration des équations différentielles d'un ordre quelconque entre deux variables.
Thursday, February 5	27	Intégration de l'équation linéaire à coëfficients constants avec un dernier terme variable.
Saturday, February 7	28	Théorèmes relatifs à l'intégration des équations linéaires.
Tuesday, February 10	29	Suite des théorèmes relatifs à l'intégration des équations linéaires.
Thursday, Februrary 12	30	Intégrale générale de l'équation linéaire $\frac{d^2z}{dx^2} = ax^m z$. On en déduit l'intégrale générale de l'équation de Riccati par une intégrale définie.
Saturday, February 14	31	Ordre de l'équation finale obtenue par l'élimination entre plusieurs équations simultanées. Intégration par séries des équations différentielles.
Tuesday, February 17	32	Integration des équations aux différences partielles du premier ordre.
Thursday, February 19	33	Suite.
Saturday, February 21	34	Application des théorèmes exposés dans la séance précédente. Observation sur l'intégration des équations aux différences partielles d'un ordre quelconque.

(*Continued*)

(*Continued*)

Date	Lecture number	Subject of the lessons
Tuesday, February 24	35	Suite.
Thursday, February 26	36	Intégration des équations linéaires aux différences partielles à coëfficiens constans sans dernier terme variable.
Saturday, February 28	37	Calcul direct aux différences finies.
Tuesday, March 3	38	Calcul inverse aux différences finies.
Thursday, March 5	39	Application à divers exemples et à la sommation des séries.

Cauchy's analysis course, first-year, 1818–1819.[4]

Date	Lecture number	Subject of the lessons
Tuesday, November 3	1	Revue des diverses espèces de quantités. Sur les moyennes entre plusieurs quantités.
Thursday, November 5	2	Suite des théorèmes sur les moyennes. Considérations générales sur les fonctions, etc.
Saturday, November 6	3	Des fonctions continues et discontinues.
Tuesday, November 10	4	Valeurs singulières des fonctions dans quelques cas particuliers. Représentation géométrique des fonctions continues par des courbes. Equation de la logarithmique.
Thursday, November 12	5	Discussion des courbes. Déterminer leur inclinaison en un point quelconque.
Saturday, November 14	6	Suite de la leçon précédente. Des courbes représentées par les équations $y = x^{+a}$, $y = \sin x$ et $y = \cos x$.
Tuesday, November 17	7	Suite de la discussion des courbes qui ont pour équations $y = x^a$, $y = \sin x, y = 1x$, $y = \sin\frac{1}{x}$.
Thursday, November 19	8	Lorque l'on considère une fonction de plusieurs variables x, y, z, continue par rapport à chacune d'elles, si ces variables convergent vers les valeurs $X, Y, Z, \ldots, f(x, y, z, \ldots)$ convergera vers la limite $f(X, Y, Z, \ldots)$. De

[4]Archives E. P., XII C7, Registre d'instruction, 1818–1819.

(*Continued*)

Date	Lecture number	Subject of the lessons
		la continuité des fonctions composées d'une seule variable. Fonctions symétriques. Usage de ces fonctions pour la résolution des équations linéaires.
Saturday, November 21	9	Des fonctions alternées et de leur usage pour la résolution des équations linéaires.
Tuesday, November 24	10	Interpolation.
Thursday, November 26	11	Application des principes établis dans la leçon précédente.
Saturday, November 28	12	Déterminer les fonctions $Q(x)$ qui satisfont à l'une des équations $Q(x+y) = Q(x) + Q(y)$, $Q(x+y) = Q(x)Q(y)$, etc.
Tuesday, December 1	13	Déterminer la fonction $Q(x)$ qui satisfait à l'équation $Q(x+y) + Q(x-y) = 2Q(x)Q(y)$.
Thursday, December 3	14	Sur les séries convergentes. Valeur de $e = 1 + \frac{1}{1} + \frac{1}{1.2} + \cdots$
Saturday, December 5	15	Règle sur la convergence des séries.
Tuesday, December 8	16	Développement du binôme dans le cas d'un exposant quelconque.
Thursday, December 10	17	Développement des exponentielles et des logarithmes.
Saturday, December 12		Point de leçon (M. Cauchy indisposé).
Tuesday, December 15	18	Notions sur les maxima et minima des fonctions entières.
Thursday, December 17	19	Suite.
Saturday, December 19	20	Fin de la théorie des maxima et minima des fonctions entières
Tuesday, December 22	21	Sur les expressions imaginaires. Exprimer les sinus et les cosinus d'arcs multiples en fonction des puissances des sinus et cosinus d'arcs simples, et réciproquement.
Thursday, December 24	22	Tout polynôme est décomposable en facteurs réels du second degré.
Saturday, December 26	23	Suite de la leçon précédente.
Tuesday, December 29	24	Résolution des équations binômes.
Thursday, December 31	25	Résolution des équations trinômes.
Saturday, January 2	26	Résolution des équations du troisième degré.
Tuesday, January 5	27	Résolution des équations du quatrième degré. Décomposition des fractions rationnelles en fractions simples dans le cas des racines réelles.

(*Continued*)

(*Continued*)

Date	Lecture number	Subject of the lessons
Thursday, January 7	28	Décomposition des fractions rationnelles, dans le cas des racines imaginaires et des racines égales.
Saturday, January 9	29	Séries imaginaires.
Tuesday, January 12	30	Des fonctions imaginaires, exponentielles, logarithmiques et circulaires.
Thursday, January 14	31	Élevation des expressions imaginaires à des puissances quelconques.
Saturday, January 16	32	Développemens en séries entières des fonctions imaginaires $\{1+x(\cos\theta+\sqrt{-1}\sin\theta)\}^{\mu}$, $\log\{1-x(\cos\theta+\sqrt{-1}\sin\theta)\}$ et des fonctions réelles $1+2x\cos\theta+x^2$, $\arctang\dfrac{x\sin\theta}{1+2x\cos\theta+x^2}$.
Tuesday, January 19	33	Suites récurrentes. Principes du calcul infinitésimal.
Thursday, January 21		Point de leçon (anniversaire de la mort de Louis XVI).
Saturday, January 23		Point de leçon (M. Cauchy étant indisposé).
Tuesday, January 26	34	Différentielles des fonctions simples et des fonctions de fonctions.
Thursday, January 28	35	Suite de la leçon précédente.
Saturday, January 30	36	Différentielles des fonctions de plusieurs variables indépendantes et des fonctions composées d'une seule variable.
Tuesday, February 2	37	Différentielles des séries convergentes et des fonctions imaginaires. Notions sur les intégrales définies.
Thursday, February 4	38	Propriétés des intégrales définies. Intégration de l'équation $dy=f(x)\,dx$.
Saturday, February 6	39	Valeurs d'intégrales définies et indéfinies.
Tuesday, February 9	40	Examen des diverses méthodes d'intégration.
Thursday, February 11	41	Emploi des expressions imaginaires dans l'intégration indéfinie.
Saturday, February 13	42	Revue des formules différentielles auxquelles les méthodes d'intégration exposées dans les séances précédentes sont applicables.
Tuesday, February 16	43	Idem.
Thursday, February 18	44	Formules de réduction pour les différentielles binômes.
Saturday, February 20	45	Suite.

(*Continued*)

Date	Lecture number	Subject of the lessons
Tuesday, February 23	46	Différentielles de divers ordres des fonctions d'une seule variable indépendante, et intégrales successives.
Thursday, February 25	47	Changement de la variable indépendante. Différentielles des fonctions implicites.
Saturday, February 27	48	Différentielles totales de divers ordres pour les fonctions de plusieurs variables indépendantes et les fonctions composées d'une seule variable. Théorème des fonctions homogènes.
Tuesday, March 2	49	Différentiation et intégration sous le signe $\int$. Conditions d'intégrabilité pour la différentiation des formules différentielles qui renferment plusieurs variables indépendantes. Intégration de ces mêmes formules lorsqu'elles satisfont à ces conditions.
Thursday, March 4	50	Théorème de Taylor.
Saturday, March 6	51	Applications du théorème de Taylor aux maxima et minima des fonctions d'une ou de plusieurs variables.
Tuesday, March 9	52	Valeur des fractions qui se présentent sous la forme 0/0. Intégration par séries.
Thursday, March 11	53	Considérations générales sur les applications du calcul différentiel à la géométrie. Tangentes aux courbes, sous-tangentes, normales et sous-normales.
Saturday, March 13	54	Contact des divers ordres. Les courbes se traversent quand le contact est d'ordre pair. Points d'inflexion. Cercles osculateurs et rayons de courbure.
Tuesday, March 16	55	Maxima et minima par la géométrie. De la courbe qui est produite par les intersections successives d'un système de droites. Le centre de courbure est le point de rencontre de deux normales infiniment rapprochées.
Thursday, March 18	56	Différentielle de l'arc d'une courbe. Equations des développées. Description de la développante par le moyen d'un fil appliqué sur la développée.
Saturday, March 20	57	Propriétés de la cycloïde.
Tuesday, March 23	58	Asymptotes, points isolés, etc.

(*Continued*)

(*Continued*)

Date	Lecture number	Subject of the lessons
Thursday, March 25	59	Coordonnées polaires. Détermination des tangentes, des rayons de courbure, développées, lorsque les équations des courbes sont exprimées à l'aide des coordonnées polaires. Des spirales, et en particulier des spirales hyperbolique, logarithmique, etc.
Saturday, March 27	60	Quadratures et cubatures.
Tuesday, March 30	61	Exemples de quadrature. Application de l'interpolation aux quadratures.
Thursday, April 1	62	Sur les limites des intégrales simples et doubles qui représentent les aires et les volumes. Du plan tangent à une surface courbe.
Saturday, April 3	63	Rectification des courbes, et quadratures des surfaces courbes.

Appendix III

Selection of Letters from Cauchy to Various Persons (1821–1857)

Letter from Cauchy to Libri, February 3, 1821, Biblioteca Moreniana of Florence, Fonds Palagi, File 431, Insert 76:

Paris, February 3, 1821

Sir,

Several days ago, I received and read with great interest the study *Sur la Théorie des Nombres* that you were kind enough to send me.[1] I particularly noted the ingenius method by which you solve the equation

$$z^n = ax^n + bx^{n-1} + \cdots + px + q$$

over the integers. However, at the same time, it would seem to me that one should be able to substitute for this method a general principle that supplies the solution for a rather broad class of indeterminate equations. The confidence that you have been kind enough to place in me is evidence that I may here expand on this principle.

Suppose, then, that two variables x and z are related to a certain equation

$$f(x,z) = 0. \tag{1}$$

One attempts to find a function u of these variables that, as regards the given relation, can never assume any numerical value less than a certain limit U, as long as the variables are positive. If the function u is rational and if, at the same time, its coefficients are integers, then this function should, for positive values of the variable, be equivalent (up to the sign) to one of the integers between 0 and U. By successively letting it assume the values of these numbers, one obtains new equations that determine one or several

[1] The Académie received Libri's paper on January 22, 1821, and Cauchy presented a verbal report on the work at the following meeting (*O.C.*, 2, **15**, *Annexes documentairs*, p. 519).

systems of values for the variables x or z. Among these systems, those that constitute the positive values yield all the possible solutions, over the integers, of the given equation.

Let us consider, for example, that from the given equation one obtains for the values z or z^2, or $z^3, \ldots$, a function of x with rational coefficients and a series ordered in terms of descending and negative powers of x. Then, one can take for u a multiple of the difference between z or z^2, or $z^3, \ldots$, etc. and the integral function that was just mentioned.

For the sake of greater clarity, I will present here the application of the general method given above to the solution of some particular equations.

1st Problem: To find the positive values of x or of z that satisfy the equation

$$z^2 = x^4 + x^3 + x^2 + x + 1. \tag{2}$$

Solution: From Eq. (2), one obtains

$$z = x^2\left(1 + \frac{1}{x} + \frac{1}{x^2} + \frac{1}{x^3} + \frac{1}{x^4}\right)^{\frac{1}{2}}$$

and

$$z = x^2 + \frac{x}{2} + \frac{3}{8} + \frac{a}{x} + \frac{b}{x^2} \cdots, \ a, b, \ldots \text{are integers.}$$

For u, we can take a multiple of the expression

$$\frac{3}{8} + \frac{a}{x} + \frac{b}{x^2}, \text{etc.} = z - x^2 - \frac{x}{2}$$

and express u by

$$u = 2z - 2x^2 - x.$$

One sees that u ought to remain less that the limit 2. Thus, $u = 0$ or $u = +1$. For $u = 0$, we have $x = 0, z = 0$, and for $u = 1$, we have $x = 3, z = 11$. Such, then, are the positive solutions of equation (2).

2nd Problem: To find the positive solutions of the equation

$$z^4 + 8z = x^4 + 21x. \tag{3}$$

Solution: Let us assume $u = z - x$. The values of u are determined by the equation

$$(u + x)^4 + 3 = x^4 + 21x,$$

and its maximum for the formula $\frac{du}{dx} = 0$, which reduces to

$$4(x+u)^3 + 3 = 4x^3 + 21,$$

or, what is the same thing,

$$(x+u)^3 - x^3 = \frac{9}{2}.$$

Then, always,

$$u < (9/2)^{\frac{1}{3}} < 2.$$

This means that either $u = 0$ or $u = 1$; $u = 0$ gives $x = 0$, $z = 0$, and $u = 1$ gives $x = 1$, $z = 2$.

3rd Problem: To find the positive solutions of

$$(z^2 - x)(z + x) = 5 + 4x. \tag{4}$$

Solution: In this case, we can set $u = z^2 - x$. Then, $u < 4$. Since $5 + 4x$ is odd, u cannot be even. Thus, $u = 1$ or $u = 3$. Moreover, the value $u = 1$ does not yield any solutions. But, $u = 3$, $z = 1$, $z = 2$. The investigation of negative solutions is the same as the one of positive solutions, only the signs of the variables are to be changed. Consider also that in a large number of cases it is possible to solve indeterminate equations in more than two variables by principles similar to the ones that we have just discussed. I close by asking that you be assured of my personal regards.

Augustin Cauchy

Letter from Cauchy to Libri, May 18, 1821, Biblioteca Moreniana of Florence; Fonds Palagi, File 431, Insert 76

Paris, May 18 1821

I have received the letter that you were so kind as to write to me and with which you sent me a new work on prime numbers. It will be a true pleasure to make myself familiar with this study and to transmit it, as you desire, to the Académie Royale des Sciences. Please be sure, I pray you, of my highest regards.

Augustin Cauchy

Letter from Cauchy to Libri, March 28, 1828, Biblioteca Moreniana of Florence, Fonds Palagi, File 431, Insert 76

Paris, March 28, 1828

My dear Count,

It has always been a special pleasure to answer your kind letters. But, my poor health or the multitude of things that I must do always come up and do not permit me to respond. Today, I have a few moments of leisure, and I hasten to take advantage of it in order to tell you that I have fulfilled the commission with which you charged me. I was delighted to nominate you as correspondent of the Académie des Sciences.[2] I also thank you for the works that you have been so kind as to send me at various times, and I pray that in exchange you will accept the first fourteen fascicles of my *Exercices de Mathématiques* along with a study on the applications of the calculus of residues to questions in physics. I will be very flattered if this letter should be of some interest to you. M. De Bure, my father-in-law, will be responsible for sending it, as well as the various booklets of *Exercices*, to you.

I pray you be assured, my dear Count, of my highest regards,

A. L. Cauchy

Letter from Cauchy to Libri, July 21, 1829, Biblioteca Moreniana of Florence, Fonds Palagi, File 431, Insert 76

Paris, July 21, 1829

Sir,

I have received the interesting studies that you were so kind as to send me. I will be delighted to read them, and I beg you to accept all my thanks in this matter. I regret that the person who was responsible for sending you *Exercices* for me has only sent you the first seventeen fascicles; and, in order to make up for this, I have just addressed to M. Freddoni the installments up to no. 39, as well as several studies that were printed separately. I hope that they will be worthy of your interest.

The formulas that are contained in the 39th fascicule of *Exercices* and that had to do with the torsions of rectangular elastic rods have quite recently been

---

[2] Libri was presented for a position as correspondent in September 1826 and December 1827. However, he was not elected until December 31, 1832.

confirmed by the experiments of one of our most able physicists.[3] The numbers supplied by the theory and observation agree to an extraordinarily high degree. I eagerly seize this occasion to pray that you will be assured of my highest regards.

A. L. Cauchy

Letter from Cauchy to the president of the Académie des Sciences, July 14, 1831, Arch. Ac. Sci., Cauchy File

Geneva, June 29, 1833

Mr. President,

I should like to be able to present to my colleagues the study that I now have the honor of addressing to you. However, the precarious condition of my poor health has again obliged me to extend my leave of absence for a while; and I must ask you to be so kind as to offer the Académie des Sciences the work to which I refer. This paper is an extract from a more extended study that has as its aim the investigation of various analytical methods; this work has been published in volume 60 of the *Bibliothèque Italienne*. By means of methods that are analogous to those, I developed lectures on differential calculus and its main analytical applications and also on the calculus of variations. Later, I will be honored to offer the Académie a second study that will present a new application of the calculus of residues. Please be assured anew, etc.

Your very humble and obedient servant,

A. L. Cauchy

Letter from Cauchy to the president of the Académie des Sciences, March 25, 1833; Bibl. Inst. M663 H*

Mr. President,

Permit me to ask you to be kind enough to offer the Académie, on my behalf, the second part of my lithographed memoir. I hope that geometers will not consider as unimportant, in this second part, the applications of this new calculus, which I call the *calculus of limits*, to celestial mechanics, because it serves to determine the limits of the errors in the series expansions of implicit or explicit functions of one or several variables.

In the addition to this work, and in subsequent works that I will have the

---

[3] The reference is to Savart.

honor of addressing to the Académie, I will present some new developments on this topic and will show how, by means of this new calculus, it is possible to determine the errors in the series expansions of functions that represent the integrals of a system of differential equations, such as, for example, those that give rise to the theory of planetary motion. Be assured, I beg you, of my highest regards, etc.

A. L. Cauchy

Letter from Cauchy to Comte de l'Escarène, July 18, 1833, Musée National de l'Éducation, Rouen, Ms. A 10822-1

My dear Count,

In agreement with your Excellency's views, I have told M. de Collegno of the reasons that obliged me to hasten my departure. Many obstacles arose: on Saturday, I conclude my last examination, and I am booked for the Geneva post for Monday. But, an incident came up that obliged me to let the very excellent M. d'Olry into our secret. I must cross Switzerland and Bavaria, and I am carrying some mathematics books with me. It is very important that I do not find myself forced to stop before crossing into Bavarian territory. This compelled me to ask your Excellency to tell M. d'Olry of the position in which I now find myself and to send him a few lines that I include in this letter. Also, please accept assurances of the respectful and inalterable devotion of

Your very humble and very obedient servant,

A. L. Cauchy

Turin, July 18, 1838

I beg your Excellency be kind enough to send to M. d' Olry my passport, which is included herein. I should like to have your response as soon as possible.

Copy of the official letter from Baron de Damas:

Toeplitz, June 22, 1833

Sir,

You could be very helpful in educating the Duc de Bordeaux, and the King has asked me to write to you. His Majesty would like, if it is possible, that you immediately begin to work with my pupil; you will be responsible for his instruction in the sciences, a task that has heretofore been the responsibility of M. Barande.

M. Barande was removed for reasons that have greatly affected us; but, it did not compromise either his honor or his considerateness. I owe you this explanation because I do not want to give grounds for any unjust suspicions that might come up at this time.

Please be assured of my steadfast attachment and special consideration.

Baron de Damas

A second letter, filled with expressions of goodwill for me, concludes as follows:

I do not know what your personal position is, but I do know your sentiments. The heir of our King needs your services, and it is my duty to request them of you.

Sir, please be assured of the deep attachment that I swore to you in times better than these and that will not change.

Baron de Damas

P. S. The King desires that you should travel first to Prague and from there to the countryside where his Majesty resides. As soon as possible, please write to me telling the route you will be taking and the prospective date of your arrival.

The King will leave Toeplitz during the first days of July to return to Buschtiehrad near Prague. His Majesty has good reason to be pleased with his stay here: the waters do him good.

A. L. Cauchy
Professor of Sublime Physics at the University of Turin

Letter from Cauchy to Count de l'Escarène, September 24, 1834, Musée National de l'Éducation, Rouen, Ms. A 10822–2

Buschtiehrad near Prague, September 24, 1833

Moved by the recognition of your kindness, as well as that of the King of Sardinia, I impatiently await your letter at the very moment when the question that was raised here several weeks ago shall have been completely solved. Unfortunately, we are still operating on a provisional basis, so that it is impossible to say definitively whether I will remain here or whether I will return to Turin to teach the courses that his Majesty has entrusted to me. Be all that as it may, the newspapers spread many errors and falsehoods about us. The young heir of our King bears himself in a magnificent way, and his sentiments, which are well worthy of a son of Saint-Louis, are those that you will find expressed in a brochure which was published in Prague and of which I am enclosing three copies. I beg you to be kind enough to present the first copy to His Majesty the King of Sardinia as a testimonial of my gratitude, my devotion, and my respect; the second is to be presented to Count de la Tour; the third copy is for you, as is a new assurance of the respectful devotion that I feel toward Your Excellency.

Your very humble and obedient servant,

A. L. Cauchy

P. S. Please give my respects to Madame Countess de l'Escarène and also remember me to persons who have been kind enough to take an interest in me, particularly M. le Chevalier d'Olry and R. P. Grassi. If I have not returned to Turin by November 1, it would seem to me that R. P. Grassi could make very fine suggestions to M. de Collegno as to the selection of a teacher for the courses or for part of the courses that I was to teach. Both Father Moigno and Father Lachaise would be excellent candidates. Father Moigno is perfectly well acquainted with all areas of mathematical physics. M. d'Olry will receive from Switzerland either one or several copies of the first edition of my brochure.

Letter from Cauchy to Moigno written on June 12, 1837, Notebook in the Sorbonne Library, Ms 1759 bis

June 12, 1837

My dear Abbé,

I only received your letter of May 17 this morning, and I am taking advantage of the few remaining minutes to reply to it. Thus, I will not have time today to examine the objections that MM. Sturm and Liouville have raised against the theorem that concludes my 1833 study,[4] as to whether or not it will be necessary to make any changes or modifications. However, I want to look into this closely when I have the time. Far be it from me to think that I am infallible, but what really grieves me is that MM. Sturm and Liouville should imagine that I would try to claim the glory for proofs that they published earlier. It should be remembered that in my study on the theory of waves, I found and presented a theorem that M. Fourier had given earlier in a work that had not been published and the existence of which I was completely unaware; but when I later found out about it, I hurried to insert in the *Bulletin de la Société Philomatique* an article in which I acknowledged M. Fourier's prior

[4] *Calcul des Indices des Fonctions*, lithograph was published in Turin and reprinted in *J.E.P.* **15**, 25th cahier 1837, pp. 176–226 (*O.C.*, 2, **1**, pp. 416–466).

discovery.[5] Similarly, when M. Sturm published his theorem on the number of real roots, I was the first to congratulate him; and I did not write a word mentioning that I had published, on this same subject, a work that had been approved by the Académie and published in the *Journal de l'École Polytechnique*.[6] Meanwhile, M. Poisson gave a report that verified, as to an equation of arbitrary degree, I was the first to have developed methods by which it is possible to find rational functions with coefficients whose signs show the number of real roots between given limits. Finally, in the 1833 study, I cited M. Sturm's theorem. I am as sorry as your are, and I am very astonished, that you have not been able to find in Paris a copy of the paper and of the article that was published in the *Gazette du Piedmont*.[7] That by no means proves that they cannot be found there; and, in particular, I remember that, at the moment when I distributed to the Académie the article that was published in the *Gazette du Piedmont*, one of my colleagues told me he regretted being unable to read the article in Italian. In the meanwhile, I requested that the study be deposited with the *Bibliothèque de l'Institut*. Perhaps someone borrowed it and did not return it. That is all I can think of. At least, however, it can be found in the hands of scholars here in Italy, especially in Turin, Milan, and Modena. The only copy that I have in my hands is now in a two-volume work containing various studies that were lithographed in Turin. While I am waiting to see if you can find the paper, I will have a copy made for you and send it. You will see the elementary proofs that you are interested in, especially for the case that the equation

$$\mathscr{I}_{x_0}^{X}((u)) + \mathscr{I}_{x_0}^{X}\left(\left(\frac{1}{u}\right)\right) = \frac{1}{2}\left(\frac{\mathscr{I}U}{((x))} - \frac{\mathscr{I}u_0}{((x))}\right),$$

where u is a rational fraction in x and u_0, U its values corresponding to $x = x_0$ and $x = X$. It is from this equation that I deduced M. Sturm's theorem in the study of 1833.

I am waiting with the greatest impatience for the news of the presentation to the Académie and of the printing of the three letters, each 15 or 16 pages, that I sent to you. These letters were dated May 6, May 13, and May 18. They contain results that are as important as the proof of the theorem given at the end of the 1833 paper.[8] I do not understand that the objectives raised by M. Sturm should have suggested to you the thought of delaying the presentation of the three letters. In fact, it is just one more reason for letting them appear right away, unless you found some obvious errors, which you could then have

---

[5] See Chapter 7, p. 113.

[6] *J.E.P.*, **10**, 17th cahier, 1815, pp. 457–558 (*O.C.*, 2, **1**, pp. 170–257). See Chapter 3, p. 39.

[7] *Gazetta piemontese*, **113**, September 22, 1832, p. 620. The text is reproduced by A. Terracini in 'Cauchy a Torino', Rendiconti del Seminario Matematico dell Universita e del Politecnico di Torino, **16**, 1957, pp. 178–179, but it does not appear in the *Oeuvres Complètes* of Cauchy.

[8] See *O.C.* 1, **4**, pp. 48–81.

eliminated. However, I do not think that they can be shortened. The theorems stated are, for the most part, given with such clarity and precision that it will be very easy for you to verify them; and, for those that might possibly constitute a problem, I myself have taken care to make a corresponding note. The only thing I fear is that this delay will give other persons the time to publish, before I do, the proof of my theorem—and you can very well imagine how disconcerting that would be, particularly since M. Libri and other geometers have pressed me to send my proofs, which I did.

What I am requesting of you, then, is that you publish my letters as quickly as possible, showing the date that they were postmarked. Please communicate the following note to the Académie.

Note: The fundamental theorem on the integration of differential equations stated in a letter of January 29 can be immediately deduced from the 1st theorem stated in the letter of May 2 and the principles discussed in the lectures for the second-year course at the *École Polytechnique* (see the copy in the *Bibliothèque de l'Institut* of the first eleven lessions).[9] I will explain this in more detail in another letter.

Please let me hear from you and be assured of my respect and unchanging attachment.

Yours,

A. L. Cauchy

Letter from Cauchy to Moigno written December 15, 1837, *Mélanges* notebook of Madame de Pomyers, p. 107

Dear Abbé

I am honored to send to you a new note on the theory of light; it will be followed by several others.[10] In waiting until I had had these various notes published, I request that they, as well as the preceding ones, be inserted in the *Comptes Rendus*—or, if that should not be possible, that they be lithographed or hand copied, as was done with some of my lectures at the Collège de France. In this way, the expense will be small, and a copy can immediately be deposited at the Bibliothèque de l'Institut after it has been properly noted in the minutes of the Académie's meeting.

---

[9] On these lectures, see Chapter 5, p. 80.

[10] The reference here is to the 1837 study *Recherches sur la théorie de la lumière*, the first part of which is entitled '*Polarisation rectiligne*'. The Study is in the *Mélanges* notebook. It has not been published.

Letter to Catalan, July 25, 1839, Bibliothèque Générale de l'Université de Liège, Correspondence of Catalan, Ms 1307C

My dear Friend and Colleague

You promised to come to Sceaux one day to dine with us, and I am asking today if you would like it to be on next Tuesday, following the Académie meeting. We dine at six o'clock, and I am looking forward to this opportunity. As we should like M. Dirichlet to be with you, and since I do not know his address, I am now mentioning it to you and am asking that you be kind enough to forward the enclosed invitation to him.

Your good friend,

Baron Augustin Cauchy

Sceaux, July 25, 1839
Rue de Voltaire, No. 49

Letter to Dirichlet, July 25, 1839, Staatsbibliothek preussischer Kulturbesitz, Berlin, Nachlass Dirichlet

Dear Friend and Colleague,

I am now asking that you please fulfill the promise that you made to me the other day by coming to dine with us at Sceaux on next Tuesday at six o'clock. M. Liouville should also be there. I hope that this does not present you any inconvenience and that you will be assured that I will receive your acknowledgment with the greatest pleasure.

Yours truly,

Baron Augustin Cauchy

Rue de Voltaire, No. 49

Letter from Cauchy to Moigno, October 26, 1842, Bibliothèque du Centre des Fontaines, Chantilly, Moigno file

My dear Abbé

I have again thought about what we said yesterday and have become more and more convinced that in free space there are no knots or bulges—at least so that a reflected ray shall not interfere with an incident ray. For instance, if one places himself behind a wall and if the sound reaches the ear of the observer through an opening made in the wall, so that there is no reflection from neighboring walls, there will be no knot, properly speaking, at which the intensity of the second becomes zero. There will only be, as I said at the last session, diffraction of the sound waves, and the points corresponding to the maxima and minima are obviously located on parabolas whose parameters form an arithmetical progression. I should like you to tell me tomorrow if this absence of knots in free space is confirmed by experiments and, in particular, if this agrees with the experiments of Monsieur M.[11]

I would also like you to bring me last Wednesday's copy of *Le National* tomorrow, as well as the medal claimed by Alicia. Finally, I cannot tell you too often how much I believe in what was not replied to in last week's articles. The objections that could be made fell flat in view of the actual facts. We count on you tomorrow. If it should be impossible for you to come, you can send me, but without the wrapper, the article from the *National.*

Yours,

Baron Augustin Cauchy

Sceaux, October 16, 1842

Letter to Saint-Venant, February 6, 1843, fonds Saint-Venant, Archives E.P. not classified

Sir,

In a note that you sent to the Académie this morning, you mentioned the last of the notes that I inserted earlier in the *Comptes Rendus* of the meeting before the last session.[12] I said that the definition of pressure that you adopted led precisely to the formulas that I first gave, if I am not mistaken, in Volume III of *Exercices de mathématiques.* I added that my note was only an extract from a study which should soon appear in the *Exercices d'analyse* ⋯ (?)[13] If in the *Compte Rendus* of the session before last I did not give the proof even of the proposition recalled below, it was simply that I was afraid that the article would become too long.

On balance, this proof was precisely based—and I make this remark today—

---

[11] The reference here is undoubtedly to Savart (see *C.R. Ac. Sci.*, **15**, pp. 761–762 and *C.R. Ac. Sci.*, **15**, pp. 815–816).

[12] 'Note sur les pressions supportées, dans un corps solide ou fluide, par deux portions de surface trés voisines, l'une extérieure, l'autre intérieur à ce même corps', *C.R. Ac. Sci.*, **16**, January 23, 1843, p. 151 (*O.C.*, 1, **7**, p. 252).

[13] 'Mémoire sur les dilatations, les condensations et les rotations produites par un changement de forme dans un système de points matériels', *Ex. An. Phys. Math.*, **2**, pp. 302–330 (*O.C.*, 2, **12**, pp. 343–377).

on the consideration of the groups of molecules that are situated, pairwise, at the extremities of parallel and equal lines.[14]

The proof that you gave in your notes uses the same argument. I observed, as you do, that the groups of molecules that are situated pairwise, as I just said, are always found on a right or oblique cylinder.

I should like to know if this proof is already found in the paper you presented to the Académie at today's session. In this latter case, my note would be pointless, and I would withdraw it. In any case, write a few lines to me here in the country where I am now residing, that is, in Sceaux, on the Rue de Voltaire, No. 41, Department de la Seine.

Yours truly,

Baron Augustin Cauchy

Letter to Saint-Venant, July 31, 1845, fonds Saint-Venant, Archives E.P. not classified

Sceaux, July 31, 1845

Sir,

Yesterday, I presented a study to the Académie that had been initialled by the Permanent Secretary and included a considerable number of formulas and theorems on geometry.[15] I discussed them with several of my colleagues, who thought them to be new discoveries; I intended to present the main results to the Académie itself next Monday. However, toward the end of the last session, I had an opportunity to speak with M. Lamé, and I found out from him that M. Binet is presently in possession of one of your papers that probably contains certain results that are similar to mine.[16] Accordingly, yesterday, I unsuccessfully sought to get in touch with M. Binet who, as it turned out, was not at the Académie. He had left for an outing in the country and, I am told, will be out of town for several days. Nevertheless, I should still very much like to know if there is any connection between your formulas and some of those that I have come up with; and, thus, in order to satisfy myself on this point, I am forced to ask for your assistance. There are, in particular, two points that I hope can be clarified.

---

[14]'Mémoire sur les pressions ou tensions intérieures, mesurées dans un ou plusieurs systèmes de points matériels que sollicitent des forces d'attraction ou de répulsion mutuelles', *C.R. Ac. Sci*, **16**, February 6, 1843, p. 299 (*O.C.*, 1, 7, pp. 252–260).

[15]'Mémoire sur de nouveaux théorèmes de géométrie et, en particulier, sur le module de rotation d'un système de lignes droites menées par les divers points d'une directrice données', *C.R. Ac. Sci.*, **21**, July 30, 1845, p. 273, and August 4, 1845, p. 305 (*O.C.*, 1, **9**, p. 253).

[16]Probably 'Mémoire sur les lignes courbes non planes', *J.E.P.*, **18**, 30th cahier, 1845, pp. 1–76.

In my study, I established the main properties of two systems of curves traced out simultaneously on an arbitrary surface. These curves include, as a special case, those that M. Lamé found as orthogonal lines, as well as a considerable number of others. They lead to curved coordinates formed by the intersection of three systems of curved surfaces. From this, in order to arrive at the various results I obtained, I used new properties of curves with double curvatures and the special case of the *modulus of rotation* of a system of lines drawn through various points of a given directrix. This modulus of rotation is simply the quotient obtained by dividing an infinitely small arc of the directrix into the infinitely small angle formed by two straight lines drawn from the endpoints of the arc. The inverse of the modulus of rotation is a certain radius that reduces to the radius of curvature when the system of lines is identified with the system of tangents and, when the lines are perpendicular to the osculating plane, with the radius of the flexion. Moreover, the sum of the squares of the moduli corresponding to these two cases is precisely the square of the modulus of rotation relative to the case in which the system of lines is reduced to the system of the radius of curvature, etc.

Please be kind enough to see if there is something in common between this theorem and those that you have obtained and reply to my letter as soon as possible. I will be in the country—that is to say, at Sceaux—today (Thursday) and tomorrow (Friday) as well as Sunday. On Saturday and on Monday, I will be in Paris, so that if it is your intention to come and see me in the country, then it would be better not to choose either Monday or next Saturday. Please forgive my scrawl, but I do not have time to check and correct this letter, since the mail will be picked up very shortly. I close by offering you, once more, assurance of my best wishes.

Baron Augustin Cauchy

Letter to Leverrier, November 23, 1846?, Bibliothèque de l'Institut, Ms 3710

Rue Serpente
Paris, November 23

Dear Colleague,

I have been working with determination, and I got up at six o'clock this morning to complete your calculations. It could not have gone better. But, my poor brain could not endure this sort of work. When I got ready to leave for the Institute, I became quite ill and was forced to go to bed. Nevertheless, I have quite a few things to discuss at the Académie, things that I should think will be of great interest to you; since my new paper is ready, I do not want to wait until next Monday. I fully appreciate the depth of our friendship, and so, in this extremity, I must call on you. I am not far away from either the Rue Saint-Thomas or from the Académie. You would be of particular help to me if you would allow me a few moments of conversation so that you would be able today to inform the Académie of everything that I have done. I hope that you can come.

Come... (?)

Yours,

A. L. Cauchy

Letter to Herschel, undated, 1848, Archives of the Royal Society, Ms 328

Sir,

Being called to America to direct the astronomical works at the Observatory in Georgetown, Father Vico—who is rightly respected by all true friends of science[17]—asked me to write a few lines to you on his behalf. I am now taking advantage of his request in order to refreshen your memory of me.

Allow me, then, to use this opportunity to ask you to offer in my name to the Royal Society of London my study on the theory of light. This work was published in 1836, and I presented a copy of it to you when, on your trip to Paris, I had a chance to meet with you for a few moments,[18] I am also enclosing a study, which I published in 1813, on the method for an a priori determination of the number of real positive roots and the number of real negative roots of an equation of arbitrary degree. This method furnished the first solution to the problem that has since been solved, by other means, by M. Sturm's theorem. Aside from this, it is known that the method in question and M. Sturm's theorem are found to be covered by certain more general theorems dealing with arbitrary roots—real or imaginary—that I published in 1833.

Finally, I am enclosing with the 1836 study on light several reports that were written in 1813; one of them contains the proof that I gave at that time of

[17] Franco de Vico was an Italian Jesuit astronomer. When the revolution of 1848 broke out, Father de Vico left Rome to become Director of the Georgetown Observatory. He died on November 15, 1848, in London while en route to his new appointment.

[18] See C. R. *Ac. Sci.*, **8**, May 17, 1839, p. 38, for more on Daguerre's intervention: 'M. Cauchy, who also saw M. Herschel when he [M. Herschel] passed through Paris, confirms M. Arago's remark. He adds: M. Herschel declared that the attempts made in England are, in fact, mere child's play as compared to M. Daguerre's methods. M. Talbot himself will soon come to share my opinion, for I am going to write to him and ask him to come to see these wonders'.

Euclid's theorem. The last two opuscules, which I just mentioned, are in two copies. I am asking that you will please be kind enough to offer one of the two copies to the Royal Society, and I also hope that you will be assured of my highest regards and deep respect.

Your devoted friend,

Baron Augustin Cauchy

Letter to Saint-Venant, October 5, 1850, fonds Saint Venant, Archives E. P. not classified

Sceaux, October 5, 1850

Sir,

Upon returning from a short trip to Normandy, I read the letter that you were so kind as to write me, and as you requested, I am hastening to send back to you the paper that you presented to the Académie des Sciences on August 26.[19] As to the information that you would like to have relative to M. Dubuat, I think that you can obtain it from M. Binet, who was Inspecteur des Études at the École Polytechnique in 1816 and who continued in that capacity until 1830.[20] I pray that you give my highest regards to Mme de Saint-Venant and that you will be assured of my highest estimation and my sincerest attachment.

Baron Cauchy

MMes. de Bure and Cauchy remember you fondly and ask me to bid you give Mme. de Saint-Venant their compliments and greetings.

Letter to the Dean of the Faculté des Sciences, May 14, 1852, Archives Ac. Sci. Cauchy file

Sceaux, May 14, 1852

My dear Dean,

I have received the letter that you so kindly sent me relating to the formality prescribed by Article 14 of the Constitution. You understand, I am sure, the caution and even—should it be necessary—the sacrifices that the position in which I now find myself placed imposes upon me.

---

[19]'Formules nouvelles pour la solution des problèmes relatifs aux eaux courantes' is a work for which Cauchy served as commissioner. On October 25, Saint-Venant gave a sequel to this work.
[20]Saint-Venant published the *Notice sur la Vie et les Travaux de Pierre Louis-Georges Comte du Buat* in 1865.

When, in 1830, the Duke of Orléans ascended to the ruins of a throne that he himself had worked to destroy, I renounced the three (professorial) chairs that I then held and swore to myself to remain faithful to the oaths I had sworn. Can I forget this promise, after the signal honors that Charles X bestowed on me by requesting me as teacher to the heir of Saint-Louis, Henri IV, and Louis XIV?

If, in 1849, I did not hesitate to subscribe to the requests of certain of my colleagues who pressed me to return to a career in teaching, it was not because (and that is surely well known) I had any illusions about the dangers of the anarchist doctrines and theories that threaten to bring desolation to our homeland and overwhelm society everywhere in Europe. But, the political oath came to be abolished, and I was happy to think that I might once more be of use in guiding youth along toward the harmless study of a science that has ever been the delight of my life. My efforts in this respect should be easier yet under a government whose head has so loudly proclaimed his deep desire to see growing and flourishing among us that true science that enlightens minds without corrupting hearts.

Sir, I deeply regret the fact that the impossibility to subscribe to the oath should separate me from my honorable colleagues and oblige me to renounce my heartfelt hopes of working alongside them in serving my country in a field in which I am able to do some good.

Please be assured, Sir, etc.

Augustin Cauchy

P.S. Please let me know if I should still hold my class session in mathematical astronomy next Monday morning.

Appendix IV

Unpublished Documents on Cauchy's Two Candidacies to the Collège de France (1843 and 1850–1851)

I. The Election of 1843

Letter from Liouville to Letronne June 6, 1843 (Archives of the Collège de France C. XII, Liouville 1 A)

Paris, June 9, 1843

My Dear Mr. Administrator:

I am honored to ask you to present me as a candidate for the vacant chair in mathematics at the Collège de France, a vacancy created by the death of your venerable colleague, M. Lacroix. The members of the Committee are of the opionion that the works I have published as a geometer and the students I have taught as a professor entitle me to their consideration. I am no less certain that I am worthy of such consideration on account of my enthusiasm and scholarly standards. However, putting aside any discussion of matters not having to do with the science, I nevertheless think that I ought to state here clearly that if the Committee should choose M. Cauchy, I would—far from regarding myself as having been hurt by that choice—be the first to applaud it. Fairness to the superior merit of a fellow scholar is, at least to my thinking, a sacred duty that one should observe at all times, and I can entertain no base thoughts of personal interest and advancement that run contrary to that duty.

Mr. Administrator, I am

Your very humble and obedient servant,

J. Liouville

Letters from Cauchy to Letronne, June 11, 1843 (Archives of the Collège de France, B-11, Mathématiques b – 1)

1. My Dear Mr. Administrator:

You yourself can bear witness to the promptness with which I sacrificed my own candidature out of the desire to avoid knowingly giving anyone the slightest pretext for seeing my efforts and aims in a bad light. However, today it appears that the delicacy of this way of proceeding has not been appreciated in full by everybody, contrary to what I had hoped. This situation, of course, no longer allows me the liberty of remaining on the sidelines, for there are certain points at which the very honor of the contestants—when the dispute is of a scientific nature—will not allow them to withdraw from the fray. Under this assumption, then, Mr. Administrator, I now find myself obligated to reenter the lists. Accordingly, I request that you please be kind enough to read to the honorable professors of the Collège de France not only the present note (which I hope you will be kind enough to do in any event) but also the enclosed letter. I also hope that the honorable professors will show me some indulgence relative to the formalities that I ought to have satisfied by recognizing that time did not allow me the customary opportunity of visiting each professor separately. Please be assured, Mr. Administrator, that I have the honor of being

Your very humble and obedient servant,

Augustin Cauchy

2. Dear Mr. Administrator:

If the honorable professors of the Collège de France believed that my having devoted my entire life, over the past forty years, to the study and teaching of the physical and mathematical sciences should entitle me to lay claim to the chair that became vacant upon the death of M. Lacroix, I would be happy to think that I can still be of use, as a professor of differential and integral calculus, both to science and to France, my homeland. The honorable professors may be of the mind that those works of mine whose aim is the advancement of infinitesimal analysis and the courses that I taught over all those years not only at the École Polytechnique but also at the Collège de France, courses which, I might add, were attended by many students, who today successfully occupy chairs in mathematics in various countries and are distinguished members of the Institut de France as well as various learned societies all over Europe—give me special claim to a favor whose value I now realize.

I hasten to seize this opportunity to assure you, Mr. Administrator, that I am honored to be

Your very humble and obedient servant,

Augustin Cauchy

Letter from Cauchy to Letronne, undated, probably a few days prior to June 18, 1843 (Archives of the Collège de France, B-II, Mathematics b-1)

Mr. Administrator:

The honorable professors of the Collège de France already know that I am presenting myself as candidate for the chair that became vacant upon the death of M. Lacroix. My long years of work—the courses that I have taught not only at the École Polytechnique and at the Faculté des Sciences but also at the Collège de France, courses which were taken by outstanding scholars, both French and foreign, and even by members of the Institut—justify, I should think, my request, and they would appear to give me some title to such a flattering distinction. The high regard and goodwill that the honorable professors of the Collège de France have shown me, in particular, those professors who are themselves in the physical sciences or in mathematics and the publicly expressed esteem for my works, both by word of mouth and in writing, not only by masters of the science whose loss sadden us all—scholars such as MM. Lagrange, Laplace, Legendre, and Ampère—but also by scholars now at the Collège de France—all this allows me to hope for success. However, one objection has arisen that might give reason for doubt to the professors. It might be, I am told, that my election may not be acceptable to the government. However, I am morally convinced that the contrary is true. Today, I believe that I am able to assure the professors that, in case they should favor me, there will be no serious problems to fear from the government. Moreover, I can assure them, if that be necessary, that the instructional program in mathematics at the Collège de France will not be subjected to any interruption.

Please be assured, Mr. Administrator, that I am honored to be

Your very humble and obedient servant,

A. L. Cauchy

Letter from Libri (the addressee is unknown, but was probably Letronne), undated, probably June 17, 1843 (Archives of the Collège de France, Libri file)

Sir and Very Respected Colleague:

Allow me to add a few words to the official letter that I had the honor of addressing to you the other day. M. Thénard has urged me to ask you to state, on my behalf, to the Assembly of Professors—in case this is necessary—that if I should be elected I will never become involved with nor accept any function or position that would require me to be out of Paris, even temporarily, during the school year....

I forgot to tell you in my first letter that before substituting for M. Lacroix, I had filled in for M. Biot. I was a substitute Professor at the Collège de France for eleven years. I ceased filling in for M. Biot, who paid me 2000 francs per year, only in order to undertake (gratuitously) M. Lacroix's course, who could not afford to pay a replacement.

An absurd rumor is now afoot, a rumor that has it that the *Jesuits are pleased by my candidature.* This hateful trick seems to have made a big impression on M. Michelet. I need not tell you, Sir, that nothing whatever can make me give up my struggle against the Jesuits. My second letter is going to appear on the 15th of this month. If you could make it clear to everybody that if I am rejected, the Jesuits will cry victory in their journals, then you will have done me an outstanding service. Moreover, Sir, I ask you to please use this statement as you think best in the eyes of your colleagues.

It is important to everyone that the election should be held as soon as possible. If then, there is a second session held for the voting, I ask you to promptly get an understanding with your colleagues so that this second session can take place one day next week, preferably before Thursday, if at all possible. Sunday should be rejected outright because nowadays everybody goes out of town on that day....

G. Libri

Letter from Liouville to Letronne, June 19, 1843 (Archives of the Collège de France, C. XII, Liouville, 1 B)

Mr. Administrator:

I am deeply humiliated, both as a man and as a geometer, by what took place yesterday at the Collège de France. Accordingly, from now on it will be impossible for me to teach at that institution. Please receive and accept my resignation as a substitute Professor there.

I have the honor, Sir, of being

Your very humble and obedient servant

J. Liouville

Letter from Cauchy to J. B. Dumas, June 20, 1843 (Arch. Ac. Sci. Libri file)

June 20, 1843

My Dear Mr. President:

I hasten to send to you what you honored me by requesting, namely, a readable copy of the reflections that were presented to the Académie last Monday. I take joy in the fact that at last, thanks to you, the question is going to be squarely put and that there will be an official answer. After all the testimonies of respect and esteem that my colleagues so kindly gave on my behalf, it would be fitting neither to the Académie nor to me if I were to renounce my candidature; for, at this time the battle is purely a scientific one. This is, of course, the reverse of the situation wherein it would be convenient neither for me nor the Académie that I should consent to remain a candidate if the battle were on political grounds; in such a case, I would certainly withdraw. As to the rest, I shall always have the same confidence and moral certitude that I dared to express on Monday before the Académie. I am well convinced that as soon as the honorable Minister reads my observations and remarks and has a clear account of the present state of things (and, in particular, of the fact that it has now come to pass that the chair in political economy at the Collège de France has now been bestowed to a foreigner, a person who is not even naturalized), as soon as learns the details of what a respected jurist consulted by his colleagues of the Académie des Inscriptions et Belles-Lettres said at the Institut a few days ago, he will not exclude a French geometer from a chair in mathematics at the Collège de France by subjecting him, and him alone, to a formality that has long since fallen into disuse at the Collège de France so that it was decided to name as professors persons who are found in this obvious incapability to fill positions. Thus, my dear Mr. President, thanks to you, the Académie on the one hand and I on the other, will both know what we have to do. Thanks to you, the Académie will officially know whether or not I have been defeated by a political burden that will never allow me to play either an active part in a learned body or function as a professor of mathematics. Those things are for me, I swear, a true delight; in a way of speaking, they are my element and my life. I hope you will not be condemned to come to declare this incapacitating burden before the Académie. I myself cannot conceive of how it would be possible to even frame such a principle in any event without being forced by the very logic of the situation to exclude me from the Académie itself.

I say no more. Your words and your thoughts will be much more easily understood than mine; I am so certain of it that I beg you, my dear Mr. President, to accept my thanks in advance with my best wishes.

Yours truly,

A. L. Cauchy

This June 20, 1843

Cauchy's statement to the Académie on June 19, 1843, together with Libri's letter (Archives Ac. Sci., Libri file; rough draft in the Bibliothèque des Fontaines, Chantilly, Moigno file)

Académie des Sciences. Meeting of June 19, 1843. A letter written by the Minister of Public Instruction was read by the president of the Académie des Sciences. Afterward, M. Cauchy took the floor and expressed the following thoughts and reflections.

The letter that you have just heard imposes on me a duty to state the position I take relative to the kind of competition opened by this letter to the mathematicians in the Académie. My task is to explain what must be thought, what I myself think of the hypothesis, which is accepted by certain persons and rejected by others, that I am a candidate to fill the vacancy created by the death of M. Lacroix. I am pleased that I have nothing to say on this subject that will not be satisfactory to the most sensitive susceptibilities, nothing to say that will not be appropriate to the rejection of a subject that is such a source of annoyance and disunion among scholars and, in particular, among the members of the Académie. My disinterest, the moderation of my wishes, and the frankness and loyalty of my character are so well known that the Académie need have no fear that I shall say anything that any person can complain of, or that I will say anything within these walls or even outside of these walls that could be offensive to anyone.

As I had withdrawn to the country, I occupied myself by responding to the confidence that the Académie itself had placed in me by appointing me president of the committee responsible for evaluating works submitted in the competition for the Grand Prix de Mathématiques, and also being deeply involved in the calculations that this examination required, I happened to attend the session before the last at the Académie. I learned from the mouth of my respected colleagues that there would be a question of replacing M.

Lacroix. Several of them even pressed me to apply. I admit that I was not insensitive to hopes of once more being useful, as a professor of mathematics, both to the friends of science and to my homeland. I can recall, with real joy, that I have seen gathered about my chair—at the Faculté des Sciences, at the Ecole Polytechnique, and at the Collège de France—outstanding scholars from all over Europe. Of these, several have since become members of this very academy. I believe my enthusiasm and my strengths will still permit me to contribute to the advancement of the mathematical sciences. I am the more disposed not to push my case myself since I know so well what the candidates think of my works, what they have written about them, and what they have recently said about them; and I was not at all surprised by the interest that, on this occasion, was shown in an old professor by the candidates, who once attended my courses at the Collège de France. Moreover, the goodwill with which I was honored by the professors at the Collège, especially by those who are involved in the physical and mathematical sciences, encouraged me to follow through on their idea. But, a single objection has arisen, an objection that might be the source of doubts and questions in the minds of certain persons. It was feared, I am told, that based on motives having nothing to do with science, the authorization to teach courses would not be accorded to me in case I should be appointed by the Collège or even by the Académie itself. I believe the contrary is true. In France, even among the most opposite parties and factions, justice has always been accorded to nobility of sentiments and to open, loyal, honest conduct. We understand that sense of steadfastness in the face of misfortune by which one imposes great sacrifices on himself rather than knowingly give the slightest pretext for reproach or for the most blame. I recall the anecdote often told about Professor Scarpa.[1] I also think of the words addressed by one of the members of the Académie des Inscriptions et Belles-Lettres by a Minister when the question arose over awarding the chair in political economy at the Collège de France to a foreign scholar.

Am I under any illusions myself? I think not. I know only one thing, one thing that certain other persons assume. Some people—in spite of my entreaties and earnestness, in spite of a letter that I wrote and gave my thoughts on these matters—have protested against the generally accepted hypothesis. Be that as it may, in spite of my desire to be able—at least as long as my strength will allow—to be of service as a professor of mathematics to the advancement of science in France, my homeland, I will state once and for all that I will only consent to my name being placed on the list of candidates for a chair in

[1] Antonio Scarpa (1747–1832) was a famous Italian surgeon. In 1796, when the Italian republic was established, Scarpa, who was devoted to the monarchy, refused to sit on the Committee of the Juniori or to take a loyalty oath, and no one bothered him at all. As to the foreign scholar referred to, the reference is to Count Rossi (1787–1848), who was appointed to a chair in political economy in 1832 following the death of J. B. Say.

mathematics at the Collège de France if there are no serious nonscientific obstacles against my candidacy whatsoever.

Letter from Sturm to J. B. Dumas, undated, undoubtedly a few days before July 3, 1843 (Arch. Ac. Sci., Libri file)

Mr. President:

The two most senior members of the geometry section were of the view that any section called upon to designate candidates for a vacant chair ought be bound simply to making known the names of those persons who present themselves (as candidates for the vacancy), if they are members of the Académie. Monsieur Lam̃e and I thought that we should cede on this point, although the principle seems debatable to us.

However, we reserved the right to expose to the Académie our personal and individual opinion as regards the incompetence of the only candidate who presented himself to the section. M. Libri has shown by his public teaching that he has little or no talent as a professor. It is to be feared that his reputation as a geometer and as an academician has suffered because of it. Students do not attend his classes, even though the professor has made them shorter. These facts are publicly known.

We truly regret the voluntary or forced withdrawal of those two eminent geometers, MM. Cauchy and Liouville. Both of these scholars were, at first, presented as candidates for the chair in question and both were unquestionably qualified. Since, in addition to them, several other geometers with known success in teaching might be interested, we have no recourse except to declare our formal opposition to M. Libri's candidature. Our protest has no goal except for the advancement of the mathematical sciences and those who cultivate them. In concluding, we think it should be pointed out to the Académie that, by its very own report, the section does not present M. Libri to the Académie's vote, it merely announces that M. Libri is presenting himself.

M. Lamé, being absent, has authorized me, by a letter (now) at may office, to express his views to the Académie. They completely agree with my own.

C. Sturm

II. The Election of 1850–1851

Letter from Cauchy to the Administrator of the Collège de France, November 18, 1850 (Archives of the Collège de France B-II, Mathématiques, b-7)

My Dear Mr. Administrator and Honorable Professors:

Some of you have been of the opinion that it is once again time to allow the voice of a certain person to be heard at the Collège de France, a person who in 1817 was called by a brilliant professor—a true master of the science—to teach courses that, for many years, were assiduously studied by some of the outstanding scholars who today are members of various academies throughout Europe.

Some of you have been of the opinion that it is fitting to thus satisfy the wish expressed on various occasions, and quite recently, too, by young professors as well as by French and foreign scholars, who with such enthusiasm studied the course on celestial mechanics that I offered at the Faculté des Sciences. These scholars have also wanted to see developed in a special course the new methods that I have been working on for 36 years, the solution of various problems in mathematical physics and astronomy.

The sincerity and goodwill with which these thoughts were expressed to me, and the sense of urgency with which they have been communicated over the past few weeks by respected professors whom I hardly know, have convinced me that I should follow these suggestions.

My former and present achievements are too well known by the friends of science to need elaboration here. However, if it should be necessary that they be recalled, I can call on the testimony not only of geometers and physicists who presently hold positions among you, but also on the testimony of those who would like your recommendations and votes.

Be assured then, Gentlemen, of the value I now attach to making myself ever more worthy of having the title of your future colleague, a title that some of you were kind enough to honor me with as you shook my hand.

Paris, November 18, 1850

Augustin Cauchy

Dear Mr. Administrator:

I am honored to submit my candidacy for the chair in mathematics at the

Collège de France. I should also like on this occasion to ask for your kind support and that of your colleagues.

My reasons for this decision are stated in the note that is enclosed and is addressed to the professors, and I hope you will be kind enough to read it to them.

I pray you please be assured, Sir, of my best wishes.

Sceaux, November 18, 1850

Augustin Cauchy

Letters from Quatremère to Binet (Archives Ac. Sci. Binet file)

1. Letter of December 3, 1850

Sir and Dear Colleague:

I have just learned from M. Sédillot that during the meeting in which M. Letronne was appointed Professor of Archaeology at the Collège de France there were 20 votes cast, with M. Letronne receiving 10 voices since 10 blank ballots were found in the urn. On this basis, it was decided that the candidate had, in fact, received a majority of the votes. Thus, the Collège has adopted a position that is contrary to the one that I had always assumed it to have. It follows from this situation that M. Cauchy, having 11 votes as against 10, with one blank ballot, should have been recognized as the Collège's candidate. I have just written to M. Barthélémy-Saint-Hilaire in this sense and to protest that I was led into error by adhering to the procedure that has always been followed at the Académie des Inscriptions et Belles-Lettres.

My dear Colleague, I hope that you will accept my expressions of regret along with my highest regards and sincere devotion.

Quatremère

2. Letter, undated probably on December 5, 1850, or December 6, 1850

Dear Sir and Learned Colleague:

I am honored to communicate to you the reply that was sent to me by the honorable Administrator of the Collège. As you can see, he circumvented the problem, but did not even try to resolve it. The example having to do with M. Pélouze seems not to be well selected; for, if my memory serves me correctly, M. Pélouze received the totality of the votes cast. Accordingly, his election cannot really offer a basis for any discussion. I hasten to submit to you the draft of a letter that we can put before the Minister. I urge you to read over this rough draft and make whatever changes you think fitting. You may also discuss it with M. Cauchy.

Dear Colleague, please accept my expression of deep attachment.

Quatremère

M. Cauchy is leaving my house; he should see you in the morning.

Letter from Desgranges, Quatremère, and Binet to the Minister of Public Instruction, December 6, 1850 (Archives of the Collège de France, C XII, Liouville, 5A)

Dear Mr. Minister:

We are honored to call your attention to a situation whose seriousness you will readily appreciate. It is a situation presenting a very real problem that demands a sure and invariable solution for the future as much as for the present.

The chair in mathematics at the Collège de France is vacant, and the professors were called together to designate the candidate who was to be presented to the selection of the government. There were only two persons who sought this position, M. Cauchy and M. Liouville, and both are members of the Académie des Sciences. There were twenty-two votes cast, and on the first ballot, M. Cauchy received eleven, while M. Liouville received ten, there being one blank ballot cast. The honorable administrator was of the view that as neither candidate had received an absolute majority of the votes, it would be necessary to hold a second balloting. Needless to say, no objections were raised at the time against this procedure, for there was not, within the memory of any of us, a comparable situation. Accordingly, a new vote was held, and this time M. Liouville received twelve votes and thus, in effect, became the Collège's candidate.

Several of us subscribed, without qualms, to this decision, which appeared to be quite regular. Nevertheless, some felt it important to know if a comparable situation had ever arisen, and if so, whether the Collège de France had followed the procedure that we used. Here are the results of our investigations into this matter. At the meeting of November 27, 1837, the Collège was called on to present a candidate for the chair in archaeology, and 21 professors were assembled. One of these, however, declared that he would not vote, so that the number was reduced to 20, and the majority was fixed at 11. On the first vote, M. Letronne obtained 10 votes and his competitor 8, with two blank ballots. In this situation, it was decided that as the number of voting professors had been reduced to 18, M. Letronne had obtained the required majority. This situation, as can readily be seen, is identical to the one that has just arisen.

In our desire to adhere, in our statements as well as in our conduct, to exact, scrupulous, and impartial principles, we continued our inquiries by checking the minutes and records of the Collège to convince ourselves that in other cases a different procedure had not been followed. As far as we could determine, nothing similar has arisen under the regulation of 1828. And this, to speak frankly, was quite astonishing; for, it can hardly be presumed that so grave and serious minded a body should, in identical situations, follow a legal rule that is diametrically opposite to the one adopted in an earlier, similar case. It is thus demonstrated that by the customs of the Collège de France the presence of blank ballots [in a vote], by reducing the number of voting members, ought to reduce, proportionally, the size of the majority required for the presentation of a candidate.

This principle, which has been permanently adopted by the Académie des Sciences, has been solemnly proclaimed by the Institut in a meeting of the five academies. Finally, a decision by the Conseil d' État, which was approved by the Emperor and thus carried the force of law, by the date of January 25, 1807, prescribed in a formal manner that in electoral proceedings blank ballots were to be meaningless and that the majority was to be based on the actual number of valid votes cast.

By all the facts that have come to our attention, Mr. Minister, we are strongly persuaded that M. Cauchy obtained a majority on the first ballot, that the second ballot should be regarded as null and void, and M.Cauchy's right to the title of candidate designated by the Collège should be recognized.

Please be assured, Mr. Minister, of our high and respectful regards.

Signed: Alix Desgranges, Quatremère, and Binet, Professor of Astronomy.
December 6, 1850

Letter from the Administrator to the Minister of Public Instruction, December 10, 1850 (Archives of the Collège de France, C XII, Liouville, 3)

Dear Mr. Minister:

I think it fitting that I send you certain information on the last two meetings of the assembly and supplement the minutes that I have transmitted to you.

The Collège presented M. Liouville as its candidate for the chair in mathematics. Two ballots were needed to decide the question.

At this first meeting, nobody spoke out against setting aside the first ballot, as it appeared to everyone that the rule required that the blank votes be taken into account and that the majority be based on the number of members present at the deliberations and not on the number of votes cast.

At the following meeting, however, a discussion did take place relative to the point determining whether or not blank ballots ought to be excluded, and this time it was again understood that according to the terms of the regulation, a majority could only be formed on the basis of the number of members who were present.

We thought that all discussion on this matter would thus be unnecessary. But, yesterday the debate was reopened, although only incidentally. Some professors who had voted for M. Cauchy recalled that in 1837, in the election of M. Letronne, the blank ballots had been disregarded as null and void; they further claimed that it was fitting to do so in the present situation. The reply to them was that in the election of M. Beudant the blank ballots had been counted, and that from these two opposite precedents it was impossible to draw any clear-cut guidelines. Moreover, it is contended that although the regulation may indeed have been violated in the first instance, all efforts should be made to avoid violating it a second time. Moreover, as to its formal terms, there can be no possible doubt regardless of what the customary practices may be as concerns other learned bodies or other deliberative bodies in general.

Mr. Minister, I would not bother you at all with a discussion of these debates, seeing that no trace of them is to be found in the minutes of the meeting, and that they therefore did not result in a formal proposition, if I did not fear that certain efforts would be connected with the discussion. A remonstrance is in the process of being drawn up and is to be signed by the professors who voted for M. Cauchy, in the hope of placing in the balance a weight that, to my thinking, ought not be there. I have thus sought, Mr. Minister, to warn you of these efforts in order that you may more fully appreciate them.

One person at the Collège has even proposed to me an express change on matters concerning the actual facts that I have just reported to you. The decision that gave rise to the protestation has not been weakened at all, but rather retains its full strength. One may later modify the regulation or its interpretation, but art. 24 is perfectly clear; at the meeting where the candidates presented themselves and at the following meeting, this article has been unanimously understood to have the meaning that I understood it to have, then and now. By common consent, the confrontation that took place here yesterday can in no manner have the character of a new decision.

I have thought it my duty, Mr. Minister, to bring this matter to your attention, since I am entrusted with the responsibility of applying the regulation. I should be remiss in this obligations imposed on me were I to have kept quiet, and it is only to protect the Collège's deliberative process against ill-conceived attacks that I have thought it my duty to write this letter.

Please be assured, Sir,...

Letter from Cauchy to the Minister of Public Instruction (Arch. Nat. F^{17} 15333)

My Dear Mr. Minister:

The problem has been submitted to you of determining precisely who the true candidate is designated by the Collège de France for the chair in mathematics. It is stated in writing that you now have before you that this question comes down to the following: *does a blank ballot have the power to annul and destroy a majority*?

General opinion, logic, custom, the bylaws of the Collège de France, of deliberative assemblies, of academics, and finally even the very law itself (see the *Bulletin des Lois*, 46th Series, no 104, Law 2178) all seem to point to the resolution of this question in this sense: a blank ballot is a null and void ballot.

On the other hand, this question cannot be left undecided lest there be serious inconvenience at the time in which the Académie is called on, in its own turn, to present a candidate for the chair in mathematics.

Mr. Minister, I am certain that you do not want that I—a man who has devoted 37 years to the advancement and teaching of the mathematical sciences (as evidenced by the papers that you requested from me and that are, today, before you) and who, beginning in 1819, has had as students some outstanding scholars who are now members of various academies throughout Europe—should find myself excluded from a career as a professor on account of a blank ballot.

In the meanwhile, if I present myself to the votes of members of the Académie des Sciences, it is because the scholars, the lawyers, and the professors, who, after having supported my candidature, committed an involuntary error by believing in the power of a blank ballot, seem today to be generally convinced that my right to the title of candidate designed by the Collège de France cannot be questioned.

Mr. Minister, I urge you to have the extreme kindness to tell me if I am right or wrong in my belief; and if, in your view, a blank ballot can annul and destroy a majority. If you have any doubts in this respect, it would be easy, it seems to me, to throw light on this point since—at this very moment—you have at your side on the Higher Council of Education scholars, magistrates, and lawyers capable of clarifying this serious question.

Please be assured, Mr. Minister, of my high regards and consideration.

December 14, 1850

Augustin Cauchy

Letter from the Minister of Public Instruction to the Administrator, December 21, 1850 (Archives of the Collège de France, C XII, Liouville, 5B)

Dear Mr. Administrator:

I have just received a statement of protest dated December 6 from three of the professors at the Collège de France. The protest has to do with what transpired in the meeting of the Assembly of Professors that was called to consider the presentation for the (professorial) chair in mathematics and resulted in M. Liouville being designated as the candidate of the Collège.

I am pleased, Mr. Administrator, to transmit to you a copy of the letter.

I received, at the same time, a copy of the minutes of the proceedings of the Assembly of Professors of the Collège that was held on December 8 and in which I note that the discussion had been reopened for a second time on the question of the vote for a candidate for the position of professor of mathematics and that the meeting was then adjourned. Allow me to say to you, that this new complaint, according to the statement of protest that I have received, places in question, so to speak, the legality of the result of the vote of November 15. It seems to be necessary that a formal decision be made and that the Collège de France should inform me definitively as to precisely whom it presents as its candidate.

Accordingly, I request that you, Mr. Administrator, kindly convoke the Assembly of Professors right away to deliberate the questions that have been raised by their colleagues' statement of protest.

Mr. Administrator, please be assured of my highest regards.

Parieu
Minister of Public Instruction and of Religious Affairs

Letter from the Administrator to the Minister of Public Instruction, December 23, 1850 (Archives of the Collège de France, C XII, Liouville, 6)

Dear Mr. Minister:

Before I agree to the desire that you expressed in your letter of the 21st of this month, allow me to set before you some observations that I regard as essential.

The candidate presented by the Collège de France for the chair in mathematics is M. Liouville. There can be no doubt on this point.

First of all, the minutes of the meeting confirm that this is so. After two votes had been held, the Collège concluded that M. Liouville, having obtained a majority, should be its candidate, and there was not the least complaint about it.

In the following meeting, on December 1, 1850, nothing more was said about it, only some members asked that there be a check of the circumstances of the balloting. The Assembly acted on this request, and the minutes of the meeting at which M. Liouville was presented were adopted without any changes.

Finally, at the last meeting, on December 8, there was not even a question raised as to M. Liouville or his nomination. The question was not reopened, insofar as it had been closed at the last two previous meetings, and the discussion that took place had to do only with the interpretation of the regulation in cases to come and with determining whether or not the blank ballots should be counted toward the votes defining a majority.

In my opinion, Mr. Minister, the regulation is categorical in the sense that by saying *the majority of members present, titular and honorary*, it expressly means the number of persons voting and not the number of votes cast. It is particularly on this litigious point, a point that has only to do with the future, that the debate has been focused. On the motion of M. Laboulaye, the Assembly decided to adjourn once the nomination for the position in mathematics had been made.

These, then, are the facts, Mr. Minister, and I urge you to convince yourself of their validity by setting before you the minutes of our last three meetings.

Now, 3 professors out of 23 are asking that M. Liouville's nomination be set aside as null and void and that M. Cauchy be accorded his place since he won on the first ballot, as the blank ballots are to be disregarded.

This, Mr. Minister, is a question that you and you alone must determine. If you should set aside the nomination of the Collège, there will be another meeting so that a ballot can be held at which, it would appear, no reversal will be forthcoming. If, on the other hand, you accept the reasoning of MM. Quatremère, Alix Desgranges, and Binet, there will be no need to consult the Collège at all, for M. Cauchy would be the candidate, plain and simple.

Mr. Minister, I urge you to take these matters into account, and I will await your reply before I convoke a special meeting of the professors, who, you may be certain, regard this matter as being completely closed. The only point that, in their view, remains to be settled, is to interpret the regulation so that in the future there will be no possible delays.

Please be assured, Mr. Minister, ...

Letronne

Letter from the Minister of Public Instruction to the Administrator, December 30, 1850 (Archives of the Collège de France, C. XII, Liouville, 7)

Dear Mr. Administrator:

I hasten to reply to the observations you so kindly sent to me on the 23rd of this month concerning the request that I put before you that you call a meeting of the Assembly of Professors and submit to it the protest that has been formulated by three of the professors against M. Liouville's selection as candidate for the chair in mathematics.

You seem to think, Mr. Administrator, that this complaint has only to do with future situations, that the election was completely in order relative to the time it was held as well as in its confirmity to the regulation and the majority of members present. Moreover, you added that the Assembly resolved that after the nomination had been completed there should be a determination on whether or not blank ballots should be counted in the total votes establishing a majority.

A closer look at the regulation of the Collège de France, however, has persuaded me, Mr. Administrator, that the question cannot be regarded in the light in which you are considering it. Based on the information contained in the minutes, the situation comes down to the following fact: neither applicant was properly elected as candidate of the Collège and this can be concluded from Article 24 of the regulation, which asserts:

> On the day designated, the election of the candidate shall be effected by vote. The balloting shall be repeated until one of the applicants shall have obtained an absolute majority. This majority is to be based on the

number of eligible titular professors together with the honorary professors present at the deliberation.

You must not ignore, Sir, the distinction to be made between an absolute majority and a relative majority. Since the number of eligible titular professors is 25, the true number for an absolute majority would be 13. As neither M. Cauchy nor M. Liouville obtained this number, it is obvious that neither of them can rightly be regarded as having been legally elected by the Collège de France.

There are thus grounds, Mr. Administrator, for proceeding to an election again, without going into the question of blank ballots as raised in the statement of protest. I, therefore, urge you to call a special meeting of the Assembly of Professors so as to elect, according to the sense of Article 24, a candidate for the chair in mathematics.

Please be assured, Mr. Administrator, of my highest regards.

Parieu
Minister of Public Instruction and Religious Affairs

Letter from Cauchy to the Administrator and Professors of the Collège de France, January 4, 1851 (Arch. Nat. F[17] 15333 and Archives of the Collège de France, B II, Mathématiques, b-8)

Dear Mr. Administrator and Professors:

Prompted by your kind advice, I presented myself, some weeks ago, as candidate for the chair in mathematics at the Collège de France.

A new candidacy arose on the eve of the election, and, at the meeting of November 20 last, the count of the votes on the first ballot was 11 votes in favor of my candidacy, 10 for the opposing candidate, and 1 blank.

Based on this unarguable fact, I should be able to consider myself as the candidate of the Collège de France and to present myself confidently for the vote at the Académie des Sciences. This vote, especially when there is a question of selecting between two colleagues, is generally determined by the results from the election at the collège.

However, the grievous mistake of allowing a blank ballot to have the power to negate a majority apparently robs me of the candidacy of the Collège and, in effect, that of the Académie as well.

This last consequence of an extraneous error can today no longer be set right.

Before the vote at the Académie, I vainly urged the professors and the Minister of Public Instruction to recognize this obvious error, which is contrary to the customary of procedures and bylaws of the Collège de France, as well as to those of academies and other deliberative assemblies and even to the law itself. In spite of my efforts, however, I have only been able to obtain an assertion that the question had been declared to be in doubt.

Today, I learned that it is proposed that there be an examination of the double issue of the blank ballot and the candidacy itself and that the two issues might perhaps be considered separately.

Sirs, such a decision hardly conforms to justice nor, I might add, is it worthy of you. The vote of the Collège should be free and independent, it being precisely for this reason that it is always held prior to the vote at the Académie. A new election to come at a time when the issues are no longer considered as related, and at a point at which the error committed appears to dominate the situation, ill suits you.

The matter of the blank ballot and that of the candidature, which have been closely connected to each other for quite some time, cannot now be separated without compromising the diginity of your votes.

On Monday before last, it was determined, in conformity with the law and the bylaws of the Collège de France, that the question of the blank ballot assures me the double candidacy of the Collège and the Académie.

But, to decide today, in a sense, to set aside an acquired right, to decide on the basis of a new ballot the entire question of the candidacy in a contrary sense, to subscribe to the setting aside of the blank ballots, and to set against me and me alone an exception to the general rule, to deprive me of the candidacy for the Collège after having deprived me, by a fatal blunder, of the candidacy for the Académie, this, I am sure you understand, Sirs, constitutes an obvious miscarriage of justice, an act that is contrary to the impartiality that should always control your deliberations.

If, against my expectation, the Collège should decide to separate the two questions and to call for a new vote, I hereby declare that I will withdraw from the fight; which is improper both for my age and for my character and would be completely unequal in light of a generally recognized error, the unhappy effects of which would victimize me alone. This is not the first time that I have had to take a courageous stand against an undeserved reversal. In order to justify the absence of my candidacy, I ask only that this letter should be appended to the minutes of the election meeting.

Please be assured, Sirs, of my highest regards.

Paris, January 4, 1851
A. L. Cauchy

P.S. I beg you, Sir, to be so kind as to have this letter read to the professors at their next meeting.

Cauchy's lithographed note, *Sur l'Influence Souvent Exercée par des Circonstances Étrangères à la Science dans la Solution des Questions qui Paraissaient Purement Scientifiques, et sur le Pouvoir Attribué, dans une Élection Récente à un Billet Blanc*, January 6, 1851 (Arch. Ac. Sci. January 13, 1851)

I should like here to make a few remarks on how it is that considerations having no relation whatsoever to science have frequently played a role in the solution of problems that would seem to be purely scientific in nature.

In 1839, a certain member of the Institut was called by the Bureau des Longitudes to fill a vacancy that had been created by the death of M. de Prony. However, the person to whom I now refer had sworn, in spite of some altogether undeserved misfortunes that are well known, to remain firmly attached to certain principles. These he would never desert. Thus, it happened that after several years of struggle against persons who were audacious enough to confer the title of geometer upon him without inquiring whether or not he had accepted the current political situation, the government simply decided to disregard the plain letter of the law that formally declares 'The Bureau des Longitudes shall fill its own vacancies'.

In 1843 and 1850, the chair in mathematics became vacant. Certain members of the Académie, professors at the Collège de France, invited a formal petition of one of their former instructors to apply to fill the vacancy. The most serious minded and earnest of the candidates did not dispute his qualifications or his claims of right, and some of them were even so magnanimous as to support them. Nevertheless, in 1843, the government threatened to leave the vacancy unfilled unless a candidate was chosen who had its approval.

In 1850, shortly before the election, certain changes brought about a clever scheme: a candidate who, on this occasion, had not even presented himself to be placed on the lists. Moreover, a blank ballot, which was cast on the first vote, was somehow invested with a power that had been altogether denied it under the Empire [see *Bulletin des Lois*, 4th series, no. 134, law 2178]; it was, to be sure, taken as possessing a power that, until now, it has never had in the customary procedures of the Collège de France. Also, one of the candidates then declared, to the Académie in 1843 and to the Collège de France in 1851, that he would be unable to accept the conditions imposed on him. Thus it came to pass that what at first seemed to be a purely scientific matter was given a solution which was strongly affected by political exigencies in 1843, while in 1850 and 1851, there were the combined influences of a scheme and a blank ballot.

Manuscript Sources and Bibliography

I. Manuscript Sources

A. Archives Nationales

F^{14}2148 *École nationale des Ponts et Chaussées*, 1787–1855. *Organisation, règlement, personnel, concours, élèves.*

F^{14}2187^{2} *Ponts et Chaussées.* Individual files. Cauchy file.

F^{14}2221^{2} Idem. Egault file.

F^{14}2263^{2} Idem. Lehot file.

F^{14}3374 *Mouvement des ingénieurs des Ponts et Chaussées, livre de mouvement pour l'année 1816.*

F^{14}7012 *Rapport d'une commission des Ponts et Chaussées sur le canal de l'Ourcq et ses travaux au 1er janvier 1816.*

F^{17}13551 *Procès-verbaux de l'assemblée des professeurs du Collège de France.* Minutes.

F^{17}13555 *Chaires du Collège de France. Sciences mathématiques, physiques et naturelles.*

F^{17}20356 *Professeurs à la faculté des Sciences.* Individual files. Cauchy file.

F^{1}b1*531 *Organisation des bureaux de ministère de l'Intérieur par ordre chronologique, 1792–1811.*

F^{1}dIVC^{4} *Récompenses honorifiques. Demande de Légion d'honneur pour A. L. Cauchy, 1818–1819.*

AA63(167) *Letter from L. F. Cauchy to Coulomb, 14 fructidor, an XI.*

AJ^{16}25 *Scolarité de la Faculté des Sciences*, 1821–1843.

AJ^{16}207 *Nomination de Cauchy au titre de professeur adjoint de mécanique* (1823.

AJ^{16}5120 *Registre des procès-verbaux des assemblées des professeurs de la Faculté des Sciences.* December 1821–December 1843.

AJ^{16}5126 *Pièces annexes*, 1809–1865.

CC^{2}1085 *Revue de marine.* Cherbourg.

M.C.N. CXVIII, 640, 13 Octobre 1787, *contrat de mariage entre Louis-François Cauchy et Marie-Madeleine Desestre.*

LXXIII, 1260, 4 Avril 1818, *contrat de mariage entre Augustin-Louis Cauchy et Aloïse de Bure.*

B. Archives de l'Académie des Sciences

(1) Individual files, especially those of Cauchy, Binet, Libri.
(2) Ampère's papers.
(3) G. Bertrand's collection of autographs, Cauchy file, contains several letters in Cauchy's hand.
(4) Meeting's packets (pochettes de séance) containing many manuscripts of the studies or notes of presentation. The numbers for the studies are those given in the nomenclature by R. Taton (documentary Appendix, *O.C.*, 2, **15**, pp. 589–595).

January 11, 1811 **M 1**
July 20, 1812 **M 2**
May 17, 1813 **M 4**
January 22, 1813 **M 6**
December 21, 1818 **M 20**
August 2, 1819 **M 21**
September 16, 1822 **M 26 & M 27**
January 27, 1823 **M 30**
November 17, 1823 **M 34**
August 9, 1824 **M 37**
February 14, 1825 **M 41**
February 28, 1825 **M 42**
November 7, 1825 **M 9**
December 26, 1825 **M 44**
February 13, 1826 **M 45**
February 27, 1826 **M 46 & M 47**
March 20, 1826 **M 48**
April 10, 1826 **M 49**
April 17, 1826 **M 50**
May 8, 1826 **M 51**
May 29, 1826 **M 52**
November 13, 1826 'Sur la nature des racines de quelques équations transcendantes', *Ex. Math.*, **1**, pp. 297–338 (*O.C.*, 2, **6**, pp. 354–400).
December 26, 1826 **M 55**
January 15, 1827 **M 56**
March 12, 1827 **M 59**
April 16, 1827 **M 61**
April 28, 1827 De la différenciation sous le signe, *Ex. math.*, **2**, pp. 125–140 (*O.C.*, 2, **7**, pp. 160–176).

July 19, 1827 **M 64**
October 1, 1827 **M 65** sealed message
November 5, 1827 **M 66**
December 10, 1827 **M 67**
December 17, 1827 **M68** and 'Sur un mémoire d'Euler qui a pour a titre *Nova methodus fractiones simplices resoluendi*', *Ex. Math.*, **2**, pp. 315–316 (*O.C.*, 2, **7**, pp. 363–365).
January 22, 1828 **M 69**
August 18, 1828 **M 71**, **M 71bis**, **M 71ter** sealed message
October 6, 1828 **M 72**, **M 73**, **M 74**, **M 75**, **M 78**, **M 79**
December 22, 1828 **M 75**, **M 78**
February 16, 1829 **M 79**
April 6, 1829 **M 82**
April 27, 1829 **M 82**
May 4, 1829 **M 83**
June 1, 1829 **M 85**
September 14, 1829 **M 88**
January 18, 1830 Letter from Cauchy to the president of the Académie published by R. Taton in the *Revue d'Histoire des Sciences*, 1971, **24**, pp. 123–148.
April 12, 1830 **M 94**
May 10, 1830 **M 95**
June 14, 1830 *Observations présentées par M. Cauchy.*
June 21, 1830 **M 98**
July 5, 1830 **M 101**
July 4, 1831 **M 103**
October, 22, 1831 'Note sur le versement des voitures publiques'.
March 22, 1847 Sealed message: 'Sur le théorème de Fermat'.
August 11, 1856 Sealed message: 'Note sur l'intégration des équations différentielles qui représentent les mouvements des astres dont se compose le système solaire'.

C. Archives du Collège de France

Registre des procès-verbaux de l'assemblée des professeurs.

BII. Mathématiques. Chaire de mathématiques. Biot, Libri, and Liouville files.

D. Archives de l'École Polytechnique

Registres des procès-verbaux du Conseil d'Instruction et du Conseil de Perfectionnement.
Cauchy file.
VI 2a2 Rapports de Prony.
XII C7 *Registres d'instruction.*
Fonds Saint-Venant, not classified.

E. Bibliothèque Nationale

Fourier's papers.

Ms ffr 22 516, p. 88, Note sur un mémoire de Cauchy relatif aux racines de $e^x - 1 = 0$ et à la factorisation de la fonction e^{x-1}.
Ms ffr 22 525, p. 207, Note relative aux intégrales de la théorie de la chaleur. Réponse à des objection de Cauchy.
Ms ffr 22 529, p. 127, Brouillon de lettre de Fourier aux secrétaires perpétuels de l'Académie. Réclamation à propos des fonction réciproques de Cauchy.

F. Bibliothèque de l'École Nationale des Ponts et Chaussées

Ms 1845 *Mémoire sur les roues de voiture...*, 15 février 1808.
Ms 1982: 1. *Mémoire sur les moyens de perfectionner la navigation des rivières...*, 1809. 2. *Mémoire sur les ponts en pierre.* 3. *Second mémoire sur les ponts en pierre. Théorie des voûtes en berceau.*

G. Bibliothèque de l'Institut

Ms 3710. 3 letters from Cauchy to Leverrier.
f^0 AA/38. 4 autographs by Augustin Cauchy.

H. Bibliothèque de la Sorbonne

Ms 1759. Formules sur la résolution des équations (1837).
Ms 1759 bis. Notes et lettres a l'abbe Moigno (1837).
Ms 1760. Exercices d'arithmétique (1838).
Ms 1761. Exponentielles et logarithmes (undated).
Ms 1762. Recherches nouvelles sur la lumière (1836).
Ms 1786. Résolution des équations (undated).
Ms 2057. Théorie des ondes, note XVI (1815).

I. Bibliothèque de la Faculté des Sciences de Bordeaux

Cours de G. Lespiault, ancienne cote Ms 52.

J. Bibliothèque du centre des Fontaines à Chantilly

Moigno file. *Biographie du Clergé Contemporain* **10**, l'abbé Moigno, by G. Barbier, annotated by Father Cahier and 2 letters by Cauchy.

K. Family Papers

Cauchy's personal scientific papers are now destroyed.

When Cauchy's wife, Aloïse de Bure, died in 1863, they became the property of the couple's eldest daughter, Alicia, who was married to Félix de l'Escalopier, who kept them until he died in 1909. While his estate was being settled, Honoré Champion, a bookseller, surveyed Cauchy's papers. He found a number of books filled with notes that he had put aside. With the consent of the notary in charge of the estate settlement, he wrote to Painlevé, and later to Darboux, the Permanent Secretary of the Académie des Sciences, asking their advice as to what should be done with these papers (see the letters by Honoré Champion in the Cauchy file at the Académie des Sciences and also in the archives of the Honoré Champion bookstore, Arch. Nat. AQ22). Darboux seems to have replied that the Académie would be pleased to receive any papers that had been written by Cauchy or that had to do with Cauchy. In the meantime, the papers remained in the family's possession, in the hands of Madame de Leudeville, Félix de Escalopier's youngest daughter. There were two large trunks filled with notebooks and log books that had apparently belonged to the scholar and were filled with numbers and calculations of all kinds. Not knowing what to do with these items, the Leudevilles sent them to the Académie des Sciences in 1936 or 1937. The Académie, however, immediately sent them back. The decision was then made to destroy these unwanted documents, and accordingly, everything was burned. The only items to have escaped the general destruction were the notebooks that are kept in the Sorbonne Library (See H.) and two handwritten notebooks that had been given to Madame de Pomyers by her great uncle in rememberance of her ancestors.

1. *Sur la théorie des ondes* (1815–1821).
2. *Mélanges* (1836–1837).

The correspondence between A. L. Cauchy and his family (1811–1812 and 1831–1837) was rediscovered in 1989. It is in the possession of the family de Leudeville. Those personal letters have not been used for the biography.

II. Bibliography

A. Published Works of A. L. Cauchy

We refer the reader to A. L. Cauchy, *Oeuvres Complètes,* Paris, Gauthier-Villars, 1882–1974 and especially to the documentary appendix of volume **15** of the 2nd series, pp. 582–611.

1. Printed Scientific Treatises and Pamphlets

Exposé Sommaire d'une Méthode pour Déterminer a Priori le Nombre des Racines Réelles Positives et le Nombre de Racines Réelles Négatives d'une Équation d'un Degré Quelconque, Paris, 1813 (*O.C.*, 2, **15**, pp. 11–16).

Cours d'Analyse de l'École Royale Polytechnique, lère Partie, Analyse Algébrique, Paris, 1821 (*O.C.*, 2, **3**).

Résumé des Leçons Données à l'École Royale Polytechnique sur le Calcul Infinitésimal, 1st vol., Paris, 1823 (*O.C.*, 2, **4**, pp. 5–261).
Mémoire sur les Intégrales Définies Prises entre des Limites Imaginaires, Paris, 1825 (*O.C.*, 2, **15**, pp. 41–89).
Exercices de Mathématiques, Paris, 5 vol., 1826–1830 (*O.C.* 2, **6**, **7**, **8**, and **9**).
Leçons sur les Applications du Calcul Infinitésimal à la Géométrie, Paris, 2 vol., 1826–1828 (*O.C.*, 2, **5**, pp. 5–403).
Mémoire sur l'Application du Calcul des Résidus à la Solution des Problèmes de Physique Mathématique, Paris, 1827 (*O.C.*, 2, **15**, pp. 90–137).
Leçons sur le Calcul Différentiel, Paris, 1829 (*O.C.*, 2, **4**, pp. 263–615).
Mémoire sur la Théorie de la Lumière, Paris, 1830 (*O.C.*, 2, **2**, pp. 119–133).
Mémoire sur la Dispersion de la Lumière, Paris, 1830 (*O.C.*, 2, **10**, pp. 195–220).
Résumés Analytiques, Turin, 1833 (*O.C.*, 2, **10**, pp. 9–184).
Nouveaux Exercices de Mathématiques, Prague, 1836 (*O.C.*, 2, **10**, pp. 189–464).
Sur la Résolution des Équations de Degré Quelconque, Paris, 1837 (*O.C.*, 2, **15**, pp. 448–482).
Mémoire sur une Méthode Générale pour la Détermination des Racines Réelles des Équations Algébriques ou même Transcendantes, Paris, 1837 (*O.C.*, 2, **15**, pp. 483–510).
Recueil de Mémoires sur la Physique Mathématique, Paris, 1839.[1]
Exercices d'Analyse et de Physique Mathématique, Paris, 4 vol., 1841–1853 (*O.C.*, 2, **11**, **12**, **13**, and **14**).
See also the proofs of the beginning of the *Résumé des Leçons Données à l'École Royale Polytechnique* (second year).
A. L. Cauchy, *Equations Différentielles Ordinaires, Cours Inédit, Fragment*, Paris, Études Vivantes, 1981.

2. *Lithographed Scientific Pamphlets*

See the list of the pamphlets published in the documentary appendix of the *Oeuvres Complètes, O.C.*, 2, **15**, pp. 586–588.

3. *Scientific Memoirs, Articles, and Notes in the Following Reviews and Academic Collections*

Annales de Mathématiques Pures et Appliquées of Gergonne.
Biblioteca Italiana.
Bulletin des Sciences of the Société Philomatique.
Bulletin des Sciences Mathématiques of Férussac.
Comptes Rendus Hebdomadaires des Séances de l'Académie des Sciences.
Correspondance sur l'École Polytechnique.

[1]This pamphlet contains extracts from the *Comptes Rendus Hebdomadaires des Séances de l'Académie des Sciences* (see *O.C.*, 1, **4**, no. 22, p. 112, and nos. 33, 34, 35, 36, 37, 38, 39, 40, and 41, pp. 193–311).

Journal de l'École Polytechnique.
Journal de Mathématiques Pures et Appliquées of Liouville.
Journal des Mines.
Mémoires de l'Institut de France.
Mémoires Présentées à l'Académie des Sciences par Divers Savants.

4. *Scientific Works of F. Moigno Directly Inspired by A. L Cauchy*

F. Moigno, *Leçons de Calcul Différentiel et de Calcul Intégral Rédigées d'après les Méthodes et les Oeuvres Publiées ou Inédites d'A. L. Cauchy*, **1** (*Calcul Différentiel*), **2** (*Calcul Intégral*), and **4** (*Calcul des Variations*), Paris, Bachelier, 1840, 1844, and 1861.
F. Moigno, *Leçons de Mécanique Analytique Rédigées Principalement d'après les Méthodes d'A. L Cauchy et Étendues aux Travaux les plus Récents*, Paris, Bachelier, 1868.

5. *Nonscientific Works*

See the bibliography compiled by R. Taton in the documentary appendix of the *Oeuvres Complètes* of A. L. Cauchy, *O.C.*, 2, **15**, pp 606–607. This should be completed by the lectures published in the *Bulletin de l'Institut Catholique.*

'Epître d'un mathématicien à un poète, ou la leçon d'astronomie', poem read on January 13, 1842, *Bulletin de l'Institut Catholique*, 1st instalment, pp. 14–20.
'Sur quelques préjugés contre les physiciens et les géomètres', Lecture of March 3, 1842, *Bulletin de l'Institut Catholique*, 1st instalment, pp. 43–49.
'Sur la recherche de la vérité', lecture of April 14, 1842, *Bulletin de l'Institut Catholique*, 2nd instalment, pp. 18–29.
'Motif de regrets et d'espérance', lecture of April 8, 1843, *Bulletin de l'Institut Catholique*, 3rd instalment, pp. 166–171.

B. Bibliography on the Life and Work of A. L. Cauchy

A. J. C. Barré de Saint-Venant, 'Historique abrégé des recherches sur la résistance et sur l'élasticité des corps solides', in Navier, *De la Résistance des Corps Solides*, 3rd ed., Paris: Dunod, 1864.
B. Belhoste, 'Le cours d'analyse de Cauchy á l'École Polytechnique en seconde année', *Sciences et Techniques en Perspective*, **9**, 1984–1985, pp. 101–178.
B. Belhoste, *Cauchy, un Mathématicien Légitimiste au XIXe siècle*, Paris: Belin, 1985.
B. Belhoste and J. Lützen, 'J. Liouville et le Collège de France', *Revue d'Histoire des Sciences*, **37**, 1984, pp. 255–304.
J. Ben-David, 'The rise and decline of France as a scientific center', *Minerva*, **8**, 1970, pp. 160–179.
G. Bertier de Sauvigny, *La Restauration*, Paris: Flammarion, 1963.

G. Bertier de Sauvigny, *Un Type d'Ultra-royaliste, le Comte F. Bertier de Sauvigny et l'Énigme de las Congrégation*, Paris: Les Presses Continentales, 1948.

J. Bertrand, 'La vie et les travaux du baron Cauchy, par C. A. Valson', *Journal des Savants*, 1869, pp. 205–215.

J. Bertrand, *Eloges Académiques*, **2**, Paris: Hachette, 1902, pp. 101–120.

G. Bigourdan, 'Le Bureau des Longitudes, son histoire et ses travaux de l'origine (1795) à ce jours', *Annuaire du Bureau des Longitudes*, from 1928 to 1932.

J. B. Biot, *Mélanges Scientifiques et Littéraires*, **3**, Paris: Michel Lévy, 1858, pp. 143–160.

C. A. Bjerknes, *Niels-Henrik Abel, Tableau de sa Vie et de son Action Scientifique*, Paris: Gauthier-Villars, 1885.

B. Boncompagni, 'La vie et les travaux du baron Cauchy', *Bulletino di Bibliografia e di Storia delle Scienze Matematiche et Fisiche*, **2**, 1869, pp. 1–102.

L. Bucciarelli and N. Dworsky, *Sophie Germain: An Essay in the History of the Elasticity*, Dordrecht, Boston: D. Reidel, 1980.

J. Buchwald, 'Optics and the theory of the punctiform aether', *Archive for History of Exact Sciences*, **21**, 1980, pp. 245–278.

H. Burkhardt, 'Entwicklungen nach oscillierenden Funktionen und Integration der Differentialgleichungen der mathematischen Physik', *Jahresbericht der deutschen Mathematiker-Vereinigung*, **10**, 1904–1908, especially pp. 671–745.

J. Burnichon, *La Compagnie de Jésus en France, Histoire d'un Siècle, 1814–1914*, 4 vol., Paris: Beauchesne, 1914–1919.

J. P. Callot, *Histoire de l'École Polytechnique*, Paris: Les Presses Modernes, 1958.

G. Castella, 'Documents inédits pour un projet de fonder une Académie Helvétique à Fribourg en 1830', *Revue d'Histoire Ecclésiastique Suisse*, **21**, 1927, pp. 308–313.

M. Crosland (ed.), *The Emergence of Science in Western Europa*, London: Macmillan, 1976.

A. Dahan-Dalmedico, *Les Recherches Algébriques de Cauchy*, 3rd cycle thesis, typewritten, Paris: EHESS, 1979.

A. Dahan-Dalmedico, 'Les travaux de Cauchy sur les substitutions. Étude de son approche du concept de groupe', *Archive for History of Exact Sciences*, **23**, 1980, pp. 279–319.

A. Dahan-Dalmedico, 'La mathématisation de la théorie de l'élasticité par A. L. Cauchy et les débats dans la physique mathématique française (1800–1840)', *Sciences et Techniques en Perspective*, **9**, 1984–1985, pp. 1–100.

A. Dahan-Dalmedico, 'Étude des méthodes et des "styles" de mathématisation: la science de l'élasticité', in R. Rashed (ed.), *Sciences à l'Epoque de la Révolution Française. Recherches Historiques*, Paris: Blanchard, 1988, pp. 349–442.

J. Dhombres, *Nombre, Mesure et Continu. Epistémologie et Histoire*, Paris: Nathan, 1978.

J. Dieudonné (ed.), *Abrégé d'Histoire des Mathématiques, 1700–1900*, 2 vol., Paris: Hermann, 1978.

P. Dugac, *Sur les Fondements de l'Analyse de Cauchy à Baire*, thesis, typewritten, Paris: Université Paris VI, 1978.

P. Dugac, 'Histoire du théorème des accroissements finis', *Archive Internationale d'Histoire des Sciences*, **30**, pp. 86–101, 1980.

Ch. Dupin, *Discours aux Funérailles de M. Augustin Cauchy le 25 mai 1857*, Paris: Gauthier-Villars, 1857.

J. B. Duroselle, *Les Débuts du Catholicisme Social en France (1822–1870)*, Paris: PUF, 1951.

J. B. Duroselle, 'Les "Filiales" de la Congrégation', *Revue d'Histoire Ecclésiastique*, **50**, 1955, pp. 867–891.

H. Edwards, *Fermat's Last Theorem: A Genetic Introduction to Algebraic Number Theory*, New York: Springer-Verlag, 1977.

G. Fisher, 'Cauchy and the infinitely small', *Historia Mathematica*, **5**, 1978, pp. 313–331.

A. Foucault, *La Société de Saint-Vincent-de-Paul*, Paris: Editions 'Spes', 1933.

A. Fourcy, *Histoire de l'Ecole Polytechnique*, Paris, 1828, reed. Paris: Belin, 1977.

H. Freudenthal, 'Cauchy, Augustin-Louis', *Dictionary of Scientific Biography*, **3**, New York: Charles Scribner's Sons, 1971, pp. 131–148.

H. Freudenthal, 'Did Cauchy plagiarize Bolzano?', *Archive for History of Exact Sciences*, **7**, 1971, pp. 375–392.

J. P. Garnier, *Charles X, le Proscrit*, Paris: Fayard, 1967.

C. A. Geoffroy de Grandmaison, *La Congrégation (1801–1830)*, Paris: Plon-Nourrit, 1889.

C. Gilain, Introduction to A. L. Cauchy, *Equations Différentielles Ordinaires, Cours Inédit, Fragement*, Paris: Etudes Vivantes, 1981.

C. Gilain, 'Cauchy et le cours d'analyse à l'École Polytechnique', *Bulletin de la Société des Amis de la Bibliothèque de l'École Polytechnique*, **5**, July 1989, pp. 3–145.

H. Gouhier, *La Jeunesse d'Auguste Comte et la Formation du Positivisme*, 3 vol., Paris: Vrin, 1933–1941.

J. V. Grabiner, *The Origins of Cauchy's Rigorous Analysis*, Cambridge (Massachusetts): MIT Press, 1981.

I. Grattan-Guinness, 'Bolzano, Cauchy and the "New analysis" of the early nineteenth century', *Archive for History of Exact Sciences*, **6**, 1970, pp. 372–400.

I. Grattan-Guinness, *The Development of the Foundations of Mathematical Analysis from Euler to Riemann*, Cambridge (Massachusetts): MIT Press, 1970.

T. Guitard, 'La querelle des infiniment petits à l'École Polytechnique au XIXe siècle', *Historia scientiarium*, **30**, 1986, pp. 1–61.

A. d'Hautpoul, *Quatre Mois à la Cour de Prague*, Paris: Plon-Nourrit, 1912.

T. Hawkins, 'Cauchy and the spectral theory of matrices', *Historia Mathematica*, **2**, 1975, pp. 1–29.

C. C. Heyde and E. Seneta, *I.-J. Bienaymé. Statistical Theory Anticipated*, New York: Springer-Verlag, 1977, pp. 71–76.

A. P. Iushkevich, *Michel Ostrogadski et le Progrès de la Science au XIXe Siècle*, Paris: Palais de la Découverte, 1966.

A. P. Iushkevich, 'The concept of function up to the middle of the 19th century', *Archive for History of Exact Sciences*, **16**, 1977, pp. 37–85.

P. Jourdain, 'The theory of functions with Cauchy and Gauss', *Bibliotheca Mathematica*, 3rd series, **6**, 1905, pp. 190–207.

P. Jourdain, 'The origin of Cauchy's conceptions of a definite integral and of the continuity of a function', *Isis*, **1**, 1914, pp. 661–703.

M. Jullien, 'Quelques souvenirs d'un étudiant jésuite à la Sorbonne et au Collège de France, 1852–1856', *Les Études*, **127**, 1911, pp. 333–348.

M. Kline, *Mathematical Thought from Ancient to Modern Time*, New York: Oxford University Press, 1972.

F. R. de Lamennais, *Correspondance Générale*, **1**–**6**, Paris: A. Colin, 1970–1978.

A. E. H. Love, *A Treatise on the Mathematical Theory of Elasticity*, 4th ed., Cambridge (England): University Press, 1927.

J. Lützen, *Joseph Liouville, 1809–1882. Master of Pure and Applied Mathematics*, New York: Springer-Verlag, 1990.

J. Mandelbaum, *La Société Philomatique de Paris de 1788 à 1835*, 3rd cycle thesis, typewritten, Paris: EHESS, 1980.

L. Menabrea, *Memorie*, Firenze, Giunti: G. Barbèra, 1971.

J. Michelet, *Journal*, **1**, Paris: Gallimard, 1959.

F. Moigno, Préface to A. L. Cauchy, *Sept Leçons de Physique Générale*, Paris: Journal 'Les Mondes', 1868.

C. de Montalembert, *Éloge de Cauchy à la Séance des Cinq Académies du 17 Août 1857*, Paris: Firmin-Didot, 1857.

J. Paguelle de Fontenay, *M. Teysseyrre, sa Vie, son Oeuvre, ses Lettres*, Paris: Poussielgue, 1882.

J. Peiffer, *Les Premiers Exposés Globaux de la Théorie des Fonctions de Cauchy*, 3rd cycle thesis, typewritten, Paris: EHESS, 1978.

M. Pensivy, 'Jalons historiques pour une épistémologie de la série infinie du binôme', *Sciences et Techniques en Perspective*, **14**, 1987–1988.

J. C. Pont, *La Topologie Algébrique des Origines à Poincaré*, Paris: PUF, 1974.

K. Rychlik, 'Sur les contacts personnels de Cauchy et de Bolzano', *Revue d'Histoire des Sciences*, **15**, 1962, pp. 163–164.

L. Sédillot, *Les Professeurs de Mathématiques et de Physique Générale au Collège de France*, Rome: Imprimerie des sciences mathématiques, 1869.

H. Sinaceur, 'Cauchy et Bolzano', *Revue d'Histoire des Sciences*, **26**, 1973, pp. 97–112.

F. Smithies, 'Cauchy's conception of rigor in analysis', *Archive for History of Exact Sciences*, **36**, 1986, pp. 41–61.

P. Stäckel, 'Integration durch das imaginäre Gebiet', *Bibliotheca Mathématica*, 3rd series, **1**, 1900, pp. 111–121.

D. and R. Struik, 'Cauchy and Bolzano in Prague', *Isis*, **11**, 1928, pp. 364–366.

R. Taton, 'Sur les relations scientifiques d'Augustin Cauchy et d'Evariste Galois', *Revue d'Histoire des Sciences*, **24**, 1971, pp. 123–148.

A. Terracini, 'Cauchy a Torino', *Rendiconti del Seminario Matematico dell'Universita e del Politecnico di Torino*, **16**, 1957, pp. 159–203.

I. Todhunter, *A History of the Theory of Elasticity and of the Strength of Materials from Galilei to the Present Time*, **1**, Cambridge (England): University Press, 1886.

C. Truesdell, 'The rational mechanics of flexible or elastic bodies, 1638–1788', in Euler, *Opera omnia*, 2, **11**, sect. 2, Zürich, pp. 7–435, 1960.

C. Truesdell, *Essays in the History of Mechanics*, Berlin, Springer-Verlag, 1968.

C. Truesdell, 'Rapport sur le pli cacheté n° 126, paquet présenté à l'Académie des Sciences dans la séance du ler octobre 1827 par M. Cauchy et contenant le mémoire 'Sur l'Équilibre et le mouvement intérieur d'un corps solide considéré comme un système de molécules distinctes les unes des autres', *Comptes Rendus de l'Académie des Sciences*, **291**, October 13, 1980, Vie Académique.

C. A. Valson, *La Vie et les Travaux du Baron Cauchy*, 2 vol., Paris: Gauthier-Villars, 1868.

E. T. Whittaker, *A History of the Theories of Aether and Electricity*, London, New York: T. Nelson, 1951.

Abbreviations

Ac. Sci.	Académie des Sciences
Arch. Hist. Ex. Sci.	*Archive for History of Exact Sciences*
Arch. Nat.	Archives Nationales
Bull. Fer.	*Bulletin des Sciences Mathématiques* of Férussac
Bull. Phil.	*Bulletin des Sciences de la Société Philomatique*
C.R.	*Comptes rendus Hebdomadaires des Séances de l'Académie*
E.N.P.C.	École Nationale des Ponts et Chaussées
E.P.	École Polytechnique
Ex. An. Phys. Math.	*Exercices d'Analyse et de Physique Mathématique*
Ex. Math.	*Exercices de Mathématiques*
J.E.P.	*Journal de l'École Polytechnique*
J.L.	*Journal de Mathématiques Pures et Appliquées* of Liouville
Mem. Ac. Sci.	*Mémoires de l'Académie des Sciences*
Mem. Institut	*Mémoires de l'Institut de France*
Mem. Sav. Etr.	*Mémoires Présentés par Divers Savants Étrangers*
O.C.	*Oeuvres Complètes* of A. L. Cauchy (*O.C.*, 1 is the 1st Series and *O.C.*, 2 is the 2nd series)
Rev. Hist. Sci.	*Revue d'Histoire des Sciences*

Index

Note: Subjects are italicized. References to the notes are italicized.